Magnetohydrodynamics

FLUID MECHANICS AND ITS APPLICATIONS
Volume 80

Series Editor: R. MOREAU
MADYLAM
Ecole Nationale Supérieure d'Hydraulique de Grenoble
Boîte Postale 95
38402 Saint Martin d'Hères Cedex, France

Aims and Scope of the Series

The purpose of this series is to focus on subjects in which fluid mechanics plays a fundamental role.

As well as the more traditional applications of aeronautics, hydraulics, heat and mass transfer etc., books will be published dealing with topics which are currently in a state of rapid development, such as turbulence, suspensions and multiphase fluids, super and hypersonic flows and numerical modelling techniques.

It is a widely held view that it is the interdisciplinary subjects that will receive intense scientific attention, bringing them to the forefront of technological advancement. Fluids have the ability to transport matter and its properties as well as transmit force, therefore fluid mechanics is a subject that is particulary open to cross fertilisation with other sciences and disciplines of engineering. The subject of fluid mechanics will be highly relevant in domains such as chemical, metallurgical, biological and ecological engineering. This series is particularly open to such new multidisciplinary domains.

The median level of presentation is the first year graduate student. Some texts are monographs defining the current state of a field; others are accessible to final year undergraduates; but essentially the emphasis is on readability and clarity.

For a list of related mechanics titles, see final pages.

Magnetohydrodynamics

Historical Evolution and Trends

by

S. MOLOKOV

Coventry University
UK

R. MOREAU

EPM-MADYLAM
St Martin d'Hères
France

and

H.K. MOFFATT

University of Cambridge
UK

 Springer

A C.I.P. Catalogue record for this book is available from the Library of Congress.

ISBN 978-1-4020-4832-6 (HB)
ISBN 978-1-4020-4833-3 (e-book)

Published by Springer,
P.O. Box 17, 3300 AA Dordrecht, The Netherlands.

www.springer.com

S C I E N C E
QA
920
· M1346
2007

Printed on acid-free paper

COST is an intergovernmental European framework for international cooperation between nationally funded research activities. COST creates scientific networks and enables scientists to collaborate in a wide spectrum of activities in research and technology. COST activities are administered by the COST Office.

11.0

Contents

Preface

Magnetohydrodynamics (MHD) is concerned with the flow of electrically conducting fluids in the presence of magnetic fields, either externally applied or generated within the fluid by inductive action. Its origin dates back to pioneering discoveries of Northrup, Hartmann, Alfvén, and others in the first half of the twentieth century. After 1950, the subject developed rapidly, and soon became well established as a field of scientific endeavour of great importance in various contexts: geomagnetism and planetary magnetism, astrophysics, nuclear fusion (plasma) physics, and liquid metal technology.

This volume surveys both the historical evolution of the field and some of the current trends. It is based on a workshop on the History of MHD organised at Coventry University, UK, 26–28 May 2004, by the working group on "High Magnetic Fields" within the European network "Magnetofluiddynamics" (COST Action P6). It contains contributions by the workshop participants, supplemented by several additional invited papers in order to provide more comprehensive coverage of the recent trends. It also includes reminiscences of scientists who worked during the period of pioneering discoveries in the field (1950s and 1960s), together with photos of at least some of the pioneers of the subject.

Topics covered in this volume include dynamo theory and experiment, astrophysics, plasmas, high magnetic fields, turbulence, and electromagnetic processing of materials. Other topics such as magnetoconvection, magnetic reconnection, and tokamak plasmas are not included, simply because to do justice to these important topics would have required a book of unmanageable proportions.

Judging by the vitality of the field as evidenced by this volume, we believe that MHD still poses challenges of great fundamental, as well as practical, importance, and that the prospects for its continuing vitality are bright.

We gratefully acknowledge the willing cooperation of all participants of the workshop and contributors to this volume, the financial support of the European Cooperation in the field of Scientific and Technical Research (COST), the help of Svetlana Aleksandrova in organising the event, the advice of Leo Bühler on the sometimes painful process of conversion between various pieces of software, and last but by no means least, the patience and understanding of the publishers (Springer).

Coventry, Grenoble, Cambridge   *The Editors*
January 2006

Part I

Dynamo, Astrophysics, and Plasmas

How MHD Transformed the Theory
of Geomagnetism

Paul Roberts

Department of Mathematics and Institute of Geophysics and Planetary Physics,
University of California at Los Angeles, Los Angeles, CA 90095, USA
(roberts@math.ucla.edu)

Summary. The main magnetic field on the Earth is generated by, and has been maintained throughout Earth's history by, a fluid dynamo operating in the Earth's electrically conducting core. The author gives his personal view of how understanding of this 'geodynamo' grew during his lifetime, and he includes recollections of some of the scientists involved. The remarkable evolution of the subject from simple applications of electromagnetic theory to today's sophisticated magnetohydrodynamic theory is outlined. The importance of Coriolis forces in core MHD is not fully appreciated even today, but it transforms MHD into an essentially different subject that is briefly reviewed here. Proposals are made to give it its own special name.

1 Early days

As this is a meeting about the *history* of magnetohydrodynamics (MHD), it seemed to me, when I was preparing my talk, that it might be appropriate to include some reminiscences[1]; I expected a fair number of people in the audience to be too young to have interacted with the founders of the subject but who might be interested to hear snippets about them, and as an old fogey I am in a position to oblige. But it seems I am wrong: the room is full of old fogies. Nevertheless I'll continue as planned.

When was MHD born? The answer to this question is subjective. It seems to me that no study that explores only one side of the interrelationship between the magnetic field and the motion of the conductor can truly be said to be a part of MHD. Thus Faraday's famous experiment [1] of 1832 on Waterloo bridge, that was intended to detect tidal motion in the Thames by electromagnetic (=em) induction, is not an MHD experiment, since the dynamical effect of the field on the motion is negligible and is not included. The same can be said of many early attempts to understand the geomagnetic field, as we shall see.

---

[1] Sergei Molokov has encouraged me to be equally informal too in this written account of my talk, which I nearly subtitled "on falling off chairs"; read on!;

S. Molokov et al. (eds.), Magnetohydrodynamics – Historical Evolution and Trends,
3–26. © 2007 *Springer.*

Fig. 1. Hannes Alfvén

We have already heard at this meeting of exciting experiments in liquid metals in a tradition that go back to Hartmann and Lazarus [2] in 1937. These involve both sides of the MHD relationship, but to my mind they are not really MHD experiments for they do not exhibit the main feature of MHD. The interaction of conducting fluids and magnetic fields gives rise to a completely unexpected phenomenon, the Alfvén wave. That discovery [3], which was a triumph of theory over experiment, was made during World War II. As I see it, this marks the birth of MHD.

Perhaps it is time to start reminiscing. I first met Alfvén at a symposium in Saltsjöbaden near Stockholm in 1956 and, during the time I still worked in Newcastle, he visited there a couple of times. I remember two things about him. First, his colleagues held him in great awe. Second, he ate an apple in a most unusual way. He removed all the skin, where I'm told all the goodness lies, and then he ate *all* the rest including the pips in the core, which I've been told contain small amounts of prussic acid.

The Alfvén wave could have been discovered even before Maxwell introduced displacement currents. How had everyone else missed it? It was said that, as a research student, Ferraro had suggested to his supervisor, Sydney Chapman, that it might be interesting to look for wave motions, but Chapman had told him it would be a waste of time. So Ferraro discovered not the wave but (in 1937) his law of isorotation [4], which is not, according to my definition, an MHD phenomenon. True or false, this is a cautionary tale for all research students. Listen to your supervisor by all means, but do not necessarily take his/her advice. Your own instincts may be better!

Fig. 2. Walter Elsasser; look at the blackboard!

Another great thing Alfvén did in explaining his wave was to give a physical argument [3] in support of his "frozen flux theorem", which I shall call his first theorem:

Magnetic field is "frozen" to a perfect conductor as it moves.

Of course, we do not encounter perfect conductors much, but the theorem is very helpful in visualizing MHD processes whenever the *magnetic Reynolds number* is large. You will remember that this number is

$$R_m = \mathcal{L}\mathcal{V}/\eta, \quad \text{or} \quad R_m = \tau_\eta/\mathcal{T}, \tag{1}$$

where $\mathcal{L}$, $\mathcal{T}$ and $\mathcal{V}$ are typical length, time and velocity scales, $\eta = 1/\mu\sigma$ is magnetic diffusivity, μ is permeability, σ is electrical conductivity (SI units), and

$$\tau_\eta = \mathcal{L}^2/\eta \tag{2}$$

is the *electromagnetic diffusion time*. If we take $\mathcal{L} = 10^6$ m and $\eta = 2$ m^2/s as appropriate for the Earth's core, τ_η is of order 10^4 years, and $R_m = 100$ for $\mathcal{V} = 10^{-4}$ m/s (as suggested by the "westward drift" of the field). But $\mathcal{L}$ is necessarily small in laboratory experiments with liquid metals and therefore R_m is small too. The frozen field picture is then not so useful; indeed, some say it is distinctly unhelpful. The magnetic field acts more like an anisotropic friction.

To detect an MHD wave it must be seen to "cross the apparatus". The wave travels with the Alfvén speed

$$\mathbf{V}_A = \mathbf{B}/\sqrt{(\mu\rho)}, \tag{3}$$

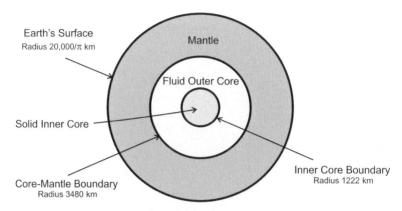

Fig. 3. Interior of the Earth

where **B** is the magnetic field and ρ is density. It crosses the apparatus in a characteristic time of $\tau_A = \mathcal{L}/\mathcal{V}_A$ but, as it does so, the induced current system that travels with it decays ohmically in a characteristic time of τ_η. The condition that the wave is detected is $\tau_A < \tau_\eta$, i.e., the *Lundquist number*,

$$Lu = \mathcal{L}\mathcal{V}_A/\eta, \tag{4}$$

must be "large enough", and generally it isn't in the laboratory.

In neutral Sweden, MHD got off to a fine start and it was several years after the war before the rest of the world caught up. Meanwhile, Lundquist wrote the first paper on magnetostatics in 1950, a famous review in 1952 [5], and attempted to demonstrate Alfvén waves in the laboratory using mercury as the working fluid [6]. Lehnert used liquid sodium instead. His method of disposing of used sodium led to an amusing incident that shows his skill as a conciliator rivals his prowess as a scientist. He relates this tale elsewhere in this book.

2 Early pioneers: Cowling, Elsasser, Bullard, and Chandrasekhar

The famous physicist, Walter Elsasser, who had emigrated to the United States in the 1930s for obvious reasons, began in 1939 to take an increasing interest in why the Earth is magnetic [7]. It may be appropriate here to remind you of the Earth's internal structure; see Fig. 3. Elsasser's studies led him ever deeper into MHD culminating in 1955/56 with a two part review [8], but initially he made use only of em theory. In 1950 he discovered the by now well-known Elsasser variables $\mathbf{V} \pm \mathbf{V}_A$, where $\mathbf{V}$ is the fluid velocity [9].

One of his most interesting papers [10] appeared in 1946, the second of a series on "Induction effects in terrestrial magnetism". This shows that he had

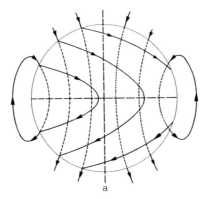

Fig. 4. The ω-effect

by now realized that rotation is very important in the MHD of the core. He got close to introducing the important dimensionless measure of the relative sizes of Lorentz and Coriolis forces in a rotating system:

$$\Lambda = \mathcal{B}^2/\mu\rho\eta\Omega = \mathcal{V}_A^2/\eta\Omega, \tag{5}$$

where Ω is the angular velocity. I christened this the *Elsasser number* in his honour and the name has stuck. Although he did not write down his number, he did introduce the important scale $\mathcal{B}_E$ based on $\Lambda = 1$, namely

$$\mathcal{B}_E = \sqrt{(\mu\rho\eta\Omega)}, \tag{6}$$

which he estimated as about 12 gauss for Earth's core, i.e., much the same value as we would accept today, taking $\rho = 10^4$ kg/m^3. I will return to the Elssaser number later.

Elsasser was also the first to draw attention to the importance of the toroidal field in geomagnetism, the existence of which had previously been unsuspected since it is trapped in the core (assuming that the electrical conductivity in and above the mantle is negligible). One expresses the magnetic field, **B**, as

$$\mathbf{B} = \mathbf{B}_T + \mathbf{B}_P, \tag{7}$$

$$\mathbf{B}_T = \mathbf{\nabla}\times(T\mathbf{x}), \qquad \mathbf{B}_P = \mathbf{\nabla}\times\mathbf{\nabla}\times(S\mathbf{x}), \tag{8}$$

where $\mathbf{B}_P$ is the 'poloidal field' that escapes to the Earth's surface where we view it, and $\mathbf{B}_T$ is the 'toroidal field'; here **x** is the radius vector from the geocenter. Elsasser pointed out, I think for the first time, that zonal motions easily create toroidal field from poloidal field, a process I later named 'the ω-effect.' With the help of the frozen flux idea, one can visualize the lines of force of an axisymmetric poloidal field, $\overline{\mathbf{B}}_P$, being stretched out along lines of latitude by a zonal shear, $\overline{\mathbf{V}}_T$ (Fig. 4). In this way a zonal $\overline{\mathcal{B}}_T$ is created of

Fig. 5. Thomas Cowling (*left*); Subrahmanyan Chandrasekhar (*right*)

order $R_m \times \overline{\mathcal{B}}_P$, where R_m is defined using the zonal velocity $\overline{\mathcal{V}}_T$. The belief quickly developed that

$$\overline{\mathcal{B}}_T/\overline{\mathcal{B}}_P \sim R_m, \qquad \overline{\mathcal{V}}_T/\overline{\mathcal{V}}_P \sim R_m \qquad (9)$$

are large; unseen is to be believed! If $R_m = O(100)$ (see above), $\overline{\mathcal{B}}_P = 3$ gauss, implies $\overline{\mathcal{B}} \approx \overline{\mathcal{B}}_T$ is about 300 gauss, the value used in making estimates below. For example, $\overline{\mathcal{B}} \approx 300$ gauss gives $\overline{\mathcal{V}}_A \sim 0.3$ m/s, so that $\varLambda \sim 100$ rather than $\varLambda \sim 1$. The concept (9) dominated core MHD until the mid-1990s, when numerical simulations did not support it [11]. One should not forget however that simulations were not, and still are not, in a geophysically realistic parameter range.

My own forays into geomagnetism started in 1951 when, as a 1st year research student I approached Herman Bondi, who had been the outstanding teacher of my undergraduate years to ask him for a topic in relativity and cosmology to work on for a Ph.D. The gist of his reply, as I recall it, was that, if he had a problem in that area to work on, he'd work on it himself. He suggested instead that I solve the dynamo problem either by generalizing Cowling's theorem [12] or by creating an example of a working dynamo.

I first met Cowling at a party at the British Embassy in Stockholm at the time of the Saltsjöbaden meeting. I found him a rather forbidding presence, especially so since he imbibed soft drinks while everyone else was getting plastered at British government expense. But I like this photograph of him (Fig. 5). A little smile plays around his lips. No doubt he was thinking at the time of all the trouble that his theorem was giving other theoreticians. Later, when I visited Leeds often to collaborate with Harold Ursell, I got to know and like "Tom" a lot better. But there is no denying that his scientific standards were high and his powers of criticism even higher. I am not the first to note that Cowling's theorem is a negative result:

Axisymmetric magnetic fields cannot be maintained by a dynamo.

When I later got to know Subrahmanyan Chandrasekhar, popularly known as "Chandra", it seemed to me that he felt that Cowling's exacting standards sometimes had a stultifying effect on the scientists. He propounded to me what he called Alfvén's second theorem:

Given Cowling, ∃ no theoretical astrophysics.

I never could discover whether Alfvén really had said that or whether this was a product of Chandra's impish sense of humour but, given its formal mathematical flavour, I suspect the latter [13].

While on the subject of Cowling's theorem, I cannot resist another story. Einstein and Elsasser maintained contact with each other after they had both moved to the new world. At one meeting, Einstein (who apparently had a long standing interest in the origin of the Earth's magnetism [14]) asked Walter for full details about his progress in solving the problem. Walter explained dynamo theory, as it stood in the early 1940s. He explained that it was still not known whether dynamos could function in simple bodies like spheres. He explained Cowling's theorem. "Enough" said the great man, "If dynamos do not work in such a simple case, they will not work at all". (I have been unable to remember or discover who told me this amusing tale, but I suspect it does grave injustice to an outstanding physicist.)

To go back to my own unfortunate experiences, I spent a year failing to generalize Cowling's theorem and also failing to find an example of a working dynamo. The nearest I came to it was something like Herzenberg's 1958 model [15], but I got lost in the maths. I was therefore very happy to accept Bondi's advice, which was to stop hitting my head against the wall, as the idea might not work anyway. A lost opportunity? Maybe. A lost year? Not entirely. I'd learned some MHD, and what did a year matter? At age 22 one is going to live forever. One pleasant part of the year was getting acquainted with that lovable character "Teddy" Bullard who was ultra supportive of research students perhaps partly because, as a research student himself, he had encountered less than tolerant treatment from Rutherford who had told him never to darken the doors of his (Cavendish) laboratory again. Bullard even invited me, a lowly research student, to be a guest in his home at the National Physical Laboratories in Teddington so that we could have more time to discuss the geodynamo. At that time he was Director of the NPL, in a position to drive its computing section crazy attempting to solve a kinematic dynamo model with the very primitive electronic computers then available [16].

My failure caused me to change direction and supervisor. In 1952, I started to work on other geomagnetic problems under the direction of Keith Runcorn, and I made my first positive contribution to geomagnetism, which I would

Fig. 6. "Teddy" Bullard (*left*); Keith Runcorn (*right*)

claim was actually the first success of MHD in the subject. At that time, DAMTP did not exist and contact between geophysicists and applied mathematicians, and between geophysicists and engineers, was non-existent. Apart from Raymond Hide (or "Spike", as he was known then), nobody I met seemed to know anything whatever about MHD. Most geophysicists thought that the sources of the geomagnetic field must lie in the upper 200 km of the core. The argument was based on the "skin effect": a field changing on a timescale of $\mathcal{T}$ penetrates a *solid* conductor only to a depth of order

$$d_\eta = (\eta \mathcal{T})^{1/2}. \tag{10}$$

For $\mathcal{T} = 10$ years, a typical secular variation timescale, $d_\eta = 30$ km; for $\mathcal{T} = 1000$ years, $d = 200$ km. So all the sources had to be near the top of the fluid core.

Even Elsasser [10] swallowed this, as did Lowes and Runcorn [17]. In 1952, I pointed out to Keith that the argument was not convincing essentially because the Lundquist number based on a poloidal field strength of 3 gauss is large: $Lu > 1000$ for $\mathcal{L} = 10^6$ m. Thus, I claimed, Alfvén waves would carry the secular variations from deep within the core to its surface with essentially negligible dissipation [18]. Keith almost fell out of his chair and then told me that I "must have made a mistake". Of course, "imitation is the sincerest form of flattery", and Runcorn flattered me by including my insight into the published account [19] of an address he gave at a meeting of the American Geophysical Union in 1953, and he acknowledged the contribution I had made in a characteristic way: "It is a pleasure to record my gratitude to my colleagues in the Department of Geodesy and Geophysics of Cambridge University for much discussion and for permission to report on the results of their work prior to publication."

In reality, no permission was sought, from me at least, but who needs to seek permission from a lowly research student? My idea may seem to be rather obvious nowadays, but it was not so in 1952, and (after all) it *was* my idea and not his. So it rankled, and still rankles a half a century later (as you see!) [20].

In 1954, after my Ph.D., I became a postdoc at Yerkes Observatory of the University of Chicago where I got to know Chandra, whose early work on white dwarfs later led to a shared Nobel prize. When I visited, he was working frenetically on hydrodynamic and hydromagnetic stability, the subject of his later book, but he was vitally interested in MHD and dynamo theory. He was the first person to emphasise to me that a marginal kinematic dynamo might be an overstable solution of the em equations (i.e., one that oscillates sinusoidally in time) rather than the steady, "exchange of stabilities" solution that people such as Cowling and Bullard had so far been seeking.

After returning to England for military service, I moved to Newcastle but attempted little geomagnetism and even less MHD. What I did was mostly in partnership with Spike [21]. One valuable lesson he taught me was how to deal with adversity in research. Prior to that time, I would blame my stupidity if I failed to reach an objective. Though he was seldom in such a position, Spike would rebound quickly, merely opining that the objective itself was more difficult than he had originally anticipated.

I returned "permanently" to Yerkes in 1961 where Chandra was now working away frenetically (apparently the only way he ever worked) on the stability of rotating self-gravitating fluids. At the time, Keith Stewartson was on sabbatical at the Army Research Center in Madison and we started collaborating [22]. This gives me an opportunity to remind you of some of the outstanding work Keith did in MHD. With his strong background in aeronautical theory, a natural problem for Keith to analyse was flow over a "wing" with an aligned magnetic field (Fig. 7).

Others looked at this problem too, and had generated solutions that, as in the non-magnetic case, had a wake following the wing. Keith realized however that, when the Alfvén number

$$A = V/V_A \qquad (11)$$

(sometimes called the 'magnetic Mach number') is less than 1, the solution is fundamentally different as the wing signals its presence upstream through Alfvén radiation. "Obvious", we might tend to say today but, when he spoke about it at the Williamsburg meeting [23] in 1960, it was still controversial and hotly disputed.

During this period, Stewartson challenged me to solve a problem on MHD duct flow concerning the steady flow of conducting fluid down an insulating pipe in the presence of a large transverse magnetic field. He wanted to

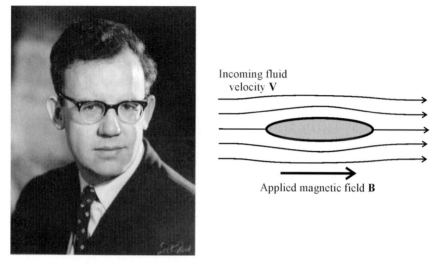

Incoming fluid
velocity **V**

Applied magnetic field **B**

Fig. 7. Keith Stewartson (*left*); aligned flow (*right*)

determine the structure of the singularities in the Hartmann layer that occur at the points where the applied field is tangential to the walls. I eventually rose to his challenge [24], and it was Stewartson's turn almost to fall out of his chair. (He also challenged me to solve the corresponding singularity structure for the Ekman layer, but that was too tough for me.)

I returned to Newcastle in 1963 and began to get interested in geomagnetism and MHD again. I started a project with Stan Scott that achieved some notoriety and recently some controversy [25]. It was based on MHD and Alfvén's theorem and argued that the secular variation could mostly be explained as the advection of field "frozen" to fluid motions beneath a boundary layer at the core surface.

3 Stanislav Braginsky, father of geomagnetic theory

In 1965 I lunched with Bullard at some meeting or other and he asked me if I'd looked at "extraordinary claims" made by "some Russian or other" about kinematic dynamos. That Russian was Stanislav Braginsky. At first sight, his paper [26] seemed obviously wrong. How could the mathematics of his nearly symmetric kinematic dynamos possibly simplify in the remarkable way he claimed? I spent a long time trying to evaluate this work. I was not helped by the notorious terseness of Soviet papers, caused by the strict length limits imposed by Soviet autocracy (because of the well-known shortage of paper caused by the death of trees in Russia!). It was the kind of situation where it was hard to re-derive results even when they are known to be correct. One marvelled at the pioneer who had faith to continue along such a complicated

Fig. 8. Stanislav Braginsky

path without really knowing for sure where it would lead. When I realized the paper was correct, it was my turn almost to fall out of my chair! Later, Andrew Soward [27] found an alternative and more fundamental route to Stanislav's result, but it is hardly simpler!

It would take me too long a digression to describe this work here. Suffice it to say that it is based on a simple idea: the magnetic Reynolds number, R_m, of the zonal core motion is large (see §2). Therefore even a small deviation from axisymmetry should be enough to defeat Cowling's theorem. Stanislav developed an asymptotic solution of the induction equation in powers of $R_m^{-1/2}$, in which (9) holds, and in which the asymmetric components $\mathbf{V}'$ and $\mathbf{B}'$ of the velocity and magnetic field are of order $R_m^{-1/2}\overline{\mathbf{V}}_T$ and $R_m^{-1/2}\overline{\mathbf{B}}_T$, respectively. Since $\mathcal{B}'/\overline{\mathcal{B}}_P = \mathrm{O}(R_m^{1/2}) \gg 1$, it appears at first sight that this solution is geophysically irrelevant, but Stanislav was able to show that the largest terms in the expansion of $\mathbf{B}'$ could not leave the conducting core; he demonstrated that $\mathcal{B}'_{\mathrm{external}} = \mathrm{O}(R_m^{-1}\mathcal{B}'_{\mathrm{internal}}) = \mathrm{O}(R_m^{-1/2}\overline{\mathcal{B}}_P)$, consistent with the small inclination of the magnetic dipole to the geographic axis.

In this work [26], Stanislav provided the first really compelling mathematical support for Eugene Parker's concept [28] of a "Γ−effect", though he did not have in mind em induction by turbulent motions as 'Gene had. I had heard' Gene talk about his ideas at Yerkes in 1955, but I was slow to realize their significance. At the time it seemed to me that he had merely replaced one equation (in three space dimensions) that I could not solve by another equation (in two space dimensions) that I could not solve either. Parker's work was taken further by Steenbeck, Krauze and Rädler [29]. They clarified *the α-effect* (Parker's Γ-effect), invented *helicity* (or "Schraubensinn" as they called it – roughly "screwiness"), explored the relationship between the two,

Fig. 9. 'Gene Parker

and created a new subject: *Mean Field Electrodynamics* (MFE). They provided a mathematical and computational model of solar activity in terms of a 'dynamo wave', much as Parker had envisaged [30].

Stanislav [31] combined his asymptotic analysis with numerical integrations, performed in his own machine code on the computers available to him in the USSR, which were even more primitive by today's standards than those used in the West at that time. This combined analytic–numeric approach to the kinematic geodynamo may fairly be said to be the first successful solutions.

Stanislav also generalized his first asymptotic theory. In [26] he had assumed that the α-effect was created by a single asymmetric planetary "wave", brought to rest by choice of reference frame. In an even more complicated generalization [32], he demonstrated that the $\alpha-$effects of an arbitrary number of such planetary waves, each having its own angular velocity about the symmetry axis, are additive. This paper illustrates how the idea of a dynamo operating by means of planetary waves had already taken root in Stanislav's mind.

The astrophysical impact of the work [29] of the Potsdam trio has been considerable. The geophysical impact of a companion paper, in which again turbulent motions in the core are held responsible for dynamo action, has been much less for what I believe are the following reasons.

Let us contrast convection in the Sun and in the Earth's core. In the former an abundant energy supply drives wildly turbulent motions. In the latter the question has always been, "Where on Earth (sorry!) does the core get enough energy to maintain the dynamo against ohmic losses?" The magnetic field in the Sun is highly intermittent. Violent motions bring flux tubes together and reconnect them with the release and dissipation of magnetic energy.

Fig. 10. The Potsdam trio; from left to right, Max Steenbeck, Karl-Heinz Rädler, and Fritz Krause

Correspondingly, the timescale of magnetic activity is greatly shortened. Instead of τ_η being of order 10^{11} years, as one might expect if one used a molecular or η, it is of order 10 years, suggesting that a turbulent diffusivity, η_T, about 10^{10} times greater than the molecular η is relevant. Models of the solar activity cycle, routinely assume that all diffusivities in the solar convection zone are of order 10^9 m^2/s. In contrast, one cannot expect such vigorous turbulence in the Earth's core, which is struggling to make do with the little energy it has available. There is absolutely no evidence that the timescales of the geomagnetic field are greatly shortened. An analogue of the regular activity cycle on the Sun is the irregular polarity reversal of the geomagnetic field. Palaeomagnetism assures us that it takes of order 10^4 years for the Earth to switch its polarity and this is about what is suggested for τ_η with the estimated molecular η. Theoretical discussions of turbulent induction link the turbulent α-effect to the turbulent diffusivity, both being proportional to the Reynolds number defined by the small scales, i.e., if the turbulent diffusivity η_T is small, so is the turbulent α-effect. This may be a simplistic way of stating the situation, since it omits the obvious influence of Coriolis forces and density stratification, but (to my mind) it distils the essence of the matter. Although it would be a brave man that said that turbulence plays no part in the generation of the large-scale geomagnetic field, it seems to me that such a statement would be nearer the truth than saying the opposite, that the main geomagnetic field is mainly produced by small-scale motions in the core. It is the large-scale motions that are responsible, and Andrew Soward was able to define a helicity for these and to relate it, in Braginsky's large R_m analysis, to his α.

I hope I have not given the impression, as Andrew Soward has told me I have on past occasions, that I believe turbulence is unimportant in the core. The Earth's core is a liquid metal, having a magnetic Prandtl number,

$$Pr_m = \nu/\eta, \tag{12}$$

of about 10^{-6}. Thus core turbulence has to supply the deficiencies of molecular processes by transporting large-scale momentum and heat. Turbulence therefore plays a crucial role in core MHD, though not in em induction processes. The significance (and anisotropy) of core turbulence was recognized in another paper [33] that Stanislav wrote in 1964, his "annus mirabilis". This contained several other important ideas too and, although it is hard to choose, I would rank this as his most important paper; I shall refer to it henceforth as "the 1964 paper". Later developments on core turbulence can be found in [34, 35].

At the time Stanislav wrote the 1964 paper [33], his estimate of η for the core was about 3 times what is accepted today and it seemed that thermal buoyancy could not provide the energy needed to offset ohmic dissipation. (Recall here that the greater the σ, the larger the $\mathcal{B}_T$ for the given, observed $\mathcal{B}_P$, and therefore the bigger the ohmic dissipation.) Stanislav came up with a new mechanism [33, 36]: compositional buoyancy. Jack Jacobs had cogently argued [37] that the Earth's solid inner core had been created by the solidi- fication of the iron-rich fluid alloy originally filling the core, and further that the inner core surface is, even to this day, continually moving slowly upwards as freezing continues. Verhoogen [38] had recognized that the concomitant release of latent heat would assist core convection. Stanislav made two impor- tant points, first that the light components of the alloy, the "admixture", would be preferentially released during freezing and would help stir the core too; second, he pointed out that such *compositional convection* would be thermo- dynamically much more significant than *thermal convection*, which is limited by (essentially) the Carnot efficiency. Others provided mathematical backing for these ideas [39] (and still others have received credit for them). Ideas of convective efficiency now dominate discussions of planetary dynamos, e.g., see [40]. The idea that core turbulence would mix entropy as effectively as it mixes the light constituents released from the inner core led Stanislav to the anelastic approximation. It is interesting to contrast his concise derivation of that approximation with those of earlier authors.

Another important idea in the 1964 paper [33] concerned the effects of rotation and buoyancy on Alfvén waves. This topic was not completely new though its application to the Earth's core was novel. I will digress again. Alfvén discovered his waves during attempts to explain the solar activity cycle;

see [18]. His colleague Walén developed [41], and Alfvén endorsed, a theory of "whirl rings", which resembled circular flux tubes. They visualized that these were produced near the Sun's energy producing core and travelled upwards as Alfvén waves, riding on the general poloidal field of the Sun. When Lehnert was a visitor to Yerkes Observatory in 1953 he analysed Alfvén waves in highly rotating systems and he found that they are replaced by slow waves, now often called 'Lehnert waves' in his honour [42]; they are sometimes also called 'MC waves' (see below). They are highly dispersive. The whirl rings would therefore not preserve their identity as they rose from deep within the Sun to its surface. I heard that Lehnert was not as thrilled by his discovery as you or I might have been. He was greatly in awe of Alfvén and did not know how to break the news to him.

The characteristic timescale of the Lehnert waves is

$$\tau_{slow} = \Omega \mathcal{L}^2 / \mathcal{V}_A^2, \tag{13}$$

which is about 200 years, for $\mathcal{L} = 10^6$ m and $\mathcal{B} = 100$ gauss ($\mathcal{V}_A = 10$ cm/s). Since

$$\tau_{slow} = \Lambda^{-1} \tau_\eta, \tag{14}$$

it is also, for $\Lambda = 100$, roughly the timescale over which ohmic dissipation would obliterate the waves.

Stanislav's had built his kinematic theory [26,32] on the dynamo action of waves, and he needed an explanation of how these waves are generated and maintained. So in the 1964 paper [33] he undertook a simple (Cartesian) analysis similar to Lehnert's but with the crucial difference that buoyancy forces were also included. This led him to the concept of the MAC wave, where M = Magnetic, A = Archimedean (buoyancy), and C = Coriolis. This acronym is mainly significant for what it omits: inertial and viscous forces (although a pressure gradient is, as always, significant). To improve geophysical realism, he later studied MAC waves in spherical geometry [43]. He argued that the geomagnetic secular variation is a manifestation of slow planetary MAC waves; see also Hide [44]. Buoyancy feeds energy to the waves and prevents their disappearance; the waves create the non-local α-effect that maintains the geomagnetic field. It's a nice idea, though a difficult one to develop theoretically; doubtless it will be pursued more in the future.

Andrew Soward joked once that the Alfvén velocity is that velocity with which no wave travels in the Earth's core! Certainly rotation changes everything. He would, however, make an exception of the torsional oscillations (TO) which are a kind of Alfvén wave travelling across the geostrophic cylinders which are coupled together by the s-component of the prevailing poloidal field (Fig. 11); here (s, ϕ, z) are cylindrical coordinates with Oz parallel to $\mathbf{\Omega}$. In a highly rotating body of fluid like the Earth's core, it is convenient to divide the fluid motion into geostrophic and ageostrophic (non-geostrophic) parts: $\mathbf{V} = \mathbf{V}_G + \mathbf{V}_N$. The geostrophic part, $\mathbf{V}_G$, is in the zonal (ϕ) direction, and is defined, for each geostrophic cylinder $\mathcal{C}(s)$, by the average of V_ϕ over

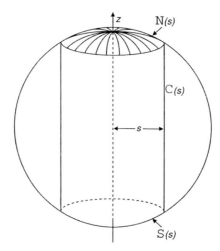

Fig. 11. Geostrophic cylinders; for simplicity, the inner core is ignored

$\mathcal{C}(s)$:

$$V_G(s,t) = \frac{1}{4\pi z_1} \int_{-z_1}^{z_1} dz \int_0^{2\pi} d\phi \ V_\phi(\mathbf{x}, t), \tag{15}$$

where $z_1 = \sqrt{(R^2 - s^2)}$. The significance of the division $\mathbf{V} = \mathbf{V}_G + \mathbf{V}_N$ is that the Coriolis forces associated with $\mathbf{V}_G$ can be absorbed into the pressure gradient, i.e., it is ineffective. This means that the inertial force, which has (compared with the Coriolis force) a negligible effect on the (ageostrophic) MAC waves, determines the fate of the geostrophic motions in much the same way as for an Alfvén wave. Stanislav [45] founded the theory of TO too and, based on some observation and analysis, these oscillations, which have a decadal timescale, have been detected [46].

This brings me to Stanislav's current work, which is motivated by a perplexing property of the geomagnetic field: the dipole varies on a decadal timescale. This cannot be explained by the TO, because these are axisymmetric and leave the dipole unaffected. Large-scale MAC waves create a dipole variation, but only on a timescale that is much too long. Stanislav currently attributes the dipole variation to MAC waves in a stably stratified "ocean" at the top of the core, about 80 km deep and having a density that is about 0.9999 times that of the fluid below it [47]. For such small-scale MAC waves, (13) gives τ_{slow} in the decadal range. It is fair to say that the geophysical community at large is not yet convinced of the existence of such an ocean. Another of Stanislav's ideas, "Model-Z" [48], has also not won general acceptance either. Nevertheless, Stanislav has been right before......

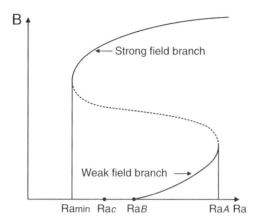

Fig. 12. Sketch of postulated bifurcation diagram

4 Epilogue: a question

I have sketched out a scenario, largely due to Stanislav Braginsky, of the MHD of the Earth's core and its consistency with what can be gleaned from past and present data on the geomagnetic field.

I will conclude by riding one of my favourite hobby-horses: the way that Coriolis forces transform traditional MHD into almost an unrecognizable form. I have touched on this topic several times already but another striking manifestation of this is the strong field dynamo which is illustrated here by a sketch (Fig. 12) that I was showing in my talks even in the 1960s. (Apologies to other old fogeys here present!) Consider the planar dynamo operating in a Bénard layer rotating about the vertical. This was originally studied by Steve Childress, Andrew Soward, and Yves Fautrelle [49], but recently it has been the focus of computations by Chris Jones in association with myself and Jon Rotvig [50]. This is the context in which my statements have so far the best theoretical backing. Figure 12 is a sketch of the $(Ra, \mathcal{B})$−plane of solutions; Ra is the Rayleigh number and Ra_c marks the onset of non-magnetic convection; if Ra is gradually increased from this state, the motions grow in magnitude and ultimately, at $Ra = Ra_B$ become strong enough to maintain a magnetic field. As Ra is further enhanced, $\mathbf{V}$ and $\mathbf{B}$ increase too and the *weak field branch* arises, on which the Hartmann number $Ha = \mathcal{BL}/\sqrt{(\mu\rho\eta\nu)}$ is O(1), i.e., on which $\Lambda = O(E)$, where $E = \nu/\Omega\mathcal{L}^2$ is the Ekman number, which is of order 10^{-14} if we assume that $\nu = 10^{-6}\mathrm{m}^2/\mathrm{s}$ is a molecular viscosity and of order 10^{-8} if we assume that ν is a turbulent viscosity of order 1 m^2/s. Ultimately, at the asymptote $Ra = Ra_A$, there is runaway field growth to the *strong field branch* where the Elsasser number, Λ, is O(1). Also evident on the strong field branch are subcritical solutions, i.e., solutions that exist at even smaller Rayleigh numbers than those at which convection can first occur. The Earth appears to be operating a dynamo on the strong field branch.

This is but one of the striking ways in which the Coriolis force transforms MHD. I believe it is an essentially different subject and deserves a different name and acronym. So I conclude this presentation by asking my audience (and readers) what that name and acronym should be. In the past I suggested 'RMHD' standing for 'rotating MHD' but that has been appropriated both by 'reduced MHD' and by 'relativistic MHD'. I recently proposed CMHD, standing for 'Coriolis MHD', but Stanislav has put forward an attractive alternative: 'MACHD', which is specially appropriate for the Earth's core, as it emphasises the central importance of Coriolis forces. (One could even take this idea further by reserving MAHD for the MHD of liquid metals in general, while using 'MHD' for the entire field, including plasma MHD.) What do you think?

Acknowledgements. I am grateful to Stanislav Braginsky and 'Gene Parker for providing pictures of themselves to use in this report. I also thank the Royal Institute of Technology, Stockholm, for supplying a photograph of Alfvén. Johns Hopkins University (the Eisenhower Library) gave permission to reproduce the photograph of Elsasser. The pictures of Bullard, Cowling, Runcorn, and Stewartson, which appeared in the biographical memoirs of the Royal Society, are reproduced by arrangement with Geoffrey Argent. Photo of Chandrasekhar courtesy AIP Emilio Segrè Visual Archives. The help of the editor, Sergei Molokov, is gratefully acknowledged. Stanislav Braginsky offered advice on an earlier draft.

5 Supplementary notes

1. Faraday M (1832) Experimental researches in electricity, Second series, Philos Trans R Soc Lond 122:163–194
2. Hartmann J (1937) Hg-Dynamics I. Theory of the laminar flow of an electrically conductive liquid in a homogeneous magnetic field. Det Kgl Danske Vid Sels Mat-Fys Medd XV(6):1–27
   Hartmann J, Lazarus F (1937) Hg-Dynamics II. Experimental investigations on the flow of mercury in a homogeneous magnetic field. Det Kgl Danske Vid Sels Mat-Fys Medd XV(7):1–45
3. Alfvén H (1942) Existence of electromagnetic-hydrodynamic waves. Nature 150:405–406
   Alfvén H (1943) On the existence of electromagnetic-hydrodynamic waves. Arkiv f Mat Astron o Fys 29B(2):1–7
4. Ferraro VCA (1937) Non-uniform rotation of the sun and its magnetic field. Mon Not R Astr Soc 47:458–472
5. Lundquist S (1952) Studies in magneto-hydrodynamics. Arkiv f Fysik 5:297–347
6. Lundquist S (1949) Experimental investigations of magneto-hydrodynamic waves. Phys Rev 76:1805–1809
7. Elsasser W (1939) On the origin of the Earth's magnetic field. Phys Rev 55:489–498

8. Elsasser W (1955) Hydromagnetism. I. A Review. Am J Phys 23:590–609
    Elsasser W (1955) Hydromagnetism. II. A Review. Am J Phys 24:85–110
9. Elsasser W (1950) The hydromagnetic equations. Phys Rev 79:183
10. Elsasser W (1946) Induction effects in terrestrial magnetism. Part II. The secular
    variation. Phys Rev 70:202–212
11. For example see
    Glatzmaier GA, Roberts PH (1997) Simulating the geodynamo. Contemp Phys
    38:269–288.
    On p. 114 of
    Braginsky SI (1991) Towards a realistic theory of the geodynamo. Geo-
    phys Astrophys Fluid Dynam 60:89–134,
    Stanislav points out that an important question needs addressing: the stability
    of MAC waves (see §3). Stanislav speculates that, if the set of developed MAC
    waves in the core is unstable, MAC wave turbulence might result. He points
    out that this could well be the main mechanism that transports large-scale heat
    and light admixture in the core and might also create a nonlocal $\alpha-$effect that
    reduces $\mathcal{B}_T/\mathcal{B}_P$
12. Cowling TG (1933) The magnetic field of sunspots. Mon Not R Astr Soc
    94:39–48
13. Chandra's picture (Fig. 5) does not do him justice; he comes across as a some-
    what haughty individual, quick to take offense at anything he could interpret as
    a criticism. I invariably found him supportive. He had a nice sense of humour,
    which I can illustrate by an amusing incident involving his colleague, Gerard
    Kuiper. One of Kuiper's ideas had featured in the national press and his young
    son rushed round to Chandra's office and demanded Chandra's opinion about it.
    "Well", admitted Chandra, "I' m not really an astronomer. . . " "Yes! yes!" inter-
    rupted Kuiper' s son, "That's what my father says. . . ". Chandra used this to
    tease Kuiper unmercifully whenever the occasion arose. Chandra also had a fund
    of amusing anecdotes, most of which unfortunately I have forgotten. I liked one
    particularly, as it nicely illustrates the pecking order of mathematics. Appar-
    ently the famous algebraist, Hermite, was a staunch conservative who sided with
    the government in the infamous treatment of Dreyfus that rocked France in the
    late nineteenth century. Hermite's favourite student, Hademard, was a liberal
    who sided with the victim. When Hermite heard of this he broke into a furi-
    ous denunciation of Hademard finishing with the most damning criticism of all:
    "And I hear that he has now taken up analysis!"
    My time as a research associate with Chandra (1954–1955) was refreshing,
    rewarding and highly educational, a revelation in fact. Chandra was ultra hard-
    working; his students claimed the initial 'S' of his first name stood for 'Super-
    man'. He even had us working on the morning of Christmas Day, 1954. He also
    set himself high standards: he told me once that he disliked writing letters of
    reference because it was so hard to be objective especially when he liked the
    person concerned; at least 5 drafts were necessary, he said. (But that was noth-
    ing new: he claimed that all his scientific work went through 5 drafts too.) His
    fairness shone through in his biography of Arthur Eddington:
    Chandrasekhar S (1983) Eddington, the most distinguished astrophysicist of his
    time. Cambridge University Press, Cambridge.
    Who, reading this monograph, would suspect that this man had treated Chandra
    in a rough and humiliating way? At various times he had encountered colour

prejudice even amongst fellow scientists but he did not show pain or anger:
"Goodness gracious", he would say, "fancy that!" One could sometimes sense his
displeasure, however, by pejorative remarks he made about their understanding
of scientific issues.

Chandra could be stubborn. I believed (and still believe) that Eddington's
famously imaginative explanation of Cepheid variability did not come "out of
the blue", but was inspired by Rayleigh's explanation of Rijke's tube a few years
earlier; see p. 231 of

Strutt JW (1896) The theory of sound, vol 2. 2nd Edition. Cambridge University
Press, Cambridge [Dover edition, 1945]

Chandra rejected my speculation "out of hand". He was also strongly opposed
to asymptotic methods, which he seemed to feel were mathematically "dirty".
Perhaps he heeded too much Abel's famous dictum: "Divergent series are the
inventions of the devil, and it is shameful to base on them any demonstration
whatever". In such supping as he did with the devil, he used a very long spoon!
Three examples come to mind from his book

Chandrasekhar S (1962) Hydrodynamic and hydromagnetic stability. Oxford
University Press, Oxford

On page 104, he discusses his $Ra \propto T^{2/3}$ law; on page 177, he does the same for
his $Ra \propto Q$ law; and in §32 he examines rotating convection for zero Prandtl
number Pr. If he had been willing to entertain asymptotic methods, he would
have quickly discovered why the first term in the $Ra \propto T^{2/3}$ law performs so
badly for no slip boundaries: the second term in the asymptotic expansion for
$T \to \infty$ is $O(T^{7/12})$, i.e., it is almost as big as the first! See

Roberts PH (1965) On the thermal instability of a highly rotating fluid sphere.
Astrophys J 141:240–250

The strange result for $Pr = 0$ has been elucidated by

Zhang K, Roberts PH (1997) Thermal inertial waves in a rotating fluid layer:
exact and asymptotic solutions. Phys Fluids 9:1980–1987

The corresponding MHD problem is addressed in

Roberts PH, Zhang K (2000) Thermal generation of Alfvén waves in oscillatory
magnetoconvection. J Fluid Mech 420:201–223

14. Merrill RT, McElhinny MW, McFadden PL (1983) The magnetic field of the
    Earth. Academic Press, San Diego

    On p. 17 it is reported that, in 1905 shortly after writing his special relativity
    paper, Einstein described the problem of the origin of the Earth's magnetic field
    as being one of the most important unsolved problems in physics.

15. Herzenberg A (1958) Geomagnetic dynamos. Philos Trans R Soc Lond
    A250:543–585

16. Bullard EC, Gellman H (1954) Homogeneous dynamos and terrestrial mag-
    netism. Philos Trans R Soc Lond A247:213–278

17. Lowes FJ, Runcorn SK (1951) The analysis of the geomagnetic secular variation.
    Philos Trans R Soc Lond A243:526–546

18. I had unwittingly stumbled onto the identical idea that had led Alfvén to the
    discovery [3] of his waves:

    Alfvén H (1943) On sunspots and the solar cycle. Arkiv f Mat Astron o Fys
    29A(12):1–17.

    He was convinced that sunspots were produced by magnetic activity deep within
    the Sun, and discovered his waves when seeking an agency that would carry this

activity to the surface; see also [41] below. In this paper he also rediscovered Ferraro's law of isorotation [4].

19. Runcorn SK (1953) The Earth's core. Trans Am Geophys Un 35:49–63
20. I provide here an example of the old adage that academics can hold a grudge longer than any other segment of society. I did (after I'd safely got my Ph.D. of course) eventually tell Keith how upset I was. He was always an easy man to talk to, and he took it well, promising that he'd acknowledge my priority the next time he wrote on the subject so I later accepted a position in his department at Newcastle. I did not discover how Runcorn honoured his promise until 1994, when I saw p. 485 of
    Runcorn SK (1990) Geophysical observations over the last few millenia and their implications. In: Stephenson FR, Wolfendale AW (eds) Secular, Solar and Geomagnetic Variations in the Last 10,000 Years. Kluwer, Dordrecht, The Netherlands: 478–488
    I told Keith, when I met him next at the 1994 AGU meeting in San Francisco, that I was not satisfied, and he proposed that we discuss matters over luncheon. There, he admitted that the notion that magnetic disturbances could reach the core surface from deep within the core was not his, but claimed that it had made him realise that this also meant that velocity disturbances could do the same, and that he deserved credit for that. (Apparently one Siamese twin can fail to notice the existence of the other!) But I am being boring. "A storm in a teacup", you'll say. I should pay heed to the advice of the Greek sage, Chilo, "De mortuis nil nisi bonum." It's a bit late for the 'nil' but I can finish with a 'bonum'! I did once see Runcorn perform a completely disinterested act: after our luncheon, he gave a panhandler a dollar bill. If I'd have been sitting in a chair at the time, I would have fallen off it!
21. Hide R, Roberts PH (1960) Hydromagnetic flow due to an oscillating plane. Rev Mod Phys 32:799–806
    Hide R, Roberts PH (1961) The origin of the main geomagnetic field. Phys and Chem of the Earth 4:25–98
    I believe Hide was the first person to appreciate the significance of the tangent cylinder, the special geostrophic cylinder (§3) that touches the inner core at its equator:
    Hide R (1953) Some experiments on thermal convection in a rotating fluid. Dissertation, University of Cambridge, UK;
    reported in
    Hide R, Roberts PH (1962) Some elementary problems in magneto-hydrodynamics. Adv Appl Mech 4:215–316
22. I got vicarious pleasure from detecting a vitiating error in a paper by my first research supervisor; see
    Stewartson K, Roberts PH (1963) On the motion of a liquid in a spheroidal cavity of a precessing rigid body. J Fluid Mech 17:1–20
23. Proceedings of the Symposium on Magneto-Fluid Mechanics. Rev Mod Phys 32:693–1032 (1960)
24. Roberts PH (1967) Singularities of Hartmann layers. Proc R Soc Lond A300:90–107
25. Roberts PH, Scott S (1965) On analysis of the secular variation. 1. A hydromagnetic constraint: Theory. J Geomagn Geoelectr 17:137–151

Love JJ (1999) A critique of frozen-flux inverse modelling of a nearly stationary geodynamo. Geophys J Intern 138:353–365

26. Braginsky SI (1964) Self-excitation of a magnetic field during the motion of a highly conducting fluid. J Exptl Theoret Phys CCCP 47:1084–1098 [translated in Sov Phys JETP 20:726–735 (1965)]

27. Soward AM (1972) A kinematic theory of large magnetic Reynolds number dynamos. Philos Trans R Soc Lond A272:431–462

28. Parker EN (1955) Hydromagnetic dynamo models. Astrophys J 122:293–314. His use of Γ ('G' for generation!) has more to recommend it than α, but the latter description has stuck

29. Many of the early papers written by the East German group have appeared in translation:
    Stix M, Roberts PH (1971) The turbulent dynamo. NCAR Technical Report IA-60.
    This report has recently become available electronically. See also
    Krause F, Rädler K-H (1980) Mean-field magnetohydrodynamics and dynamo theory. Pergamon Press, Oxford

30. Parker EN (1957) The solar hydromagnetic dynamo. Proc Nat Acad Sci USA 43: 8–14

31. Braginsky SI (1964) Kinematic models of the Earth's hydromagnetic dynamo. Geomag Aeron 4:732–747 [translated in Geomag Aeron, 4:572–583 (1964)]
    Amongst all the kinematics in this paper is a thread of dynamics: an equation of motion for the inner core's rotation under the em couple to which the outer core subjects it. This topic became popular again 32 years later with the prediction by Gary and myself that the inner core would rotate eastward:
    Glatzmaier GA, Roberts PH (1996) Rotation and magnetism of Earth's inner core. Science 274:1887–1891
    The accuracy of Braginsky's numerical integrations was confirmed in
    Roberts PH (1972) Kinematic dynamo models. Philos Trans R Soc Lond A272:663–703

32. Braginsky SI (1964) Theory of the hydromagnetic dynamo. J Exptl Theoret Phys CCCP 47:2178–2193 [translated in Sov Phys JETP 20:1462–1471 (1965)]

33. Braginsky SI (1964) Magnetohydrodynamics of the Earth's core. Geomag Aeron 4: 898–916 [translated in Geomag Aeron 4:698–712 (1964)]

34. Braginsky SI, Meytlis VP (1990) Local turbulence in the Earth's core. Geophys Astrophys Fluid Dynam 55:71–87

35. Braginsky SI, Roberts PH (1995) Equations governing Earth's core and the geodynamo. Geophys Astrophys Fluid Dynam 79:1–97
    Braginsky SI, Roberts PH (2003) On the theory of convection in the Earth's core. In: Ferriz-Mas A, Núñez M (eds) Advances in Nonlinear Dynamos. Taylor and Francis, London: 60–82

36. Braginsky SI (1963) Structure of the F layer and reasons for convection in earth's core. Dokl Akad Nauk SSSR 149:1311–1314 [translated in Sov Phys Dokl 149:8–10 (1963)]

37. Jacobs JA (1953) The Earth's inner core. Nature 172:297–300

38. Verhoogen J (1961) Heat balance of the earth's core. Geophys J R Astron Soc 4:276–281

39. Backus GE (1975) Gross thermodynamics of heat engines in the deep interior of the Earth. Proc Nat Acad Sci USA 72:1555–1558

Hewitt JM, McKenzie DP, Weiss NO (1975) Dissipative heating in convective flows. J Fluid Mech 68:721–738

As far as I know, the first person to draw attention to the issue of Carnot efficiency in compressible convection was Bullard in

Bullard EC (1949) The magnetic field within the Earth. Proc R Soc Lond A197:433–453

apart, that is, for the Eddington's theory of Cepheid pulsation mentioned above

40. Labrosse S (2003) Thermal and magnetic evolution of Earth's core. Phys Earth Planet Interiors 140:127–143

41. Walén C (1944) On the theory of sunspots. Arkiv f Mat Astron o Fys 30A(15):1–87

Walén C (1944) On the theory of sunspots. Arkiv f Mat Astron o Fys 31B(3):1–3

42. Lehnert B (1954) Magnetohydrodynamic waves under the action of the Coriolis force. Part I. Astrophys J 119:647–654

For similar dispersive effects of Coriolis forces on compact disturbances in the Earth's core, see

St Pierre MG (1996) On the local nature of turbulence in Earth's outer core. Geophys Astrophys Fluid Dynam 83:293–306

Siso-Nadal F, Davidson PA (2004) Anisotropic evolution of small isolated vortices within the core of the Earth. Phys Fluids 16:1242–1254

43. Braginsky SI (1967) Magnetic waves in the Earth's core. Geomag Aeron 7:1050–1060 [translated in Geomag Aeron 7:851–859 (1967)]

44. Hide R (1966) Free hydromagnetic oscillations of the Earth's core and the theory of the geomagnetic secular variation. Philos Trans R Soc Lond A259:615–650

45. Braginsky SI (1970) Torsional magnetohydrodynamic vibrations in the Earth's core and variations in day length. Geomag Aeron 10:3–12 [translated in Geomag Aeron 10:1–8 (1970)]

Using the more recent estimate of the electrical conductivity, Stanislav reconsidered in

Braginsky SI (1970) Oscillation spectrum of the hydromagnetic dynamo of the Earth. Geomag Aeron 10:221–233 [translated in Geomag Aeron 10:172–181 (1970)],

the consequences of the 1964 paper and of the 1967 and 1970 papers cited above. He summarises what is known about the secular variation spectrum up to periods of 10^4 years, and he expresses the hope that this spectrum will become as useful in determining the structure of the geodynamo as optical spectra are in elucidating the atom!

46. Jault D, Gire G, Le Mouël J-L (1988) Westward drift, core motions and exchanges of angular momentum between core and mantle. Nature 333:353–356

47. Braginsky SI (1993) MAC-oscillations of the hidden ocean of the core. J Geomagn Geoelectr 45:1517–1538

48. Braginsky SI (1975) Nearly axially symmetric model of the hydromagnetic dynamo of the Earth, I. Geomag Aeron 15:149–156 [translated in Geomag Aeron 15:122–128 (1975)]

Braginsky SI (1978) Nearly axially symmetric model of the hydromagnetic dynamo of the Earth. Geomag Aeron 18:340–351 [translated in Geomag Aeron 18:225–231 (1978)]

49. Childress S, Soward AM (1972) Convection-driven hydromagnetic dynamo. Phys Rev Lett 29:837–839

Soward AM (1974) A convection-driven dynamo I. Weak-field case. Philos Trans R Soc Lond A275:611–651

Fautrelle Y, Childress S (1982) Convective dynamos with intermediate and strong fields. Geophys Astrophys Fluid Dynam 22:235–279

50. Jones CA, Roberts PH (2000) Convection-driven dynamos in a rotating plane layer. J Fluid Mech 404:311–343

Rotvig J, Jones CA (2002) Rotating convection-driven dynamos at low Ekman number. Phys Rev E66:056308

Early Magnetohydrodynamic Research in Stockholm

Bo Lehnert

Alfvén Laboratory, Royal Institute of Technology, SE-100 44 Stockholm, Sweden
(bo.lehnert@alfvenlab.kth.se)

1 Introduction

Electric discharges in gases were subject to a rather limited research effort during the first three decades of the twentieth century. There was, however, a turning point when Hannes Alfvén [1,2] started his studies on electromagnetic phenomena in astrophysics at the end of the 1930s, and especially after his discovery of magnetohydrodynamic waves in 1942.

Ever since Alfvén received his professorship in 1940 at the Royal Institute of Technology in Stockholm, he continued an intensified research with his collaborators on the behavior of electrically conducting media in a magnetic field. In his spirit the research was whenever possible carried out in close connection between theory and experiments. Two essential features originating from Hannes Alfvén's earlier results formed the basis of the phenomena to be investigated, namely the concept of "frozen-in" magnetic field lines in a highly conducting medium, and the magnetohydrodynamic waves where these field lines were pictured as elastic strings in a dynamic process. During this early period of magnetohydrodynamic (MHD) research in Stockholm to be described here, the studies were mainly concentrated on MHD flow, the characteristic dimensionless numbers of MHD phenomena, MHD waves, heat convection in a magnetized electrically conducting layer, and the idea of a self-exciting dynamo. In addition to this, there were important investigations at the laboratory on gaseous discharges with application to the aurora, and on motions of charged particles in electromagnetic fields. In the 1950s Alfvén's department consisted of about 20 persons out of whom about a half were active in research and the rest in technical and administrative duties. The laboratory had a small workshop, and also a large magnet for generating a steady almost homogeneous magnetic field with a strength up to 1 T within a volume of about 0.01 m^3 (Fig. 1).

S. Molokov et al. (eds.), Magnetohydrodynamics – Historical Evolution and Trends,
27–36. © 2007 *Springer.*

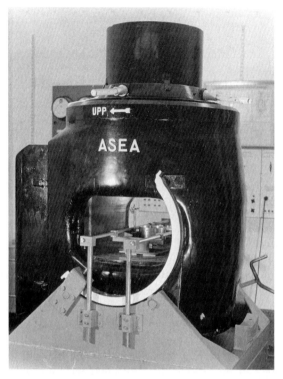

Fig. 1. The electromagnet at Alfvén's Department of Electronics at the Royal Institute of Technology in Stockholm. An almost homogeneous magnetic field up to about 1 T can be generated in the gap. The experimental arrangement is placed on a fixed stand. The magnet can be turned around a horizontal axis, such as to generate a magnetic field which forms any angle with the vertical direction

2 Magnetohydrodynamic flow

The magnet was used in simple qualitative demonstrations of MHD phenomena. In a glass vessel filled with mercury, a horseshoe-formed piece of copper was dipped into the liquid and moved across the magnetic field. Polarized electric currents were then generated which coupled the motion to a large part of the mercury body, thus demonstrating the occurrence of a strong MHD drag.

An investigation of pressure-driven MHD channel flow was first reported by Hartmann and Lazarus in 1937 [3,4]. Later the author [5] studied theoretically an equivalent case of Couette-like flow of a conducting viscous liquid placed between two parallel walls moving in opposite directions, and where there is a perpendicular magnetic field. In both cases the magnetic field and the resulting induced electric currents where shown to have a strong influence on the velocity profile, by forming boundary layers with strong velocity gradients and an enhanced drag force.

In a simple experiment with mercury in the spacing between two rotating concentric cylinders [5] the mutual torque was measured in presence of an axial magnetic field. There were drag effects both from the mid-parts and from the end regions of the cylinders. The qualitative conclusion was that, at least in certain cases, the magnetic field suppresses the turbulence which occurs at sufficiently large angular velocities.

With the main purpose of finding out whether there may also exist situations in which an imposed magnetic field can give rise to non-laminar motion in the form of regular vortices or even turbulence, an experiment was designed [6] as shown in Fig. 2. In a shallow tray filled with mercury are placed a copper disc A and two concentric copper rings C and D. A and D are at rest, whereas C rotates round a vertical axis at a constant rate of one fifth revolution per second. In a stationary unmagnetized state the whole mass of liquid rotates slowly due to viscous coupling between neighboring layers. This is shown in Fig. 3a where the motion of the surface has been made visible with grains of sand, illuminated from the side and photographed with an exposure time of 1/5 s. Figure 3b further shows the image of a grating as seen in the mercury surface which is used as a mirror. Deformations of the surface due to a fluid motion would give rise to corresponding local deformations of the image. Thus, Fig. 3a indicates a slow and diffuse motion due to the viscous drag, and Fig. 3b shows that the surface is practically plane in the absence of a magnetic field. This situation is radically changed when applying a strong vertical magnetic field of about 0.43 T, as demonstrated by Figs. 3c and d. Thus Fig. 3c shows that the fluid is kept at rest in the layers above the immovable disc A and ring D, whereas it rotates at full speed in the layers above the rotating ring C. This is a clear demonstration of frozen-in magnetic field lines. It also supports Ferraro's isorotation law [7], according to which the layers of a mag-

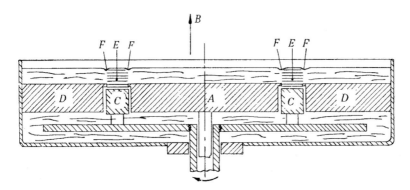

Fig. 2. Apparatus demonstrating the frozen-in magnetic effect, as well as an instability caused by a magnetic field, in the flow of a mercury layer. The copper ring C rotates round the vertical axis, whereas the disc A and the ring D are at rest, in a tray filled with mercury. The mean diameter of C is 7 cm, and the mercury layer has a thickness of 6 mm. (From [6].)

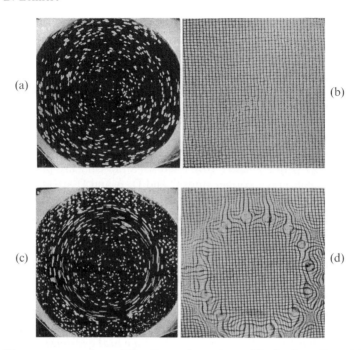

Fig. 3. The mercury surface of Fig. 2 seen from above. The basic motion at the free surface has been indicated with grains of sand, as seen in (a) and (c). The deformation of the surface is revealed by the image of a grating, as seen when the surface is used as a mirror in (b) and (d). The magnetic field is vertical, having zero strength in (a) and (b), and the strength 0.43 T in (c) and (d). (From [6].)

netized and ionized body will be forced to rotate at the same angular speed at every point on a magnetic field line. From Fig. 3d it is further seen that vortex streets are formed above both the inner and outer edges of the ring C. Consequently this presents an example in which an imposed magnetic field can give rise to a vortex motion. In the present case the frozen-in condition enforces the velocity gradient of the basic flow to become large above the edges of the ring C, and this leads in its turn to an instability driven by the velocity gradients.

The final period of decay of homogenous magneto-turbulence in an externally imposed homogenous magnetic field was studied theoretically by the author [8]. It was found to develop pronounced anisotropic properties where turbulence elements with finite wave numbers in the direction of the field are strongly damped, as also being observed. For a fluid of mercury, the decay has a time dependence $\exp(-t/\tau)$ with the decay time τ given by

$$1/\tau = 2\left[\nu k^2 + k_z^2 V^2/k^2(\lambda - \nu)\right]. \tag{1}$$

Here k is the total wave number, k_z its component in the direction of the magnetic field, V is the Alfvén velocity, ν the kinematic viscosity, $\lambda = 1/(\mu\sigma)$, σ the electrical conductivity, and μ the absolute magnetic permeability. An introduction of an angular velocity of rotation, inclined to the magnetic field, further destroys the symmetry and modifies the damping effects. The influence of the magnetic field on the damping can in certain situations be counteracted by the Coriolis force.

3 Characteristic dimensionless numbers

The basic MHD equations can be written in a dimensionless form where every quantity $Q = Q_c Q'$ is replaced by its characteristic value Q_c and the corresponding dimensionless variable Q'. In this way a system of equations for Q is obtained which includes characteristic dimensionless numbers as coefficients. The latter then consist of combinations of the various characteristic values Q_c. With each of the characteristic dimensionless numbers being fixed, a solution of the equations then generates an entire class of geometrically similar configurations in space and time. These numbers were found to have the forms [5], [9]:

$$R_1 = v_c/V, \qquad R_2 = v_c L_c/\lambda, \qquad R_3 = v_c L_c/\nu, \tag{2}$$

$$S_1 = p_c/\rho v_c^2, \qquad S_2 = \phi_c/\rho v_c^2, \tag{3}$$

$$S_3 = v_c L_c/\kappa, \qquad S_4 = v_c^2/c_v T_c, \tag{4}$$

where $V = B_c/(\mu\rho)^{1/2}$ is the Alfvén velocity with ρ as the mass density, κ is the thermometric conductivity, c_v is the specific heat at constant volume, and $B_c, v_c, L_c, p_c, \phi_c, T_c$ are the characteristic values of the magnetic field strength, fluid velocity, length scale, fluid pressure, gravitation potential, and temperature, respectively. Of these numbers the new first deduced ones [5] were the Alfvén Mach number R_1 and the magnetic Reynolds number R_2, whereas the conventional Reynolds number R_3 and the rest are earlier known parameters. A combination of R_1 and R_2 leads to the Lundquist number [10]

$$R_2/R_1 = B_c L_c \sigma(\mu/\rho)^{1/2} \equiv Lu, \tag{5}$$

the magnitude of which in many cases can be taken as a measure of the strength of a MHD phenomenon.

4 Magnetohydrodynamics waves

Eight years after Alfvén's discovery of the transverse MHD mode [11], Nicolai Herlofson showed that there also exists a longitudinal mode [12] in a compressive fluid conductor. This mode is associated with a compression of the frozen-in magnetic field.

Experiments with the purpose of confirming the existence of the transverse Alfvén mode were first undertaken about 7 years after its discovery. Lundquist [13] planned and performed an investigation in which a cylindrical column of mercury was placed in a strong axial magnetic field generated by the magnet in Fig. 1. Torsional oscillations were imposed at the lower end of the column by a motor-driven disc with paddle-wheels. The amplitude and phase of the oscillations were measured at the upper free surface of the column. In this way it was shown that the whole column behaved like an elastic body for which the torsional force and motion at the bottom were transferred along the column to its top, thereby revealing a phase-shift between its ends.

The conditions for the existence of Alfvén waves in a dissipative conducting liquid can be estimated from a simple theoretical model [9]. In a frame (x, y, z) the liquid is assumed to be situated between two parallel infinitely conducting planes with the spacing d. The z-axis is chosen to be perpendicular to the planes, and a magnetic field of the strength B_0 is imposed along the same axis. A transverse MHD wave with the wave number $k = 2\pi/d$ and a perturbed magnetic field component $\mathbf{b} = (b, 0, 0)$ is then introduced into the basic MHD equations. The decay of the standing-wave perturbation is then given by

$$b(z, t) = \left\{ b_1 \exp\left[i(\lambda - \nu)k^2(\xi^2 - 1)^{1/2}t \right] \right.$$
$$\left. + b_2 \exp\left[-i(\lambda - \nu)k^2(\xi^2 - 1)^{1/2}t \right] \right\}$$
$$\times \exp\left[-(\lambda + \nu)k^2 t \right] \sin kz, \tag{6}$$

where

$$\xi = 2B_0 / \left[(\mu\rho)^{1/2} k(\lambda - \nu) \right] \tag{7}$$

with $\lambda > \nu$ and b_1 and b_2 being constants. Consequently, there is a range $\xi > 1$ for damped wave motion, and another range $\xi \leq 1$ for aperiodic motion without wave phenomena. The critical periodic limit $\xi = 1$ then corresponds to a Lundquist number $Lu = \pi$ when $\lambda \gg \nu$.

With linear dimensions of $L_c \cong 0.1$m and a magnetic field strength $B_c \cong 1$ T, the corresponding Lundquist numbers become $Lu \cong 1$ and $Lu \cong 38$ for mercury and liquid sodium, respectively. MHD wave motion under laboratory conditions is therefore hardly realizable with mercury, whereas strongly damped waves can become available with liquid sodium, due to its larger conductivity and lower density. An experiment with the latter medium was therefore performed some years later by the author [14]. The set-up was in principle the same as that used by Lundquist. The oscillations were excited by a motor-driven copper disc at the bottom of the column, having a smooth surface with good electrical contact with the sodium. The amplitude and phase-shift of the recorded signal U were measured by an electric probe at the top free surface. The level of the sodium was adjusted by means of a system of pipes, and in a nitrogen atmosphere. The measurements were made at a fixed torsional frequency of 30 cycles per second, with a magnetic field strength varying in the range from 0.3 to 1.0 T. The results were in satisfactory agreement with theory. Thus, at increasing magnetic field strength they showed a weakly developed resonance peak of the amplitude (Fig. 4a), and a phase-shift (Fig. 4b), both being reconcilable with deduced rather strongly damped torsional Alfvén waves propagating along the column.

The practical details of the sodium experiment were not without problems. A break in the delivered pipes made sodium leak out, thereby burning and gasifying the asbestos parts of the experimental device and sending smoke up from the basement to the second floor, while the author was supervising the fire, lying on the floor in asbestos clothes. After the experiment had been completed, a contaminated part of about half a liter of sodium was put into a can with stones and was thrown by the author into the Baltic Sea at the suburb Djursholm during a beautiful summer's evening. The burst was like that of a medium–heavy artillery; the windows in the neighboring houses were shaking and the motorboats on sea turned around. Then the police appeared, but was persuaded through a lecture on MHD not to claim for penalty, and was later as thanks sent a reprint of [14].

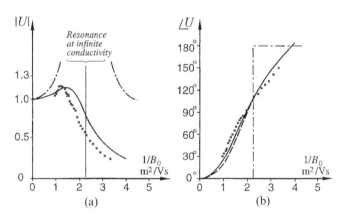

Fig. 4. The experiment on torsional MHD oscillations of a cylindrical column of liquid sodium in an axial magnetic field. Full curves present deduced behavior, and the crossed marks indicate the experimental results. The dot-and-dash curve represents the ideal case of resonance at infinite electrical conductivity. (a) The amplitude at the surface of the column, normalized to its value at infinite magnetic field strength. (b) The phase shift at the surface of the column. (From [14].)

During a stay as visitor of S. Chandrasekhar at Yerkes Observatory in the United States, the author was inspired to look into the behavior of MHD waves in rotating bodies, such as in magnetized stars. It was found that a plane Alfvén wave is split up into two circularly polarized transverse waves, traveling at velocities V_1 and V_2 in the direction of the wave normal and with moduli $|V_1| > |V_z| > |V_2|$, where $V_z = V(\cos\theta)$, V being the Alfvén velocity and θ the angle between the wave normal and the homogeneous magnetic field [15]. In a rotating medium with an immersed magnetic field, an MHD disturbance will then not necessarily travel along the magnetic field lines, as found in a further study of the group velocity [16]. These effects are caused by the Coriolis force, thus modifying the propagation of MHD waves in such cosmical bodies as the sun.

5 Heat convection

In a series of investigations, Chandrasekhar performed detailed rigorous studies on the thermal instability of a layer of fluid heated from below [17]. Among these investigations, there were studies on the inhibition of convection in an electrically conducting fluid in a magnetic field.

Experiments on a mercury layer heated from below and placed in the field of the magnet in Fig. 1 were at an early stage performed by Jirlow [18], who was a student of the author. These investigations were followed by a larger set of experiments reported by the author and Little [19], who was a visiting scientist at the institute. The technique for revealing surface motions of convective cells was the same as that applied in the experiments of Figs. 3a and c. First there were studies in which the layer was placed in a nearly vertical but slightly inhomogeneous magnetic field of varying strength in a direction along the mercury surface. Under appropriate conditions convective cells were then observed within a part of the surface, whereas there were no cells within the rest of the latter. The boundary between convection and no convection was then found to be situated where the inhomogeneous magnetic field had a local value being in agreement with the critical field strength for the onset of convection, as predicted by Chandrasekhar. A second type of experiments was undertaken with the magnet being turned around, to produce a field which was parallel to the mercury surface. Then convective cells were observed as in Fig. 5, being strongly elongated in the magnetic field direction, in agreement with the theory by Chandrasekhar.

6 The MHD dynamo

A cylindrical column of an electrically conducting fluid which is traversed by an electric current and immersed in an axial magnetic field can become subject to a kink instability, by which the column becomes distorted and

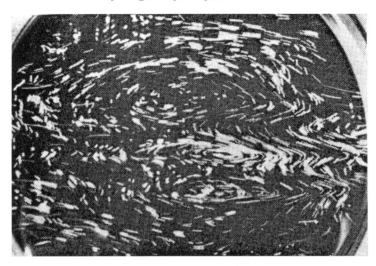

Fig. 5. Cellular convection in a layer of mercury when the magnetic field is parallel to the free surface. The cells are seen to be elongated in the field direction and extend across the entire vessel. (From [19].)

bent into a screw-shaped geometry. This instability was at an early stage proposed by Alfvén [2] to provide a dynamo mechanism for the magnetic field of stars and planets. This idea was further analyzed by Lundquist [20], who was at this stage the first to reformulate the hydrodynamic energy principle for disturbances of an equilibrium state, such as also to apply to MHD.

In the Earth's electrically conducting core a dynamo process is imagined to arise from fluid motions driven by temperature gradients. To demonstrate the dynamo mechanism, a first experimental attempt was made in a vessel containing 58 l of liquid sodium which was set into motion by a rotating disc provided with radial strips [21]. An initial poloidal magnetic field, of strength up to 0.02 T, was generated by a coil under the bottom of the vessel. In the bottom layers of the sodium radial metal strips had been placed, to inhibit a toroidal flow. Possibly occurring magnetic disturbances were recorded by an adjustable magnetic probe inside the vessel. Angular velocities of the disc up to $47s^{-1}$ were imposed on the fluid. The measurements indicated the occurrence of poloidal and toroidal induced magnetic disturbances, but no tendency of instability or of any self-excited dynamo behavior. Not until 42 years later, a similar but far more sophisticated dynamo experiment with an associated theory was successfully conducted at Salaspils near Riga in Latvia by A. Gailitis, O. Lielausis and collaborators [22]. This experiment was performed with 2 tons of liquid sodium. The rotational velocities were large enough to reach magnetic Reynolds numbers R_2 in Eq. (2) of about 10, thus being in the range of sufficiently strong MHD interaction.

References

1. Alfvén H (1950) Cosmical Electrodynamics. Oxford University Press, Oxford
2. Alfvén H, Fälthammar C-G (1958) Electromagnetic Phenomena in Cosmical Physics. Cambridge University Press, Cambridge
3. Hartmann J (1937) Math-Fys Medd 15(6)
4. Hartmann J, Lazarus F (1937) Math-Fys Medd 15(7)
5. Lehnert B (1952) Arkiv f Fysik 5:69
6. Lehnert B (1955) Proc R Soc A 233:299
7. Ferraro VCA (1937) Mon Not R Astr Soc 97:458
8. Lehnert B (1955) Quart Appl Math XII:321
9. Lehnert B (1958) In: Lehnert B (ed) Electromagnetic Phenomena in Cosmical Physics, IAU Symposium No 6. Cambridge University Press, Cambridge: 50
10. Lundquist S (1952) Arkiv f Fysik 5:297
11. Alfvén H (1942) Nature 150:405
12. Herlofson N (1950) Nature 165:1020
13. Lundquist S (1949) Phys Rev 76:1805
14. Lehnert B (1954) Phys Rev 94:815
15. Lehnert B (1954) Astrophys J 119:647
16. Lehnert B (1955) Astrophys J 121:481
17. Chandrasekhar S (1961) Hydrodynamic and Hydromagnetic Stability. Clarendon Press, Oxford
18. Jirlow K (1956) Tellus 8:252
19. Lehnert B, Little NC (1957) Tellus 9:97
20. Lundquist S (1951) Phys Rev 83:307
21. Lehnert B (1957) Arkiv f Fysik 13:109
22. Galitis A, Lielausis O, Dementev S, Platacis E, Cifersons A, Gerbeth G, Gudrum T, Stefani F, Christensen M, Hänel H, Will G (2000) Phys Rev Lett 84:4365

Dynamo Experiments

Agris Gailitis[1], Olgerts Lielausis[1], Gunter Gerbeth[2], and Frank Stefani[2]

[1] Institute of Physics, Univerity of Latvia, LV-2169 Salaspils 1, Latvia
(gailitis@sal.lv)
[2] Forschungszentrum Rossendorf, P.O. Box 510119, D-01314 Dresden, Germany
(F.Stefani@fz-rossendorf.de)

Summary. The long history of laboratory experiments on homogeneous dynamo action is delineated. It is worked out what sort of insight can be expected from experiments, and what not. Special focus is laid on the principle and the main results of the Riga dynamo experiment which is shown to represent a genuine hydromagnetic dynamo with a non-trivial saturation mechanism that relies mainly on the fluidity of the electrically conducting medium.

1 Natural dynamos

Wherever in the cosmos a large body of electrically conducting fluid is found in vigorous motion, there is also a magnetic field around the corner [1].

Geophysicists may consider it a luck to live on a planet whose liquid iron core produces a most interesting magnetic field [2]. It shows polarity reversals at irregular intervals with a reversal rate that varies from nearly zero during the Cretaceous superchron to approximately 5/Myr in the present. Some observations indicate that reversals might consist of a slow field decay and a fast field recreation [3]. There is also some evidence for a correlation of the field strength and the persistence time in one polarity [4], as well as for a possible bimodal behaviour of the dipole moment [5]. Why this is so, is still in question, despite the enormous advancements that dynamo simulations have experienced during the last decade [6].

The magnetic field of the Sun, or more precisely the magnetic field connected with sunspots, was discovered in 1908 by Hale at Mt. Wilsons observatory [7]. Actually, it was this discovery that prompted Larmor to suggest self-excitation as the source of magnetic fields of large astronomical bodies [8]. Still today, the 11-year periodicity of sunspots, their migration towards the equator (the "butterfly diagram"), and the occurrence of grand minima which are superimposed upon the main periodicity are the subject of intensive investigations [9].

However, the magnetic field of our Sun is rather moderate compared with that of other stars. The field of some white dwarfs can easily reach values of

S. Molokov et al. (eds.), *Magnetohydrodynamics – Historical Evolution and Trends*,
37–54. © 2007 *Springer*.

100 T, and even fields of 10^{11} T have been ascribed to some anomalous x-ray emitting pulsars [10].

The large-scale fields of spiral galaxies, which are closely correlated with the optical spiral pattern, have typical values in the order of 10^{-9} T [11]. But magnetic fields can spread far beyond the galaxies that created them. Only recently it was shown that the energy stored in the magnetic field within the "lobes" of free-standing giant radio galaxies can reach huge values. It has been estimated that up to 10% of the gravitational infall energy are being recycled back into the intergalactic medium by the accreting supermassive black holes [12].

It should be pointed out, however, that magnetic fields are not only passive by-products of fluid motion in the cosmos. Quite contrary, they seem to play an active role in cosmic structure formation by virtue of the *magnetorotational instability* (MRI) which is responsible for rendering (hydrodynamically stable) Keplerian flows unstable and ensuring sufficient angular momentum transport in accretion discs [13].

2 What to expect from experiments, and what not

Not one of the quoted natural systems can be put into a *Bonsai form* to be studied in laboratory. Taking the Earth dynamo as a striking example, it is not possible to actualize all of the dimensionless numbers in an equivalent experimental set-up. The Ekman number – the ratio of viscous forces to Coriolis forces – of the earth outer core is approximately 10^{-15}. A liquid sodium experiment of 1 m radius would have to rotate with 10^8 (!) rotations per second in order to reach this number.

So what, then, can we learn from dynamo experiments?

First of all, it is certainly legitimate to verify experimentally that homogeneous dynamos work at all. Theoretically, kinematic dynamo action has been proved for a large variety of velocity fields or some related turbulence parameter. But still there were open questions, in particular regarding the role of turbulence. Kinematic dynamos are governed by the induction equation for the magnetic field $\mathbf{B}$,

$$\frac{\partial \mathbf{B}}{\partial t} = \nabla \times (\mathbf{v} \times \mathbf{B}) + \frac{1}{\mu_0 \sigma} \Delta \mathbf{B} \quad \text{with} \quad \nabla \cdot \mathbf{B} = 0,$$

under the assumption of a given velocity field $\mathbf{v}$. The constants μ_0 and σ denote the magnetic permeability of the free space and the conductivity of the liquid. The evolution of the magnetic field in Eq. (1) is controlled by the relative importance of diffusion and advection. For zero velocity the magnetic field will decay within a typical time $t_d = \mu_0 \sigma l^2$, with l being a typical length scale of the system. On the other hand, the advection can lead to an increase of $\mathbf{B}$ within a kinematic time $t_k = l/v$. The ratio of the two timescales is the famous magnetic Reynolds number, $R_m = \mu_0 \sigma l v$, that rules the evolution of the magnetic field. Depending on the flow pattern, typical values of the critical R_m are in the range of 10^1–10^3. For the best liquid metal conductor, sodium, the product of conductivity and magnetic permeability is approximately 10 m^2/s. Thus, to get an R_m of 100, the product of length and velocity has to reach 10 m^2/s. It is this large value, in combination with the technical and safety problems in handling sodium, that makes hydromagnetic dynamo experiments so costly.

Assume now that a dynamo works and a magnetic field starts to self-excite. How can the exponential field growth be stopped, how does the dynamo saturate? To answer this *second* question, we have to abandon purely kinematic theory and consider dynamically consistent dynamo models which include the back-reaction of the magnetic field on the flow field. The details of this mechanism certainly depend on the concrete form of the flow, in particular on the degrees of freedom that the moving medium has. Dynamically consistent dynamos have to satisfy, in addition to the induction equation for the magnetic field $\mathbf{B}$, the Navier–Stokes equation for the velocity $\mathbf{v}$,

$$\frac{\partial \mathbf{v}}{\partial t} + (\mathbf{v} \cdot \nabla)\mathbf{v} = -\frac{\nabla p}{\rho} + \frac{1}{\mu_0 \rho}(\nabla \times \mathbf{B}) \times \mathbf{B} + \nu \Delta \mathbf{v} + \mathbf{f}_{\text{drive}} ,$$

where ρ and ν denote the density and the kinematic viscosity of the fluid, μ_0 is the permeability of the vacuum, and $\mathbf{f}_{\text{drive}}$ symbolizes a driving force.

Third, besides its influence on the large-scale flow, the magnetic field back-reaction may also change the turbulence structure of the flow. Sometimes this effect is considered the most important one that dynamo experiments may help to understand, as they provide an interesting test-case for MHD turbulence models. Those models, once validated, could gain reliability when applied to such hard problems as magnetic field generation in the Earth's core, say. But "turbulence model validation" sounds much easier than it is in reality. Even the simple flow measurement in liquid sodium is a problem in its own right, let alone the measurements of all sorts of correlation functions which might be important for the validation of turbulence models.

Intimately connected with the issue of turbulence modification is a *fourth* topic related to the destabilizing role magnetic fields can have on flows. Typically dynamo experiments and experiments on the magnetorotational instability are of similar size, making it worth to hunt for new instabilities in the presence of (self-excited or externally applied) magnetic fields.

A *fifth* issue that could possibly be addressed by dynamo experiments has to do with the distinction between steady, oscillatory, and reversing dynamos. We will come back to this point in the conclusions.

It seems worth to mention a *sixth* issue which we would like to coin *inverse dynamo theory*. In the sense of determining the tangential velocity at the core-mantle-boundary (using the frozen flux approximation) inverse dynamo theory has a long tradition in geomagnetism going back to the early work of Roberts and Scott [14]. In the framework of dynamo experiments, however, there is a wider variety of problems in connection with optimizing flows to give a minimal critical R_m, to tailor dynamos in order to show particular spectral features, or to infer flow features from measured magnetic field information.

3 Laboratory dynamos: an overview

In this section we compile some material about past and present dynamo-related experiments. Due to the space limitations, this overview is bound to be sketchy. A more comprehensive overview may be found in [15].

3.1 Lehnert's Experiment

A liquid sodium experiment was reported by Lehnert as early as 1958 [16]. His set-up can be considered as the prototype of a number of dynamo-related experiments which will be considered later. The flow is produced by motor-driven disk, partly attached with radial strips, rotating in a 0.4 m diameter vessel containing 58 l of liquid sodium. Lehnert observed the conversion of an applied poloidal magnetic field component into a toroidal field (nowadays known as the Ω-effect), which is an important ingredient of the dynamo process, but also the (quadratic in R_m) induction of an additional poloidal field.

3.2 The first homogeneous dynamos: the experiments of Lowes and Wilkinson

At the University of Newcastle upon Tyne, Lowes, and Wilkinson had carried out a long-term series of homogeneous dynamo experiments [17, 18], a summary of which can be found in [19]. These experiments were inspired by the pioneering work of Herzenberg [20], who had given, under mild assumptions, the first existence proof for a homogeneous dynamo consisting of two rapidly rotating, and well-separated, small spheres embedded in a large sphere of the same conductivity (cf. Fig. 1a). His steady solution, which results for an angle

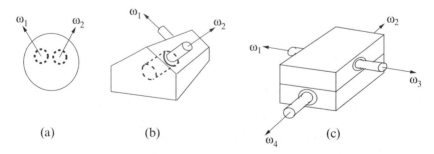

(a)   (b)   (c)

Fig. 1. The principle of the Herzenberg dynamo (a) and two of the experiments of Lowes and Wilkinson (b, c)

φ between the spin axes satisfying $90° < \varphi < 180°$, was later accomplished by on oscillatory solution for $0° < \varphi < 90°$ by Brandenburg et al. [21].

Interestingly, Herzenberg-type dynamos possess no helicity h ($h = \mathbf{v}\cdot\nabla\times\mathbf{v}$) at all. They work as dynamos because each rotor winds up a local poloidal field into a toroidal field that diffuses away and serves, in turn, as a local poloidal field for the other rotor.

In the first homogeneous dynamo Herzenberg's rotating spheres were replaced by two cylinders (each of 7 cm diameter) spinning around non-parallel axes (with 8 cm distance) in a "house-shaped" surrounding conductor (see Fig. 1b). Mercury was used as a conducting lubricator between the cylinders and the bulk [17]. After having inhibited an unfavorable current system by non-conducting layers at the top of the rotating cylinders, a steady magnetic field self-excited at rotation rates of around 1,800 rpm.

The sequence of five experiments carried out by Lowes and Wilkinson is very instructive, not only for their step-by-step improvements, but also for the continuing comparison of the resulting field features with those of the geomagnetic field [19]. Starting with a simple geometry of the rotating cylinders, which produced steady and oscillating magnetic fields, the design was made more sophisticated so that the fifth dynamo (Fig. 1c) finally permitted the observation of field reversals. In that way it was shown that a complex field structure and behaviour can be produced with comparatively simple patterns of motion.

It should be said, however, that a key point for the success of these experiments was the use of ferromagnetic materials (perminvar, mild steel) making the magnetic Reynolds number large, simply by a high relative magnetic permeability μ_r (between 150 and 250). One attempt (the third dynamo), to get self-excitation with rotating non-magnetic copper cylinders failed [19]. And, although being homogeneous, these dynamos were extremely *stiff* in that there was no freedom of the moving conductor to be deformed by the Lorentz forces. Later we will compare this stiffness with the more flexible saturation behaviour in the Riga and Karlsruhe experiments.

3.3 The "α-box"

Also in the 1960s, the concepts of mean-field dynamos and the α-effect were developed [22, 23]. Roughly speaking, the α effect means that a non-mirror-symmetric flow can induce an electromotive force that is parallel to an assumed mean magnetic field.

In order to validate this principle in experiment, Steenbeck et al. in 1967 constructed the "α-box," a system of two orthogonally interlaced channels with sodium flowing through them [24]. This experiment confirmed the α-effect to be proportional to Bv^2, which means that is independent of the flow direction whereas it reverses its sign with reversed magnetic field.

3.4 Precession

Another tradition line of dynamo-related experiments was opened up by Gans in 1970 [25]. Its background is that precession has long been discussed as a possible energy source for the geodynamo [26, 27]. Gans used a sodium filled precessing cylinder with a diameter of 0.25 m and approximately the same height. The rotation rate of the cylinder reached 3,600 rpm, and the precession rate 50 rpm. Amplifications of an applied magnetic field up to a factor of 3 were observed.

Recently, this concept has been taken up by Léorat [28] who studied a precessing flow in a cylindrical container filled with 27 l of water. Special focus was laid on influence of the precession parameter (ratio of precession to rotation rate) on the transition from laminar to turbulent flows, and on the possible dynamo capabilities of such a device. An envisioned precession experiment with liquid sodium would have the attractive features of being closed (no shafts for propellers, etc.) and having no internal constraints.

3.5 Unintended dynamos?

With the advent of fast breeder reactors it became important to know if dynamo processes could occur in the huge liquid sodium pumps where the magnetic Reynolds numbers are indeed large. Early papers on that topic indicated that this could really happen [29–31]. Recent experimental results show that no dynamo action has occurred in the Superphenix [32].

3.6 The first hydromagnetic dynamos in Riga and Karlsruhe

It was an interesting historical coincidence that, after decades of theoretical and numerical studies and years of preparation, the hydromagnetic dynamo effect was experimentally demonstrated at two liquid sodium facilities in Riga and Karlsruhe almost simultaneously at the end of 1999 [33, 34].

While the Riga experiment, which will be portrayed in more detail below, can be seen as the elementary cell of dynamos, namely a single helical flow, the Karlsruhe dynamo can be considered as a demonstration of mean-field dynamo theory.

It is not widely known that the underlying geophysical motivation, the basic idea, the mathematics and even a final formula for the critical flow-rates for a sort of Karlsruhe experiment can already be found in a paper of 1967 [35]. The idea was to substitute real helical ("gyrotropic") turbulence by "pseudo-turbulence", actualized by a large (but finite) number of parallel channels with a helical flow inside. Later, in 1975 [36], Busse considered a similar kind of dynamo which prompted him to initiate the Karlsruhe dynamo experiment.

In 1972, Roberts had proved dynamo action for a velocity pattern periodic in x and y that comprises both a rotational flow and an axial flow [37]. The α-part of the electromotive force for this flow type can be written in the form $\mathcal{E} = -\alpha_\perp (\mathbf{B} - (\mathbf{e}_z \cdot \bar{\mathbf{B}})\mathbf{e}_z)$, which represents an extremely anisotropic α-effect that produces only electromotive forces in the x- and y-directions but not in the z-direction [38].

In the specific realization of the Karlsruhe experiment the Roberts flow in each cell is replaced by a flow through two concentric channels. In the central channel the flow is straight, in the outer channel it is forced by a "spiral staircase" on a helical path. This design principle of the Karlsruhe dynamo being given, a fine-tuning of the geometric relations was carried out with the aim to achieve a maximum α effect for a given power of the pumps. Such an optimization led to a number of 52 spin generators, a radius of 0.85 m and a height of 0.7 m for the dynamo module.

The scheme in Fig. 2a depicts the central dynamo module and one of the 52 spin generator in detail. Figure 2b shows the stability diagram and the pressure increase beyond the critical flow rate. Below we will compare this rather steep pressure and power increase with the corresponding behaviour for the Lowes and Wilkinson experiment and for the Riga experiment. During its comparably short lifetime, the Karlsruhe dynamo experiment has brought about many results on its imperfect bifurcation behaviour and on MHD turbulence which are documented in [39–43].

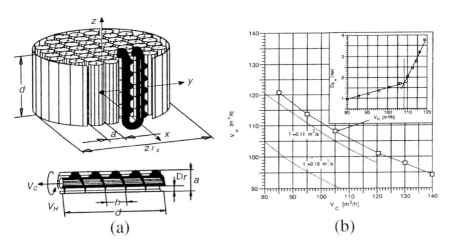

Fig. 2. (a) Sketch of the central module of the Karlsruhe dynamo and details of one of the 52 spin generators. (b) Phase diagram of the experimentally determined dynamo action as a function of the volumetric flow rates in the central channel (V_C) and in the helical channel (V_H), and its comparison with numerical results by Tilgner for different magnetic diffusivities l. The inset shows the pressure increase for increasing V_H at $V_C = 105$ m^3/s. (Figures are taken from [41]).

3.7 Maryland

D. Lathrop and his collaborators at the University of Maryland have investigated a number of liquid sodium flows with a view on their dynamo capabilities [44, 45]. The first experiment (Dynamo I), a fast rotating (up to 25,000 rpm) 0.2 m diameter titanium torus, which was heated from the outer side and cooled at the axis, was motivated by common ideas about planetary convection. In a second experiment (Dynamo II), 15 l of liquid sodium were stirred by two counterrotating propellers with up to 7,200 rpm within a 0.3 m diameter steel sphere. A third experiment (Dynamo III) was basically a spherical Couette flow of 0.6 m outer diameter.

Up to the present, in neither of these experiments was there any sign of a dynamo effect. However, in a modified Dynamo II configuration (with the propeller replaced by an inner sphere of 5 cm diameter), the Maryland group might have achieved a very important result with the detection of an instability which they identify, for some good reasons, with the magnetorotational instability [46]. Presently, a 3 m diameter spherical Couette experiment is set up at Maryland, which will reach magnetic Reynolds numbers of 700!

3.8 Cadarache

Since a couple of years, the "von Karman Sodium" (VKS) dynamo experiment has been pursued at the CEA research center in Cadarache (France). The sodium flow of the $s2^{+}t2$ type (comprising two poloidal vortices directed inward in the equatorial plane, and two toroidal vortices) is produced by two disks in a cylinder. The VKS 1 experiment was carried out with 50 l sodium in a cylinder with diameter and height of 0.4 m, using two 75 kW motors at rotation rates up to 1,500 rpm.

The results of the VKS 1 experiments, including measured inductions and turbulence data, have been published in [47–49]. No self-excitation has been achieved, although remarkable deformations of applied magnetic fields have been measured.

Presently, a second version of this experiment, VKS 2, is being tested. The volume is extended to 100 l, the available motor power can reach 300 kW, and great effort was spent in order to optimize the shape of the impellers. Even with these preparations the experiment is subject to imponderabilities, one of them being evidently the role of the rather high turbulence level on the dynamo threshold, the other being hitherto unsolved questions concerning the role of the boundary conditions in axial direction.

3.9 Madison

At the University of Wisconsin, Madison, C. Forest and his colleagues have set up a liquid sodium experiment within a 1 m spherical shell [50, 51]. Two propellers drive a flow of the same $s2^{+}t2$ type as in the VKS experiment.

The Madison dynamo experiment (MDX) is instructive on how inverse dynamo theory can be utilized in dynamo experiments. First, a lot of effort has been spent in the numerical optimization of the precise geometry of the flow. Second, much attention has been paid to the inference of the velocity field from externally measured magnetic field signals. In a recent experiment, the latter technique has brought about a surprising coincidence of the measured induced magnetic fields with the predicted ones for a given propeller speed. This gives hope for a success of the real dynamo experiments, albeit it is not a guarantee since for highly fluctuating flows the average of the product of two induction effects is not equal to the product of two averages of the individual induction effects.

3.10 Grenoble

A group of geophysicists in Grenoble has build a medium-size rotating sphere experiment which they coined "Derviche Tourneur Sodium (DTS)" [52]. The Grenoble *ansatz* relies on the concept of *magnetostrophic* equilibrium between the Coriolis forces and the Lorentz forces. Similar to Dynamo III and the 3 m sphere in Maryland, DTS is a spherical Couette experiment with 0.4 m outer diameter and 0.15 m inner diameter. A peculiarity of this experiment is that the inner sphere is made of a permanent magnet (having 0.15 T at its poles). This configuration is intended to reveal several aspects of rotating fluids under the influence of a magnetic field, including torsional oscillations, the features and instabilities of a super-rotating layer [53, 54], and several turbulence characteristics in the presence of Coriolis and Lorentz forces. First experiments have been carried out in spring 2005. Of course, the experiment is not expected to reach the critical magnetic Reynolds number for such types of flows, which has been estimated (in the non-magnetized case) to be in the order of a few thousand [55].

3.11 Perm

An interesting dynamo concept which avoids the usual large size and driving power of other facilities has been pursued in the group of P. Frick at the Institute of Continuous Media Mechanics in Perm, Russia [56]. The experiment is based on the fact that a non-stationary helical flow of the Ponomarenko type can be produced within a torus when its rotation is abruptly braked and a fixed diverter forces the inertially continuing flow on a helical path. In the final sodium experiment the torus will have a major radius of 0.4 m and a minor radius of 0.12 m. Its rotation with 3,000 rpm (which means velocities up to 140 m/s) will be braked by hydraulic brakes within 0.1 s.

Both numerical [57] and experimental studies at a gallium prototype experiment [58] indicate that self-excitation could be possible in such a configuration.

3.12 New Mexico

The $\alpha-\Omega$ dynamo experiment at the New Mexico Institute of Mining and Technology has as its physical background the magnetic field generation in active galactic nuclei and in stars [59, 60]. In its present configuration the experiment consists of two coaxial cylinders with radii of 0.15 m and 0.3 m, respectively, which are rotating with a frequency ratio of 4:1. This combination makes the experiment marginally unstable to Taylor vortices. In the present form the experiment is intended to investigate MRI; for studying dynamos a modification in the form of additional injected α-plumes has been envisioned.

4 Some lessons from the Riga experiment

Having sketched the long history of dynamo related experiments, we focus now in more detail on the Riga experiment. Evidently, this has to do with our personal involvement, but besides this we feel that this experiment represents, in particular, a genuine *hydro*magnetic dynamo, which makes the analysis of its saturation mechanism also interesting for future experiments.

4.1 Basics, theory, and numerics

The principle idea of the Riga dynamo experiment (see Fig. 3) goes back to Ponomarenko [61], who had proved that a helically moving, electrically conducting cylinder embedded in an infinite stationary conductor can show dynamo action. This simple, paradigmatic configuration was analysed in more detail by Gailitis and Freibergs [62] who found a remarkably low critical magnetic Reynolds number of 17.7 for the convective instability. By adding a back-flow, this convective instability can be rendered into an absolute instability [63]. All this early numerical work, including the optimization [64] of the main geometric relations which led to the design of the Riga dynamo (see Fig. 3b), was done with a 1D eigenvalue solver.

For refined kinematic simulations a 2D finite difference code (in radial and axial direction) was written whose main advantage is the possibility to treat velocity structures varying in axial direction, which is indeed of relevance for the Riga dynamo [65]. The magnetic field structure as it comes out of this code is illustrated in Fig. 3c.

As for the saturation regime, we tried to capture the most essential back-reaction effects within a simple model [65, 66]. Roughly speaking, this model relies on the fact that, in contrast to the axial velocity component, the rotational velocity component is only maintained by inertia. Therefore, the Lorentz force can brake this component without significantly increasing the pressure and hence the motor power. Under some assumptions, this amounts to a simple 1D differential equation for the downstream accumulating

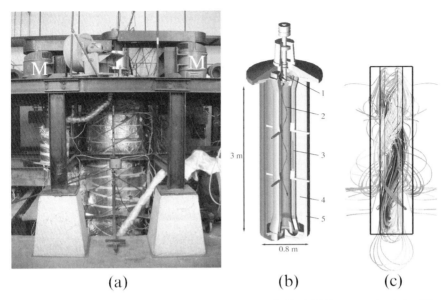

Fig. 3. The Riga dynamo experiment and its eigenfield. (a) Photograph of the Riga dynamo facility as of February 2005. M: Two motors with 100 kW each. (b) Details of the central dynamo module. 1: Propeller; 2: central helical flow region; 3: back-flow region; 4: outer sodium region; 5: thermal insulation. At one third and two thirds of the dynamo length there are four ports for various magnetic field, pressure and velocity probes. (c) Simulated magnetic eigenfield in the kinematic regime

brake of the azimuthal velocity, which can be applied both to the flow in the inner and in the back-flow cylinder.

Only recently, a more sophisticated back-reaction model has been developed in collaboration with the Delft Technical University, which basically confirmed, and slightly improved, the results of the simple model.

4.2 The experiments up to present

It should not be forgotten that a predecessor of the present Riga experiment, in which an external pump was used instead of a propeller, was carried out in St. Petersburg already in 1986 [67]. It showed a considerable amplification of an externally applied magnetic field, although self-excitation could not be achieved.

At the present facility, seven experimental campaigns have been carried out between November 1999 and March 2005. In the first campaign in November 1999, a self-exciting field was documented for the first time in a liquid metal dynamo experiment, although the saturated regime could not be reached at that time [33]. This had to be postponed until the July 2000 experiments [68]. In June 2002, the radial dependence of the magnetic field was determined by

the use of Hall sensors and induction coils situated on "lances" going throughout the whole dynamo module. In February and June 2003, first attempts were
made to measure the Lorentz force induced motion in the outermost cylinder.
A novelty of the May 2004 campaign was the measurements of pressure in the
inner channel by a piezoelectric sensor that was flash mounted at the innermost wall. In February/March 2005, a newly developed permanent magnet
probe was inserted into the innermost cylinder in order to get information
about the velocity there, and two traversing rails with induction coils were
installed to get continuous field information along the z-axis and across the
whole diameter of the dynamo. More details about these results can be found
in [15, 65, 66, 69–71], and will be published elsewhere.

4.3 Main results and their interpretation

Figure 4 shows the axial magnetic field measured by induction coils inside the
upper and the lower port close to the innermost wall during the last run on
1 March 2005. This figure might serve as an example of how the magnetic
field can be switched on and off at will, and on how its amplitude depends
on the propeller rotation rate. Comparing Figs. 4a and b, a peculiarity of this
dependence becomes visible. Whereas at the upper sensor (Fig. 4a) the field
amplitude increases from 27 mT for 2,000 rpm to 120 mT for 2,500 rpm, at
the lower sensor the corresponding increase is only from 24 to 65 mT. This is a
clear indication for a drastic change of the field dependence in axial direction
with increasing overcriticality, which in turn mirrors a significant change of
the axial dependence of the flow.

This effect can be explained by the selective braking of the azimuthal
velocity component described above, and it is also consistent with the growth
rate and frequency behaviour that is documented in Fig. 5. In this figure, the
numerical curves in the kinematic regime result from the 2D solver (slightly

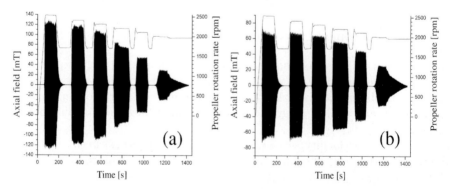

Fig. 4. Axial magnetic field and propeller rotation rate during the run 6 of the
February/March 2005 campaign measured by induction coils within the upper
(a) and the lower (b) port, close to the innermost wall

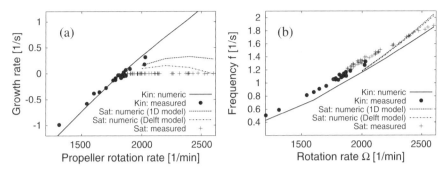

Fig. 5. Measured and computed growth rates (a) and frequencies (b) in the kinematic and saturated regime. For the saturation regime, a simple 1D model of braking and a 3D hydrodynamic model from TU Delft has been used

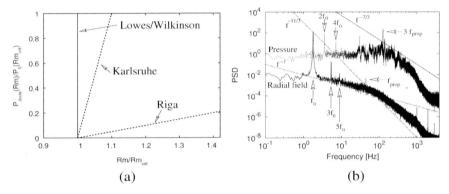

Fig. 6. (a) A rough picture of the Joule power dependence on $R_m/R_{m,\mathrm{crit}}$ for the Lowes and Wilkinson experiments, the Karlsruhe experiment and the Riga experiment. (b) Power spectral density of the radial magnetic field and the pressure

corrected by the effect of the different wall conductivity which was estimated separately by a 1D solver). In the saturated regime, we show the results of our simple back-reaction model, together with the results from a hybrid method including a 3D hydrodynamic solver from TU Delft. Evidently, the simple back-reaction model covers the most essential saturation effects.

The quoted deformation of the flow field has consequences for the dependence of the Joule power on the overcriticality. In Fig. 6a we try to compare the corresponding curves for the Lowes and Wilkinson experiment, the Karlsruhe experiment and the Riga experiment. Note that all these three curves are not very accurate: for the Lowes and Wilkinson case we refer to their paper [19], for the Karlsruhe results we simply use the pressure increase shown in the inset of Fig. 2b, and for Riga we take our own motor power measurements [65]. In stark contrast to the sharp rise for the Lowes and Wilkinson experiment, but also strongly differing from the steep increase in the Karlsruhe experiment, the Joule power dependence on the overcriticality in the Riga experiment is very flat. Quite different to the back-reaction of a rigid body, the sodium flow deforms under the influence of the Lorentz forces, and the resulting deterioration of the dynamo condition makes the growth rate drop down to zero.

In Fig. 6b we turn to some turbulent properties of the Riga experiment by showing two sorts of spectra, one for the radial magnetic field measured in the lower port in the central cylinder, 2 cm from the wall. The other spectrum results from the data of the piezoelectric pressure sensor, which is also mounted on the lower level. The main feature of the magnetic spectra is, of course, the peak at the eigenfrequency f_0. However, there is also a peak at the triple frequency $3f_0$, and even a small one at $5f_0$. Neither of these peaks is seen in the pressure spectrum. Instead, we detect here a dominant peak at $2f_0$ and some smaller peak at $4f_0$. This cascade can be easily explained by the back-reaction of different azimuthal magnetic field modes on the flow. Concerning the inertial range of the spectrum, we have plotted the $f^{-11/3}$ law for the magnetic field and a $f^{-7/3}$ law for the pressure for comparison, without claiming a perfect agreement with the measured data. Between the main field frequency f_0 and the propeller frequency f_{prop} there seems to be a region with f^{-1}, which has also been observed experimentally in [47] and numerically in [72].

5 Are we at the end, or at the beginning of dynamo experiments?

The homogeneous dynamo effect has been validated, up to present, in three experiments. The experiments of Lowes and Wilkinson represented an ingenious demonstration of Herzenberg's theory. The Riga and Karlsruhe experiments, which are based on complementary dynamo concepts, have introduced the aspect of fluidity. The Riga experiment has been shown to possess a nontrivial saturation mechanism, which relies on the deformability of the fluid flow.

For the still functioning Riga experiment, as well as for all dynamo experiments to come, a central topic will be the development of measuring techniques for all sorts of velocity, vorticity, and electromagnetic field quantities and their correlations. Hence, with the view on the validation of MHD turbulence models, we might well be at the very beginning of experimental dynamo work.

The study of rotating systems under the influence of magnetic fields has just begun. Possibly, large-scale dynamo activities may be superseded by somewhat smaller experiments using externally applied rather than self-excited magnetic fields. Those experiments are particularly important for the investigation of the magnetostrophic regime and the magnetorotational instability [73].

The last point: It seems that the geomagnetically motivated drive of Lowes and Wilkinson to construct a dynamo that shows irregular reversals has been lost somehow in the recent experimental work, which is more than under-standable with view on the grave technical problems to actualize *a working hydromagnetic dynamo at all*. But postponed is not abandoned. In an attempt to understand the essence of reversal we have studied a very simple model, a spherically symmetric mean-field α^2 dynamo with α-quenching, influenced by noise [74]. It turned out that such a simple dynamo shows already the most prominent features of Earth's magnetic field reversals, namely, asymmetry, correlation between field strength and chron duration, and bimodal field dis-tribution. Interestingly, all this can be attributed to the special magnetic field dynamics in the vicinity of the "exceptional points" of the spectrum of the non-self-adjoint dynamo operator. Hence this model could give us a general recipe on how the kinematic growth rate curve must be shaped for reversals to occur in the saturated regime. First numerical work to apply this concept to a modified Karlsruhe type dynamo and to a s2+t2 dynamo showed encouraging results [75].

References

1. Rüdiger G, Hollerbach R (2004) The magnetic universe. Wiley, Weinheim, Germany
2. Valet JP (2003) Time variations in geomagnetic intensity. Rev Geophys 41:1004
3. Valet JP, Meynadier L (1993) Geomagnetic field intensity and reversals during the past 4 million years. Nature 366:234–238
4. Tarduno JA, Cottrell RD, Smirnov AV (2001) High geomagnetic intensity during the mid-Cretaceous from Thellier analyses of single plagioclase crystals. Science 291:1779–1783
5. Heller R, Merrill RT, McFadden PL (2003) The two states of paleomagnetic field intensities for the past 320 million years. Phys Earth Planet Inter 135:211–223
6. Glatzmaier GA (2002) Geodynamo simulations – How realistic are they? Annu Rev Earth Planet Sci 30:237–257
7. Hale GE (1908) On the probable existence of a magnetic field in sunspots. Astrophys J 28:315–343
8. Larmor J (1919) How could a rotating body such as the sun become a magnet? Rep Brit Assoc Adv Sci:159–160
9. Ossendrijver M (2003) The solar dynamo. Astron Astrophys Rev 11:287–367
10. Kouveliotou C et al. (1998) An X-ray pulsar with a superstrong magnetic field in the soft gamma-ray repeater SGR 1806-20. Nature 393:235–237
11. Beck R et al. (1996) Galactic magnetism: Recent developments and perspectives. Annu Rev Astron Astrophys 34:155–206
12. Kronberg PP (2002) Intergalactic magnetic fields. Phys Today 55(12):40–46
13. Balbus SA (2003) Enhanced angular momentum transport in accretion disks. Annu Rev Astron Astrophys 41:555–597
14. Roberts PH, Scott S (1965) On the analysis of secular variation. 1. A hydro-magnetic constraint – Theory. J Geomag Geoelectr 17:137–151
15. Gailitis A et al. (2002) Colloquium: Laboratory experiments on hydromagnetic dynamos. Rev Mod Phys 74:973–990

16. Lehnert B (1958) An experiment on axisymmetric flow of liquid sodium in a magnetic field. Arkiv f Fysik 13(10):109–116
17. Lowes FJ, Wilkinson I (1963) Geomagnetic dynamo – A laboratory model. Nature 198:1158–1160
18. Lowes FJ, Wilkinson I (1968) Geomagnetic dynamo – An improved laboratory model. Nature 219:717–718
19. Wilkinson I (1984) The contribution of laboratory dynamo experiments to our understanding of the mechanism of generation of planetary magnetic fields. Geophys Surv 7(1):107–122
20. Herzenberg A (1958) Geomagnetic dynamos. Philos Trans R Soc Lond A 250:543–585
21. Brandenburg A, Moss D, Soward AM (1998) New results for the Herzenberg dynamo: Steady and oscillatory solutions. Proc R Soc Lond A 454:1283–1300
22. Steenbeck M, Krause F, Rädler KH (1966) Berechnung der mittleren Lorentz-Feldstärke $\overline{\mathbf{v} \times \mathbf{B}}$ für ein elektrisch leitendes Medium in turbulenter, durch Coriolis-Kräfte beeinflußter Bewegung. Z Naturforschung 21A:369–376
23. Krause F, Rädler KH (1980) Mean-field magnetohydrodynamics and dynamo theory. Akademie, Berlin
24. Steenbeck M et al. (1967) Der experimentelle Nachweis einer elektromotorischen Kraft längs eines äußeren Magnetfeldes, induziert durch eine Strömung flüssigen Metalls (α-Effekt). Mber Dtsch Akad Wiss Berlin 9:714–719
25. Gans RF (1971) Hydromagnetic precession in a cylinder. J Fluid Mech 45:111–130
26. Malkus WVR (1968) Precession of Earth as cause of geomagnetism. Science 160:259–264
27. Tilgner A (2005) Precession driven dynamos. Phys Fluids 17:034104
28. Léorat J: http://melamp.obspm.fr/LEORAT/ATER
29. Bevir MK (1973) Possibility of electromagnetic self-excitation in liquid-metal flows in fast reactors. J Br Nucl Energy Soc 12:455–458
30. Pierson ES (1975) Electromagnetic self-excitation in liquid-metal fast breeder reactor. Nucl Sci Eng 57:155–163
31. Kirko IM et al. (1982) On the existence of thermoelectric currents in the reactor BN-600 of Beloyarskaia nuclear power plant. Dokl Akad Nauk SSSR 266:854–856
32. Alemany A et al. (2000) Experimental investigation of dynamo effect in the secondary pumps of the fast breeder reactor Superphenix. J Fluid Mech 403:263–276
33. Gailitis A et al. (2000) Detection of a flow induced magnetic field eigenmode in the Riga dynamo facility. Phys Rev Lett 84:4365–4368
34. Müller M, Stieglitz R (2000) Can the Earth's magnetic field be simulated in the laboratory? Naturwissenschaften 87:381–390
35. Gailitis A (1967) Self-excitation conditions for a laboratory model of a geomagnetic dynamo. Magnetohydrodynamics 3(3):23–29
36. Busse FH (1975) Model of geodynamo. Geophys J R Astr Soc 42:437–459
37. Roberts GO (1972) Dynamo action of fluid motions with 2-dimensional periodicity. Philos Trans R Soc Lond A 271:411–454
38. Rädler KH et al. (2002) On the mean-field theory of the Karlsruhe dynamo experiment I. Kinematic theory. Magnetohydrodynamics 38:41–71
39. Stieglitz R, Müller U (2001) Experimental demonstration of a homogeneous two-scale dynamo. Phys Fluids 13:561–564

40. Müller U, Stieglitz R (2002) The Karlsruhe dynamo experiment. Nonl Proc Geophys 9:165–170
41. Stieglitz R, Müller U (2002) Experimental demonstration of a homogeneous two-scale dynamo. Magnetohydrodynamics 38:27–33
42. Müller U, Stieglitz R, Horanyi S (2004) A two-scale hydromagnetic dynamo experiment. J Fluid Mech 498:31–71
43. Müller U, Stieglitz R, Busse F (2004) On the sensitivity of dynamo action to the system's magnetic diffusivity. Phys Fluids 16:L87–L90
44. Peffley NL, Cawthorne AB, Lathrop DP (2000) Toward a self-generating magnetic dynamo: The role of turbulence. Phys Rev E 61:5287–5294
45. Shew WL, Sisan DR, Lathrop DP (2002) Mechanically forced and thermally driven flows in liquid sodium. Magnetohydrodynamics 38:121–127
46. Sisan DR et al. (2004) Experimental observation and characterization of the magnetorotational instability. Phys Rev Lett 93:114502
47. Bourgoin M et al. (2002) Magnetohydrodynamics measurements in the von Karman sodium experiment. Phys Fluids 14:3046-3058
48. Marié L et al. (2002) Open questions about homogeneous fluid dynamos: The VKS experiment. Magnetohydrodynamics 38:163–176
49. Petrélis F et al. (2003) Nonlinear magnetic induction by helical motion in a liquid sodium turbulent flow. Phys Rev Lett 90:174501
50. Forest CB et al. (2002) Hydrodynamic and numerical modeling of a spherical homogeneous dynamo experiment. Magnetohydrodynamics 38:107–120
51. Nornberg MD, Spence EJ, Kendrick RD, Forest CB (2006) Measurements of the magnetic field induced by a turbulent flow of liquid metal. Phys Plasmas 13:055901
52. Cardin P et al. (2002) Towards a rapidly rotating liquid sodium dynamo experiment. Magnetohydrodynamics 38:177–189
53. Dormy E, Cardin P, Jault D (1998) MHD flow in a slightly differentially rotating spherical shell, with conducting inner core, in a dipolar magnetic field. Earth Planet Sci Lett 160:15–30
54. Hollerbach R, Skinner S (2001) Instabilities of magnetically induced shear layers and jets. Proc R Soc Lond A 457:785–802
55. Schaeffer N, Cardin P (2006) Quasi-geostrophic kinematic dynamos at low magnetic Prandtl number. Earth Planet Sci Lett 245:595–604
56. Frick P et al. (2002) Non-stationary screw flow in a toroidal channel: Way to a laboratory dynamo experiment. Magnetohydrodynamics 38:143–162
57. Dobler W, Frick P, Stepanov R (2003) Screw dynamo in a time-dependent pipe flow. Phys Rev E 67:056309
58. Noskov V et al. (2004) Magnetic field rotation in the screw gallium flow. Eur Phys J B 41:561–568
59. Colgate SA, Li H, Pariev V (2001) The origin of the magnetic fields of the universe: The plasma astrophysics of the free energy of the universe. Phys Plasmas 8:2425–2431
60. Colgate SA et al. (2002) The New Mexico $\alpha\omega$ dynamo experiment: Modelling astrophysical dynamos. Magnetohydrodynamics 38:129–142
61. Ponomarenko YB (1973) On the theory of hydromagnetic dynamos. J Appl Mech Tech Phys 14:775–779
62. Gailitis A, Freibergs Y (1976) Theory of a helical MHD dynamo. Magnetohydrodynamics 12:127–129

63. Gailitis A, Freibergs Y (1980) Nature of the instability of a turbulent dynamo. Magnetohydrodynamics 16:116–121
64. Gailitis A (1996) Project of a liquid sodium MHD dynamo experiment. Magnetohydrodynamics 32:58–62
65. Gailitis A et al. (2004) Riga dynamo experiment and its theoretical background. Phys Plasmas 11:2838–2843
66. Gailitis A et al. (2002) On back-reaction effects in the Riga dynamo experiment. Magnetohydrodynamics 38:15–26
67. Gailitis A et al. (1987) Experiment with a liquid-metal model of an MHD dynamo. Magnetohydrodynamics 23:349–353
68. Gailitis A et al. (2001) Magnetic field saturation in the Riga dynamo experiment. Phys Rev Lett 86:3024–3027
69. Gailitis A et al. (2001) On the results of the Riga dynamo experiments. Magnetohydrodynamics 37:71–80
70. Gailitis A et al. (2002) Dynamo experiments at the Riga sodium facility. Magnetohydrodynamics 38:5–14
71. Gailitis A et al. (2003) The Riga dynamo experiment. Surv Geophys 24:247–267
72. Ponty Y, Politano H, Pinton JF (2004) Simulation of induction at low magnetic Prandtl number. Phys Rev Lett 92:144503
73. Rosner R, Rüdiger G, Bonanno A (eds) (2004) MHD Couette Flows: Experiments and Models. AIP Conference Proceedings:733
74. Stefani F, Gerbeth G (2005) Asymmetric polarity reversals, bimodal field distribution, and coherence resonance in a spherically symmetric mean-field dynamo model. Phys Rev Lett 94:184506
75. Xu M, Stefani F, Gerbeth G (2005) How to actualize reversals in dynamo experiments? http://www.cosis.net/abstracts/IAGA2005/00618/IAGA2005-A-00618.pdf

Mean-Field Dynamo Theory: Early Ideas and Today's Problems

Karl-Heinz Rädler

Astrophysical Institute Potsdam, An der Sternwarte 16, D-14482 Potsdam, Germany (khraedler@arcor.de)

1 Introduction

Mean-field dynamo theory has proved to be a useful tool for understanding the generation of magnetic fields in the Earth and the Sun, in stellar bodies, and even in galaxies. It provides a basis for the elaboration of detailed dynamo models of these objects. Fundamentals of this theory were developed in the 1960s of the last century in the Institute for Magnetohydrodynamics in Jena in Germany under the directorship of Max Steenbeck. The 21st of March 2004, a date close to that of the Coventry meeting, would have been his 100th birthday. Let me say first some words about his life and his contributions to various fields in physics. This will lead naturally to the early ideas of mean-field dynamo theory, to its remarkable findings and to some of the problems of its further development.

2 Max Steenbeck (1904–1981)

Born on 21st of March 1904 in Kiel in North Germany, on the Baltic coast, Max Steenbeck spent his childhood and school days there. Later he went to Kiel University, starting with chemistry but soon switching to physics. Among his teachers were Christian Gerthsen, Hans Geiger, and Walther Kossel. Under the supervision of the latter Max Steenbeck provided a thesis on energy measurements of x–rays, and on this basis he was granted a Ph.D. degree in 1928.

Already in 1927 Max Steenbeck accepted a position in the Siemens-Schuckert Company in Berlin, where he worked until the end of the Second World War in 1945. For many years he was a member of the Scientific Department and later became the director of the Rectifier Plant. His main activities lay in the physics of gas discharges and plasmas. Many of the tasks he dealt with were connected with technical applications, mainly in heavy current engineering. The scientific work of that time is documented in many publications.

S. Molokov et al. (eds.), Magnetohydrodynamics – Historical Evolution and Trends,
55–72. © 2007 *Springer.*

The most prominent ones are the two-volume monograph "Elektrische Gasentladungen, ihre Physik und Technik" by Alfred von Engel and Max Steenbeeck (1932/1934) [1] and the famous 120-page-long article "Der Plasmazustand der Gase" by Robert Rompe and Max Steenbeck (1939) [2].

In this period Max Steenbeck also designed and constructed a device for the acceleration of electrons, the "betatron", which worked successfully for the first time in 1935. Devices of that kind were later used for several purposes in medicine. An important achievement in this context was his discovery of a configuration of the magnetic field in the betatron which ensures stable paths of the electrons.

One of the remarkable activities of Max Steenbeck during the Second World War was the search and deactivation of British underwater mines, which proved to be an enormous challenge for a physicist.

At the end of the war, in May 1945, officers of the Soviet Army found Max Steenbeck in Berlin. Assuming that he, the director of an important plant, had some higher position in the Nazi hierarchy, and not believing that he was not even a party member, they took him to a detention camp in Poland. He lived there a few months in terrible conditions and was soon close to starvation. It may sound strange but it was probably the atomic bombs on Hiroshima and Nagasaki which saved his life. The Soviet government believed that its atomic program should be advanced by all possible means, including the help of scientists of the defeated Germany. So he was brought to a place near Sukhumi at the Black Sea, where other German scientists were also interned, among them the Nobel Prize winner Gustav Hertz. Living in a paradise place but with very restricted personal freedom, later joined by their families, they enjoyed reasonable working conditions for Uranium isotope separation and related topics. Max Steenbeck developed here a new type of gas centrifuge for isotope separation. Later he worked also in other parts of the Soviet Union on other research fields, e.g., on semiconductor problems. In this way he came into contact with several outstanding Soviet physicists and earned their respect. Later he was elected a Foreign Member of the Soviet Academy of Sciences.

After 11 hard years, in 1956, Max Steenbeck and also the other members of the German group were allowed to return to Germany. He decided to go to Jena in the German Democratic Republic (G.D.R.). Apart from holding a chair in plasma physics at Jena University, he was Director of the Institute for Magnetic Materials outside the University from 1956 to 1959. In 1959 he handed over the directorship to younger co-workers and established the Institute for Magnetohydrodynamics, which he directed until his retirement in 1969. One of the main ideas he had in mind for this institute was to contribute to the technical realization of energy-producing nuclear-fusion devices, which at that time seemed to be feasible in the near future. In fact the institute was a place for a variety of research work in plasma physics and magnetohydrodynamics, including that on magnetic field generation by the motion of electrically conducting fluids, which is discussed in more detail below.

Fig. 1. Max Steenbeck

In addition to these positions Max Steenbeck was the head of the Scientific–Technical Office for the Construction of Nuclear Power Stations between 1957 and 1962, which accompanied the development of the first experimental nuclear reactor in the G.D.R. He shared his time every week between Jena and this office, situated in Berlin, working very hard at both places. From 1962 to 1966 Max Steenbeck was Vice-President of the Academy of Sciences of the G.D.R., and from 1962 until his retirement in 1969, first Vice-President and later President of the Research Council of the G.D.R. In this way he continued to share his life between Jena and Berlin. Despite all his official and representation duties Max Steenbeck always found time and energy for his own detailed work on scientific topics, which he then introduced in discussions with his younger co-workers.

After his retirement Max Steenbeck followed the progress in many fields of science with great interest. His achievements were honored in several ways, including the Lomonosov Gold Medal in 1973 and the Krupp Award in 1977.

The sad experiences of his life also motivated Max Steenbeck to dedicate himself to political issues. After his retirement he continued or even intensified his activities on a political level. For example, since 1970, as the President of a corresponding committee in the G.D.R., he was involved in the International Conferences for Security and Cooperation in Europe.

In addition to a large number of papers on his research work in physics, Max Steenbeck wrote many articles and essays on general aspects of science and scientists, on the importance of basic research, on the responsibility of the scientist in human society etc. In his memoirs "Impulse und Wirkungen - Schritte auf meinem Lebensweg" (1977) [3] he recounts the very interesting events of his turbulent life in Germany and in the Soviet Union and gives some insights into political and social developments during his life.

Max Steenbeck died on 15th of December 1981 in Berlin.

3 Early ideas on mean-field electrodynamics and dynamos

3.1 The Jena findings

As already mentioned, one of the original aims of the Institute of Magneto-hydrodynamics in Jena was to contribute to the realization of nuclear fusion reactors. Initially it was not intended to do research on astrophysical or geo-physical problems. It was some kind of hobby of Max Steenbeck to think also about the question how, for example, the Sun produces and maintains its magnetic fields. He liked to discuss such questions from time to time with his younger co-workers, in particular with Fritz Krause (born 1927, mathematician) and the author (born 1935, physicist), who both had also a number of other duties in the institute. During the first years of the 1960s these discussions and the investigations stimulated by them were carried out in isolation. There was no close contact, for example, with astrophysicists who knew E. N. Parker's early works on solar magnetic fields (1955/1957) [4,5], and no easy access to the relevant literature.

Let me sketch some of the ideas which we discussed in relation to the Sun. The laws of the occurrence of sunspots, which are magnetic phenomena, and of their distribution with respect to heliographic latitude and time, suggested a general magnetic field, which is axisymmetric, consists of a strong toroidal and a weak poloidal part, and varies periodically in time. This general field has to be understood as a mean field, that is, a smoothed version of the real field or, in other words, its large-scale part. It was clear how the differential rotation of the convection zone produces a toroidal magnetic field from a given poloidal one. A dynamo seemed to be possible with a mechanism that reproduces a poloidal magnetic field from a toriodal one. Many ideas on mechanisms of this kind were discussed and investigated. Some of them were reported in a lecture by Max Steenbeck (1963) [6]. For example, a proper anisotropy of the electric conductivity relevant to mean fields could create, in addition to the electric currents which correspond to a toroidal field, others corresponding to a poloidal field. This anisotropy could be a consequence of randomly distributed conductivity inhomogeneities of the solar plasma if it is subject to the shear connected with the differential rotation. In another example, a magnetization of the conducting medium was defined by interpreting a part of the electric currents on small scales, resulting from small-scale motions, as magnetization currents. The mentioned shear of the plasma may then lead to an anisotropy of the magnetic permeability which has to be ascribed to mean fields. This anisotropy may also cause a poloidal field in addition to the toroidal one.

These ideas, the essence of which consists in assuming modified material properties of the solar plasma when considering mean fields, did however not lead to realistic models of the solar dynamo. In 1965 a new finding emerged: if on average the small-scale, convective or turbulent motions lack reflectional symmetry, that is, show helical features, then Ohm's law for the mean fields contains an electromotive force with a component parallel or antiparallel to the mean magnetic field, which has no counterpart in the original version of Ohm's law. The occurrence of this kind of electromotive force has been named the "α-effect." The lack of reflectional symmetry, which is crucial for the α-effect, occurs naturally with motions on rotating bodies, that is, under the influence of the Coriolis force. The α-effect leads in particular to the generation of a poloidal magnetic field from a toroidal one and in this way resolves the main difficulty for the construction of a solar dynamo model. It is, moreover, crucial for all dynamo theory and may lead to dynamo action even in the absence of differential rotation. This finding was reported in a paper by Steenbeck et al. (1966) [7]. More detailed results are given in the Ph.D. thesis of the author [8] (see also [9,10]) and in the Habilitationsschrift of Krause (1968) [11].

The α-effect, which was found without knowing Parker's earlier idea of "cyclonic convection" (1955/1957) [4,5], is nevertheless closely connected with it. The theory of the α-effect has to be considered as a mathematically more rigorous formulation of this idea. Considerable progress in modeling the geodynamo was achieved by S. I. Braginsky with his theory of the "nearly symmetric dynamo" (1964) [12,13]. The Γ term in his equations corresponds in a sense to the α-effect.

After establishing the fundamentals of mean-field electrodynamics, more or less detailed mean-field dynamo models for the Sun and also for the Earth and the planets were developed and investigated numerically. The first extended presentations of results were given in two papers by Steenbeck and Krause (1969) [14,15].

The Jena phase of the elaboration of the mean-field electrodynamics and dynamo theory, which was characterized by a continuous participation of Max Steenbeck, ended with his retirement in 1969. After that Fritz Krause and the author moved first to the Geomagnetic Institute in Potsdam and then to the Central Institute for Astrophysics at the same place. The first comprehensive representation of mean-field electrodynamics and dynamo theory developed so far is given in an extended article by Krause and Rädler (1971) [16].

Almost all papers on the Jena findings were written in German. This reflects the spirit of Max Steenbeck, who started his scientific carrier at a time when German was still an important language in physics and spent a large part of his life in Russian-speaking surroundings. Of course, our colleagues outside the German-language area only slowly became aware of our results. Fritz Krause and the author are very grateful to Keith Moffatt, who pointed out our results extensively in a paper (1970) [17], and to Paul Roberts and Michael Stix, who translated a large number of papers into English and compiled in 1971 a volume with these translations [18].

3.2 Experimental verifications and the danger of dynamo action in nuclear reactors

Max Steenbeck always strove for experimental verifications of theoretical findings and thought about their practical value. A short time after discovering the α-effect, already in 1967, on his initiative an experiment was carried out in the Institute of Physics in Riga, Latvia, under the directorship of I.M. Kirko. In a box with a system of channels, the "α-box," a liquid-sodium flow possessing helical features was organized. In the presence of an imposed magnetic field an electromotive force corresponding to the α-effect was indeed measured [19, 20].

There were numerous discussions between Jena and Riga colleagues on the realization of a dynamo in the laboratory, which reflects some features of the geodynamo. Already in 1967 Agris Gailitis in Riga theoretically investigated a model in which an α-effect results from a 2D flow pattern [21]. More precisely he considered a pattern in which, when referred to a Cartesian coordinate system, all three velocity components are in general unequal to zero but depend only on two coordinates. He found that, if a pattern of that kind is realized inside a spherical volume of a conducting fluid, it may indeed act as a dynamo. In this way he anticipated in some sense the dynamo with a spatially periodic flow pattern theoretically investigated by G.O. Roberts (1970) [22, 23], and also the proposal of a laboratory dynamo made by F.H. Busse (1975) [24], which has been realized later in the Karlsruhe experiment (see below).

The investigation of the first dynamo models of cosmic objects has brought some understanding of the self-excitation conditions of magnetic fields. In 1971 Max Steenbeck came up with the idea that self-excitation of magnetic fields cannot be excluded in the huge liquid-sodium loops of fast-breeder reactors. In these devices rather high magnetic Reynolds numbers are possible, and due to the construction of the pumps the flow patterns may show helical features. As a Foreign Member of the Soviet Academy of Sciences he explained this in a note to its President and pointed out that this implies a big danger for the reactor security. Stimulated by this note in 1974 a meeting took place in the reactor town of Obninsk near Moscow with Soviet reactor specialists, colleagues from the Riga magnetohydrodynamic community, Max Steenbeck and the author. Since that time attention was paid to the avoidance of self-excitation of magnetic fields in the liquid-metal circuits of nuclear reactors.

Careful magnetic measurements were carried out, for example in 1981, when the large Soviet BN600 reactor was put into operation [25,26]. The danger of self-excitation of magnetic fields in reactors has been independently recognized in Great Britain, too, as documented in papers by Bevir (1973) [27] and Pierson (1975) [28]. A series of investigations in that sense was carried out later for the French Superphenix reactor [29–31].

In 1975 Max Steenbeck proposed in a letter to the President of the Academy of Science of the G.D.R. and to several Soviet scientists in high positions to study dynamo action in a large liquid-sodium experiment. He argued that it could deliver not only valuable contributions to astrophysics and geophysics, but it would be at the same time of great importance for the reactor technology. It is interesting that he considered it necessary to have an active volume of about $10\,\mathrm{m}^3$ liquid sodium and volumetric flow rates not less than $10\,\mathrm{m}^3/\mathrm{s}$.

Only 18 years after his death, at the end of 1999, the first dynamo experiments have run successfully: the Riga experiment realizing a Ponomarenko-type dynamo [32,33] and the Karlsruhe experiment already mentioned above [34,35]. For sure, Max Steenbeck would have been highly delighted with these experiments. Incidentally, compared to the estimates he mentioned, both the active volume of sodium and the volumetric flow rates in the Karlsruhe experiment were smaller by a factor of about 3.

At the moment laboratory experiments with dynamos are under preparation at several places. A survey is given in a Special Issue of the journal "Magnetohydrodynamics" in 2002 (vol. 38, 1–2).

3.3 The solar dynamo, the geodynamo, etc.

As mentioned above, the first results for more or less detailed mean-field dynamo models for the Sun and also for the Earth and the planets were given in two papers by Steenbeck and Krause (1969) [14, 15]. These papers were the starting point for a large number of studies of that kind performed by a growing international community of colleagues. Further early results have been delivered by Deinzer et al.(1971) [36] and by Roberts and Stix (1972/1973) [37–39]. At the beginning only axisymmetric mean fields were considered. A complete analysis of mean-field dynamo models of course requires to admit also non-axisymmetric mean fields. First results of that kind were given in a paper of the author (1975) [40]. In addition to spherical dynamo models being of interest for the mentioned class of objects also others were considered in view of the magnetic fields found in galaxies, first in papers by Stix (1975) [41] and by White (1978) [42]. In between, a large number of studies has been carried out, which go beyond the kinematic consideration of dynamos, and take into account the back-reaction of the dynamo-generated magnetic field on the fluid motion; for references see, e.g. [43–45].

The early mean-field dynamo models for the Sun reflected many of the observed magnetic phenomena very well. The distribution of sunspots with respect to latitude and time could be reproduced under some assumptions on α-effect and angular velocity in the deeper layers of the convection zone, which were inaccessible to observations. So the dynamo theory served also as some kind of probe for these layers, which provided us in particular with information on the radial dependence of the angular velocity. Later, however, the conclusions drawn in this way came in conflict with the findings of helioseismology. Although there is hardly any doubt that the α-effect and differential rotation are key ingredients of the solar dynamo mechanism, many details are now again under debate [46].

Like the theory of the nearly symmetric dynamo, the mean-field dynamo theory, too, contributed much to the understanding of the geodynamo process. There are, however, several difficulties in developing a consequent detailed mean-field theory that reflects essential features of the geodynamo. One of them is the missing scale separation in the fluid flow inside the Earth's core (see below). The real progress in modeling the geodynamo came with direct numerical simulations on powerful computers; see, e.g. [47]. By the way, recently numerical results for a simple geodynamo model have been used to calculate the parameters of a traditional mean-field model with axisymmetric magnetic fields. This mean-field model indeed reproduces essential features that occur in the numerical simulations [48].

4 A critical view on mean-field electrodynamics

Let me now summarize some essentials of mean-field electrodynamics and point out a few misunderstandings or incorrect statements, which sometimes occur in the literature, as well as some open problems.

4.1 Basic concept

Consider electromagnetic processes in a homogeneous electrically conducting fluid. Assume that the magnetic field $\mathbf{B}$, the electric field $\mathbf{E}$ and the electric current density $\mathbf{J}$ are governed by Maxwell's equations and constitutive relations in the magnetohydrodynamic approximation, that is,

$$\mathbf{\nabla} \times \mathbf{E} = -\partial_t \mathbf{B}, \quad \mathbf{\nabla} \cdot \mathbf{B} = 0, \quad \mathbf{\nabla} \times \mathbf{B} = \mu \mathbf{J}, \quad \mathbf{J} = \sigma(\mathbf{E} + \mathbf{U} \times \mathbf{B}), \quad (1)$$

where μ and σ are the magnetic permeability and the electric conductivity, and $\mathbf{U}$ is the velocity of the fluid. Focus attention on $\mathbf{B}$. As a consequence of (1) it has to satisfy the induction equation

$$\eta \mathbf{\nabla}^2 \mathbf{B} + \mathbf{\nabla} \times (\mathbf{U} \times \mathbf{B}) - \partial_t \mathbf{B} = 0, \quad \mathbf{\nabla} \cdot \mathbf{B} = 0, \quad (2)$$

where η is the magnetic diffusivity $1/\mu\sigma$.

Assume further that both the electromagnetic fields $\mathbf{B}$, $\mathbf{E}$, and $\mathbf{J}$, as well as the fluid velocity $\mathbf{U}$, show in addition to variations on large scales in space and time also small-scale, e.g., turbulent variations. Then it is useful to introduce mean fields, which describe the large-scale behavior of such fields. A mean field $\overline{F}$ assigned to an original field F is defined by applying a proper averaging procedure to F. In the case of vector or tensor fields, averaging has to be restricted to their components with respect to the chosen coordinate system. Note that mean vector fields defined on the basis of a Cartesian coordinate system are therefore very different from those defined, e.g., with respect to a cylindrical or a spherical coordinate system. Various choices of the averaging procedure may be admitted. The only requirement is that the Reynolds averaging rules apply, that is,

$$\overline{F + G} = \overline{F} + \overline{G}, \quad \overline{\overline{F}G} = \overline{F}\,\overline{G}, \quad \overline{\partial_x F} = \partial_x \overline{F}, \quad \overline{\partial_t F} = \partial_t \overline{F}, \quad (3)$$

where F as well as G are arbitrary fields, and x stands for any space coordinate. With $F = \overline{F} + f$ and $G = \overline{G} + g$ this implies that $\overline{f} = \overline{g} = 0$ and further $\overline{FG} = \overline{F}\,\overline{G} + \overline{fg}$. The Reynolds rules apply exactly for statistical or ensemble averages and also, e.g., for azimuthal averages, defined by averaging over the coordinate φ in cylindrical or spherical coordinate systems, (r, φ, z) or (r, ϑ, φ). For the usual space average, defined for a given point by averaging certain surroundings of it, the Reynolds rules can be justified as an approximation if there is a clear separation of small and large scales in the spatial scale spectrum, sometimes labelled the "two–scale situation". This applies analogously to the usual time average, too.

Split now the magnetic and velocity fields according to $\mathbf{B} = \overline{\mathbf{B}} + \mathbf{b}$ and $\mathbf{U} = \overline{\mathbf{U}} + \mathbf{u}$ into mean fields $\overline{\mathbf{B}}$ and $\overline{\mathbf{U}}$ and deviations $\mathbf{b}$ and $\mathbf{u}$ from them, which are called "fluctuations" in the following. Averaging of Eq. (1) yields:

$$\boldsymbol{\nabla} \times \overline{\mathbf{E}} = -\partial_t \overline{\mathbf{B}}, \quad \boldsymbol{\nabla} \cdot \overline{\mathbf{B}} = 0, \quad \boldsymbol{\nabla} \times \overline{\mathbf{B}} = \mu \overline{\mathbf{J}}, \quad \overline{\mathbf{J}} = \sigma(\overline{\mathbf{E}} + \overline{\mathbf{U}} \times \overline{\mathbf{B}} + \boldsymbol{\mathcal{E}}). \quad (4)$$

From these equations, or by averaging (2), the mean-field induction equation

$$\eta \nabla^2 \overline{\mathbf{B}} + \boldsymbol{\nabla} \times (\overline{\mathbf{U}} \times \overline{\mathbf{B}} + \boldsymbol{\mathcal{E}}) - \partial_t \overline{\mathbf{B}} = 0, \quad \boldsymbol{\nabla} \cdot \overline{\mathbf{B}} = 0, \quad (5)$$

can be derived. Here $\boldsymbol{\mathcal{E}}$ is the mean electromotive force due to fluctuations,

$$\boldsymbol{\mathcal{E}} = \overline{\mathbf{u} \times \mathbf{b}}, \quad (6)$$

which is crucial for all mean-field electrodynamics.

For the following discussion of $\boldsymbol{\mathcal{E}}$, the velocity $\mathbf{U}$, that is, its mean part $\overline{\mathbf{U}}$ and the fluctuations $\mathbf{u}$, are considered as given. As can be concluded from (2) and (5), the fluctuations $\mathbf{b}$ are determined by

$$\eta \nabla^2 \mathbf{b} + \boldsymbol{\nabla} \times (\overline{\mathbf{U}} \times \mathbf{b} + \mathbf{G}) - \partial_t \mathbf{b} = -\boldsymbol{\nabla} \times (\mathbf{u} \times \overline{\mathbf{B}}), \quad \boldsymbol{\nabla} \cdot \mathbf{b} = 0,$$
$$\mathbf{G} = \mathbf{u} \times \mathbf{b} - \overline{\mathbf{u} \times \mathbf{b}}. \quad (7)$$

These equations imply that $\mathbf{b}$ can be considered as a sum $\mathbf{b}^{(0)} + \mathbf{b}^{(\overline{B})}$, where $\mathbf{b}^{(0)}$ is independent of $\overline{\mathbf{B}}$, and $\mathbf{b}^{(\overline{B})}$ is linear and homogeneous in $\overline{\mathbf{B}}$. This in turn leads to

$$\boldsymbol{\mathcal{E}} = \boldsymbol{\mathcal{E}}^{(0)} + \boldsymbol{\mathcal{E}}^{(\overline{B})}, \quad (8)$$

with $\mathcal{E}^{(0)}$ being independent of $\overline{\mathbf{B}}$, and $\mathcal{E}^{(\overline{B})}$ being linear and homogeneous in $\overline{\mathbf{B}}$. As can be concluded from very general arguments, $\mathcal{E}^{(\overline{B})}$ can be represented in the form

$$\mathcal{E}_i^{(\overline{B})}(\mathbf{x}, t) = \int_0^\infty \int_\infty K_{ij}(\mathbf{x}, t; \boldsymbol{\xi}, \tau) \, \overline{B}_j(\mathbf{x} - \boldsymbol{\xi}, t - \tau) \, \mathrm{d}^3\xi \, \mathrm{d}\tau \,, \qquad (9)$$

where the kernel K_{ij} is determined, apart from η and from initial and boundary conditions for $\mathbf{b}$, by $\overline{\mathbf{U}}$ and $\mathbf{u}$. Here Cartesian coordinates are used and the summation convention is adopted.

In many cases, in particular if $\mathbf{u}$ corresponds to turbulence, the kernel K_{ij} is only in some range of small $|\boldsymbol{\xi}|$ and τ markedly different from zero. This suggests to expand $\overline{B}_j(\mathbf{x} - \boldsymbol{\xi}, t - \tau)$ under the integral in a Taylor series with respect to $\boldsymbol{\xi}$ and τ,

$$\overline{B}_j(\mathbf{x} - \boldsymbol{\xi}, t - \tau) = \overline{B}_j(\mathbf{x}, t) - \frac{\partial \overline{B}_j(\mathbf{x}, t)}{\partial x_k} \xi_k - \frac{\partial \overline{B}_j(\mathbf{x}, t)}{\partial t} \tau - \cdots \,. \qquad (10)$$

In most of the traditional representations of mean-field electrodynamics, only the first two terms on the right-hand side are taken into account. Then it follows from (8), (9), and (10) that

$$\mathcal{E}_i = \mathcal{E}_i^{(0)} + a_{ij}\overline{B}_j + b_{ijk}\frac{\partial \overline{B}_j}{\partial x_k} \,, \qquad (11)$$

where

$$a_{ij} = \int_0^\infty \int_\infty K_{ij}(\mathbf{x}, t; \boldsymbol{\xi}, \tau) \, \mathrm{d}^3\xi \, \mathrm{d}\tau,$$

$$b_{ijk} = -\int_0^\infty \int_\infty K_{ij}(\mathbf{x}, t; \boldsymbol{\xi}, \tau) \, \xi_k \, \mathrm{d}^3\xi \, \mathrm{d}\tau \,. \qquad (12)$$

Assume for a simple (somewhat academic) example that there is no mean motion, $\overline{\mathbf{U}} = \mathbf{0}$, and $\mathbf{u}$ corresponds to a homogeneous isotropic turbulence. Then symmetry arguments lead to $\mathcal{E}^{(0)} = \mathbf{0}$, further to $a_{ij} = \alpha \, \delta_{ij}$ and $b_{ijk} = \beta \, \epsilon_{ijk}$, and consequently to

$$\mathcal{E} = \alpha\overline{\mathbf{B}} - \beta\boldsymbol{\nabla} \times \overline{\mathbf{B}} \,, \qquad (13)$$

where the two coefficients, α and β, are independent of position and are determined by $\mathbf{u}$. The term $\alpha\overline{\mathbf{B}}$ describes the α-effect. If in addition the $\mathbf{u}$-field is on the average mirror-symmetric, that is, all averages depending on $\mathbf{u}$ are invariant under reflexion of this field at a plane or a point, α turns out to be equal to zero. Hence the occurrence of an α-effect requires a deviation of $\mathbf{u}$ from mirror-symmetry. In the mirror-symmetric case the mean-field version of Ohm's law, that is, the last relation in (4), takes the form $\overline{\mathbf{J}} = \sigma_m\overline{\mathbf{E}}$ with the mean-field conductivity $\sigma_m = \sigma/(1 + \beta/\eta)$. The mean-field induction Eq. (5) then formally agrees with (2) after replacing there η by the mean-field diffusivity $\eta_{\mathrm{m}} = \eta + \beta$. In a wide range of assumptions β proves to be positive. Return now to the more general relation (11) for $\boldsymbol{\mathcal{E}}$. It is equivalent to

$$\boldsymbol{\mathcal{E}} = \boldsymbol{\mathcal{E}}^{(0)} - \boldsymbol{\alpha} \cdot \overline{\mathbf{B}} - \boldsymbol{\gamma} \times \overline{\mathbf{B}} - \boldsymbol{\beta} \cdot (\boldsymbol{\nabla} \times \overline{\mathbf{B}}) - \boldsymbol{\delta} \times (\boldsymbol{\nabla} \times \overline{\mathbf{B}}) - \boldsymbol{\kappa} \cdot (\boldsymbol{\nabla}\overline{\mathbf{B}})^{(s)} \; ; \; (14)$$

see, e.g., [45]. Here $\boldsymbol{\alpha}$ and $\boldsymbol{\beta}$ are symmetric second-rank tensors, $\boldsymbol{\gamma}$ and $\boldsymbol{\delta}$ are vectors, and $\boldsymbol{\kappa}$ is a third-rank tensor, which may be assumed to be symmetric in the indices connecting it with $(\boldsymbol{\nabla}\overline{\mathbf{B}})^{(s)}$. The latter is the symmetric part of the gradient tensor of $\overline{\mathbf{B}}$, with respect to Cartesian coordinates given by $(\boldsymbol{\nabla}\overline{\mathbf{B}})^{(s)}_{jk} = \frac{1}{2}(\partial \overline{B}_j / \partial x_k + \partial \overline{B}_k / \partial x_j)$. Of course, the quantities $\boldsymbol{\alpha}$, $\boldsymbol{\gamma}$, $\boldsymbol{\beta}$, $\boldsymbol{\delta}$ and $\boldsymbol{\kappa}$ can be expressed by the components of a_{ij} and b_{ijk}. The term with $\boldsymbol{\alpha}$ in (14) describes again the α-effect, which is now in general anisotropic. That with $\boldsymbol{\gamma}$ corresponds to an advection of the mean magnetic field. Thus, the effective velocity responsible for advection is $\overline{\mathbf{U}} - \boldsymbol{\gamma}$. The term with $\boldsymbol{\beta}$ can be interpreted by introducing a mean-field conductivity or a mean-field diffusivity, which are, in general, no longer isotropic. That with $\boldsymbol{\delta}$ can be included in this interpretation. Whereas the conductivity or diffusivity tensors introduced on the basis of $\boldsymbol{\beta}$ alone are symmetric, those involving $\boldsymbol{\delta}$ have also non-symmetric parts. The term with $\boldsymbol{\kappa}$ is more difficult to interpret.

As is well known, the α-effect is in general capable of dynamo action. In the absence of any shear in the mean motion there is the possibility of an α^2 dynamo. A sufficiently strong shear, e.g., by differential rotation, opens up the possibility of an $\alpha\omega$ dynamo. However, even in the absence of any α-effect, dynamos are possible due to the combination of effects described by the $\boldsymbol{\delta}$ or $\boldsymbol{\kappa}$ terms in (14) with shear. For more details on mean-field dynamos see, e.g., [43–45].

Some comments

1. It is sometimes said that mean-field electrodynamics or mean-field dynamo theory applies only under the assumption of a clear separation between the large and small scales in space and time. This is not generally correct. An assumption of that kind is of some interest in view of the Reynolds rules. For statistical or azimuthal averages, however, these rules apply independent of any such an assumption. To justify them as an approximation for the usual space average as explained above only a scale separation in space, for the time average only one in time are necessary. When working with the integral representation (9) for $\boldsymbol{\mathcal{E}}$ there is no further reason for any scale separation. Of course, the more special relations (11) or (14) are based on assumptions concerning space and time scales of $\overline{\mathbf{B}}$.

2. In some representations of mean-field electrodynamics the part $\boldsymbol{\mathcal{E}}^{(0)}$ of the mean electromotive force $\boldsymbol{\mathcal{E}}$ is ignored. This is justified if in the case $\overline{\mathbf{B}} = 0$ the fluctuations $\mathbf{b}$ decay to zero, that is, the turbulence is of pure hydrodynamic nature. It is not generally justified if also for $\overline{\mathbf{B}} = 0$ a real magnetohydrodynamic turbulence exists. Then $\boldsymbol{\mathcal{E}}^{(0)}$ vanishes only in particular cases,

e.g., if this turbulence is isotropic and therefore does not allow the definition of a vector. Possibilities of non-zero $\boldsymbol{\mathcal{E}}^{(0)}$ have been discussed, e.g., by Rädler [49] or, in particular for a case with non-zero $\boldsymbol{\nabla} \times \overline{\mathbf{U}}$, by Yoshizawa [50]. A non-zero $\boldsymbol{\mathcal{E}}^{(0)}$ may create a mean magnetic field $\overline{\mathbf{B}}$ from a state with $\overline{\mathbf{B}} = \mathbf{0}$. It loses its importance as soon as $\overline{\mathbf{B}}$ exceeds some magnitude.

3. In general, the mean-field induction Eq. (5) has to be completed by (8) and (9), and has therefore to be considered as an integro-differential equation. Only if (8) and (9) are reduced to (11) or (14) with $\boldsymbol{\mathcal{E}}^{(0)} = \mathbf{0}$, as is done in most applications, Eq. (5) has again the same mathematical character as (2). It is however important to note that (11) and (14) apply only under restrictive assumptions on the variations of $\overline{\mathbf{B}}$ in space and time.

4. Although the above derivations were done mainly in view of situations with turbulence, no specific properties of $\mathbf{u}$ were used which exclude the application of the results given here to cases in which $\mathbf{u}$ corresponds, for example, to regular flow patterns or shows space or time behaviors other than those of turbulence.

4.2 First-order smoothing and other approximations

A central problem in the elaboration of mean-field electrodynamics is the determination of the mean electromotive force $\boldsymbol{\mathcal{E}}$ for a given fluid velocity, that is, given $\overline{\mathbf{U}}$ and $\mathbf{u}$. Many calculations are based on the "second–order correlation approximation" (SOCA), sometimes also called the "first–order smoothing approximation" (FOSA). This approximation is based on Eq. (7) for $\mathbf{b}$, but assumes some smallness of $\mathbf{u}$ such that it is justified to ignore there the term $\mathbf{G}$. Consider for the sake of simplicity the case of an infinitely extended fluid with zero mean motion, $\overline{\mathbf{U}} = \mathbf{0}$. Then Eq.(7) can be solved analytically and an expression for K_{ij} in (9) and (12) can be derived,

$$K_{ij}(\mathbf{x}, t; \boldsymbol{\xi}, \tau) = (\epsilon_{ilm}\delta_{nj} - \epsilon_{ilj}\delta_{mn}) \frac{\partial G(\xi, \tau)}{\partial \xi} \frac{\xi_n}{\xi} Q_{lm}(\mathbf{x}, t; -\boldsymbol{\xi}, -\tau), \quad (15)$$

where G is a Green's function

$$G(\xi, \tau) = (4\pi\eta\tau)^{-3/2} \exp(-\xi^2/4\eta\tau), \quad (16)$$

and Q_{lm} is the second-rank correlation tensor of $\mathbf{u}$ defined by

$$Q_{lm}(\mathbf{x}, t; \boldsymbol{\xi}, \tau) = \overline{u_l(\mathbf{x}, t)u_m(\mathbf{x} + \boldsymbol{\xi}, t + \tau)}. \quad (17)$$

In order to define special cases for the calculation of $\boldsymbol{\mathcal{E}}$, a correlation length and a correlation time, λ_{c} and τ_{c}, of the velocity field $\mathbf{u}$ is introduced such that Q_{lm} is no longer markedly different from zero if ξ/λ_{c} or τ/τ_{c} markedly exceed unity. Consider then the dimensionless quantity $q = \lambda_{\mathrm{c}}^2/\eta\tau_{\mathrm{c}}$. The limits $q \to \infty$ and $q \to 0$ are called "high–conductivity limit" and "low–conductivity limit", respectively. For most astrophysical applications the high-conductivity limit is of particular interest. A sufficient condition for the validity of the second-order approximation in the high-conductivity limit reads $u\tau_{\mathrm{c}}/\lambda_{\mathrm{c}} \ll 1$, in the low-conductivity limit instead $u\lambda_c/\eta \ll 1$, where u means a characteristic magnitude of $\mathbf{u}$.

Return now to the simple example considered above, in which $\overline{\mathbf{U}} = \mathbf{0}$ and $\mathbf{u}$ corresponds to a homogeneous isotropic turbulence. A calculation of the coefficient α in (13) with the help of (12), (15), (16), and (17), delivers

$$\alpha = -\frac{1}{3} \int_{\infty} \int_0^\infty G(\xi, \tau) \overline{\mathbf{u}(\mathbf{x}, t) \cdot (\nabla \times \mathbf{u}(\mathbf{x} + \xi, t - \tau))} \, d^3\xi \, d\tau. \tag{18}$$

In the high-conductivity limit, $q \to \infty$, this turns into

$$\alpha = -\frac{1}{3} \int_0^\infty \overline{\mathbf{u}(\mathbf{x}, t) \cdot (\nabla \times \mathbf{u}(\mathbf{x}, t - \tau))} \, d\tau, \tag{19}$$

or, provided $\overline{\mathbf{u}(\mathbf{x}, t) \cdot (\nabla \times \mathbf{u}(\mathbf{x}, t))}$ does not vanish,

$$\alpha = -\frac{1}{3} \overline{\mathbf{u}(\mathbf{x}, t) \cdot (\nabla \times \mathbf{u}(\mathbf{x}, t - \tau))} \, \tau_c^{(\alpha)}, \tag{20}$$

with $\tau_c^{(\alpha)}$ determined by equating the two right-hand sides. The corresponding result for the low-conductivity limit, $q \to 0$, reads

$$\begin{aligned} \alpha &= -\frac{1}{3\eta} \int_0^\infty \overline{\mathbf{u}(\mathbf{x}, t) \cdot (\nabla \times \mathbf{u}(\mathbf{x} + \xi, t))} \, \xi \, d\xi \\ &= -\frac{1}{3\eta} \int_0^\infty \overline{\mathbf{u}(\mathbf{x}, t) \cdot (\xi \times \mathbf{u}(\mathbf{x} + \xi, t))} \, \frac{d\xi}{\xi}. \end{aligned} \tag{21}$$

Note that the integrands do not depend on the direction of ξ but only on ξ. With the vector potential ψ of $\mathbf{u}$ defined by $\mathbf{u} = \nabla \times \psi + \nabla \cdots$ and $\nabla \cdot \psi = 0$ this can be rewritten in the simple form

$$\alpha = -\frac{1}{3\eta} \overline{\psi(\mathbf{x}, t) \cdot (\nabla \times \psi(\mathbf{x}, t))}. \tag{22}$$

In the high-conductivity limit it is the mean kinetic helicity $\overline{\mathbf{u} \cdot (\nabla \times \mathbf{u})}$, in the low-conductivity limit the related quantity $\overline{\psi \cdot (\nabla \times \psi)}$, which are crucial for the α-effect. Both indicate the existence of helical features in the flow pattern and vanish for mirror-symmetric turbulence.

The sufficient conditions for the applicability of the second-order correlation approximation given above are rather narrow for most of the applications. It is basically possible to proceed from the second-order correlation approximation to approximations of arbitrarily high order with a larger range of validity. However, calculations of that kind for specific cases are extremely tedious and have been done so far only for very simple examples. Several other approaches to results with a wider range of validity have been also explored, e.g. [51], but their correctness is still under debate [52]. So the calculation of coefficients like α, γ, β, δ, and κ in a wide range of validity remains a challenge. As already mentioned, recently such coefficients have been extracted from results of direct numerical simulations [48]. This also opens up the possibility to check the results obtained analytically.

Some comments

1. Almost all dynamo models studied in relation to cosmic objects work with the α-effect. In the mean-field induction equation, often the relevant term of $\mathcal{E}$ is simply taken in the from $\alpha\overline{\mathbf{B}}$, which corresponds to the case of isotropic turbulence. It is then further assumed that α differs only by a factor from the mean kinetic helicity $\overline{\mathbf{u} \cdot (\boldsymbol{\nabla} \times \mathbf{u})}$. Even if there should be reasons to work with $\alpha\overline{\mathbf{B}}$ one should have in mind that the proportionality of α to $\overline{\mathbf{u} \cdot (\boldsymbol{\nabla} \times \mathbf{u})}$ is a specific result applying in second-order correlation approximation and high-conductivity limit only. In general, however, it is not the coefficient α in the sense of (13) and (20), but the components of the tensor $\boldsymbol{\alpha}$ which enter the mean-field dynamo equations. In the case of an $\alpha\omega$ dynamo, e.g., mainly the component $\alpha_{\varphi\varphi}$ is of interest, where φ indicates again the azimuthal coordinate. It is the trace of the tensor $\boldsymbol{\alpha}$, which in second-order correlation approximation and high–conductivity limit, is proportional to $\overline{\mathbf{u} \cdot (\boldsymbol{\nabla} \times \mathbf{u})}$. The individual components are given by other properties of $\mathbf{u}$.

2. Simplified considerations on α-effect dynamos for astrophysical objects, for which the high-conductivity limit is appropriate, have led to the opinion that any dynamo requires a non-vanishing kinetic helicity of the fluid flow. This is definitely wrong for several reasons. As mentioned above, in the low-conductivity limit the α-effect is no longer determined by the kinetic helicity. In addition, as mentioned above, there are mean-field dynamos which work in the absence of any α-effect, e.g., due to a combination of effects described by $\boldsymbol{\delta}$ and $\boldsymbol{\kappa}$ with a mean shear.

3. Concerning the dependence of the α-effect on the kinetic helicity, a dynamo proposed by G. O. Roberts [23], with a steady spatially periodic flow pattern depending on two Cartesian coordinates only, say x and y, deserves some attention. It can easily be interpreted within the mean-field concept, with averaging over x and y, and appears then as an α-effect dynamo [53]. Roberts considered a flow pattern with a non-zero helicity. It can however be shown that dynamo action and α-effect do not vanish if it is modified such that the helicity (not only the mean helicity) is everywhere equal to zero. The α-effect is instead related to a quantity of the type $\boldsymbol{\psi} \cdot (\boldsymbol{\nabla} \times \boldsymbol{\psi})$. In another interesting example of a dynamo investigated by Zheligovsky and Galloway [54] both $\mathbf{u} \cdot (\boldsymbol{\nabla} \times \mathbf{u})$ and $\boldsymbol{\psi} \cdot (\boldsymbol{\nabla} \times \boldsymbol{\psi})$ (not only their mean values) are equal to zero everywhere.

4.3 Magnetic quenching

In general, a magnetic field acts on the motion of the fluid via the Lorentz force. In this way also the mean-field coefficients like $\boldsymbol{\alpha}$, $\boldsymbol{\gamma}$, β, $\boldsymbol{\delta}$, and $\boldsymbol{\kappa}$ are influenced by this magnetic field. Assume as a simple example a turbulence which is homogeneous and isotropic in the limit of vanishing magnetic field. Then symmetry arguments show that for finite $\overline{\mathbf{B}}$

$$\mathcal{E} = \left\{ \alpha + \tilde{\alpha} \left(\overline{\mathbf{B}} \cdot (\boldsymbol{\nabla} \times \overline{\mathbf{B}}) \right) \right\} \overline{\mathbf{B}} - \tilde{\gamma} \, \boldsymbol{\nabla} \overline{\mathbf{B}}^2 \times \overline{\mathbf{B}} - \beta \, \boldsymbol{\nabla} \times \overline{\mathbf{B}} \,, \qquad (23)$$

where the coefficients α, $\tilde{\alpha}$, $\tilde{\gamma}$, and β may be considered as functions of $\mathbf{u}^{(0)}$, which means $\mathbf{u}$ in the limit of small $\overline{\mathbf{B}}$, and of $|\overline{\mathbf{B}}|$; see, e.g., [45, 55]. In the limit of small $\overline{\mathbf{B}}$, of course, α and β agree with those in (13), and $\tilde{\alpha}$ and $\tilde{\gamma}$ vanish. Simple considerations suggest that, e.g., $|\alpha|$ is reduced, or "quenched," if $|\overline{\mathbf{B}}|$ grows. Investigations on the dependence of α, $\tilde{\alpha}$, $\tilde{\gamma}$, and β on $\overline{\mathbf{B}}$ have, of course, to be done on the basis of the induction equation and the momentum balance, that is, the Navier–Stokes equation. They lead from mean-field electrodynamics to the more comprehensive mean-field magnetohydrodynamics.

Many attempts have been made to determine the dependence of quantities like α on $\overline{\mathbf{B}}$. Specific results, in particular of numerical simulations, have been interpreted in the sense of a drastic α-quenching, which would prevent magnetic fields from growing to such magnitudes as are observed in real objects, e.g., at the Sun. There is a persistent, controversial debate on this "catastrophical quenching" and mechanisms which avoid it; see, e.g., [56].

A great challenge in the further elaboration of mean-field electrodynamics, or mean-field magnetohydrodynamics, consists in establishing a reliable theory of the behavior of quantities like $\boldsymbol{\alpha}$, $\boldsymbol{\gamma}$, $\boldsymbol{\beta}$, $\boldsymbol{\delta}$, and $\boldsymbol{\kappa}$ in a regime with finite or even large $\overline{\mathbf{B}}$.

A comment

Most of the investigations on the problems addressed here start from the very beginning with the induction equation and the momentum balance. It should be noted that the general relations for the calculation of $\mathcal{E}$ and quantities like $\boldsymbol{\alpha}$, $\boldsymbol{\gamma}$, $\boldsymbol{\beta}$, $\boldsymbol{\delta}$, and $\boldsymbol{\kappa}$ delivered by the second-order correlation approximation or its generalization to higher orders do not lose their validity if $\mathbf{u}$ or the correlation tensors like Q_{lm} depend on $\overline{\mathbf{B}}$. They may be specified by inserting $\mathbf{u}$ as determined by the momentum balance including the Lorentz force. In that sense they can well be a starting point for the solution of the problems discussed.

References

1. von Engel A, Steenbeck M (1932, 1934) Elektrische Gasentladungen, ihre Physik und Technik, Vols 1 and 2. Springer, Berlin
2. Rompe R, Steenbeck M (1939) Der Plasmazustand der Gase. In: Ergebnisse der exakten Naturwissenschaften, Springer, Berlin 18:257–376
3. Steenbeck M (1977) Impulse und Wirkungen – Schritte auf meinem Lebensweg. Verlag der Nation, Berlin
4. Parker EN (1955) Hydromagnetic dynamo models. Astrophys J 122:293–314
5. Parker EN (1957) The solar hydromagnetic dynamo. Proc Natl Acad Sci 43:8–14
6. Steenbeck M, Krause F, Rädler K-H (1963) Elektromagnetische Eigenschaften turbulenter Plasmen. Sitzungsber Dt Akad Wiss Berlin Kl Math Phys Techn 1/1963

7. Steenbeck M, Krause F, Rädler K-H (1966) Berechnung der mittleren Lorentz-Feldstärke $\overline{\mathbf{v} \times \mathbf{b}}$ für ein elektrisch leitendes Medium in turbulenter, durch Coriolis–Kräfte beeinflußter Bewegung. Z Naturforsch 21a:369–376

8. Rädler K-H (1966) Zur Elektrodynamik turbulent bewegter leitender Medien. PhD thesis, Friedrich-Schiller-Universität Jena

9. Rädler K-H (1968) Zur Elektrodynamik turbulent bewegter leitender Medien I Grundzüge der Elektrodynamik der mittleren Felder. Z Naturforsch 23a:1841–1851

10. Rädler K-H (1968) Zur Elektrodynamik turbulent bewegter leitender Medien II Turbulenzbedingte Leitfähigkeits- und Permeabilitätsänderungen. Z Naturforsch 23a:1851–1860

11. Krause F (1968) Eine Lösung des Dynamoproblems auf der Grundlage einer linearen Theorie der magnetohydrodynamischen Turbulenz. Habilitationsschrift, Friedrich-Schiller-Universität Jena

12. Braginskii SI (1964) Self-excitation of a magnetic field during the motion of a highly conducting fluid. Sov Phys JETP 20:726–735

13. Braginskii SI (1964) Theory of the hydromagnetic dynamo. Sov Phys JETP 20:1462–1471

14. Steenbeck M, Krause F (1969) Zur Dynamotheorie stellarer und planetarer Magnetfelder I Berechnung sonnenähnlicher Wechselfeldgeneratoren. Astron Nachr 291:49–84

15. Steenbeck M, Krause F (1969) Zur Dynamotheorie stellarer und planetarer Magnetfelder II Berechnung planetenähnlicher Gleichfeldgeneratoren. Astron Nachr 291:271–286

16. Krause F, Rädler K-H (1971) Elektrodynamik der mittleren Felder in turbulenten leitenden Medien und Dynamotheorie. In: Rompe R, Steenbeck M (eds) Ergebnisse der Plasmaphysik und der Gaselektronik. Akademie, Berlin 2:1–154

17. Moffatt HK (1970) Turbulent dynamo action at low magnetic Reynolds number. J Fluid Mech 41:435–452

18. Roberts PH, Stix M (1971) The Turbulent Dynamo – A translation of a series of papers by F Krause, K-H Rädler and M Steenbeck. Technical Report NCAR-TN/IA-60, National Center for Atmospheric Research, Boulder, Colorado. Electronic version available in NCAR Library

19. Steenbeck M, Kirko IM, Gailitis A, Klawina AP, Krause F, Laumanis IJ, Lielausis OA (1967) Der experimentelle Nachweis einer elektromotorischen Kraft längs eines äußeren Magnetfeldes, induziert durch eine Strömung flüssigen Metalls (α–Effekt). Mber Dtsch Akad Wiss Berlin 9:714–719

20. Steenbeck M, Kirko IM, Gailitis A, Klawina AP, Krause F, Laumanis IJ, Lielausis OA (1968) Experimental discovery of the electromotive force along the external magnetic field induced by a flow of liquid metal (α–effect). Sov Phys Dokl 13:443–445

21. Gailitis A (1967) Conditions of the self–excitation for a laboratory model of the geomagnetic dynamo. Magnetohydrodynamics 3:45–54

22. Roberts GO (1970) Spatially periodic dynamos. Phil Trans R Soc Lond A266:535–558

23. Roberts GO (1972) Dynamo action of fluid motions with two-dimensional periodicity. Phil Trans R Soc Lond A271:411–454

24. Busse FH (1975) A model of the geodynamo. Geophys J R Astron Soc 42:437–459

25. Kirko IM, Mitenkov FM, Barannikov VA (1981) Observation of MHD phenomena in the liquid–metal volume of the first loop of the fast–neutron reactor BN600 of the Beloyarskaia nuclear power plant. Dokl Akad Nauk SSSR 257(4):861–863 (In Russian)

26. Kirko GE (1985) Generation and self–excitation of a magnetic field in technical devices. Nauka, Moscow (In Russian)

27. Bevir MK (1973) Possibility of electromagnetic self-excitation in liquid metal flows in fast reactors. J Br Nucl Soc 12(4):455–458

28. Pierson ES (1975) Electromagnetic self-excitation in the liquid-metal fast breeder reactor. Nucl Sci Eng 57:155–163

29. Plunian F, Alemany A, Marty Ph (1995) Influence of magnetohydrodynamic parameters on electromagnetic self-excitation in the core of a fast breeder reactor. Magnetohydrodynamics 31:382–390

30. Plunian F, Marty Ph, Alemany A (1999) Kinematic dynamo action in a network of screw motions; application to the core of a fast breeder reactor. J Fluid Mech 382:137–154

31. Alemany A, Marty Ph, Plunian F, Soto J (2000) Experimental investigation of dynamo effect in the secondary pumps of the fast breeder reactor Superphenix. J Fluid Mech 403:263–276

32. Gailitis A, Lielausis O, Dement'ev S, Platacis E, Cifersons A, Gerbeth G, Gundrum T, Stefani F, Christen M, Hänel H, Will G (2000) Detection of a flow induced magnetic field eigenmode in the Riga dynamo facility. Phys Rev Lett 84:4365–4368

33. Gailitis A, Lielausis O, Platacis E, Gerbeth G, Stefani F (2001) On the results of the Riga dynamo experiments. Magnetohydrodynamics 37:71–79

34. Müller U, Stieglitz R (2000) Can the Earth's magnetic field be simulated in the laboratory? Naturwissenschaften 87:381–390

35. Stieglitz R, Müller U (2001) Experimental demonstration of a homogeneous two-scale dynamo. Phys Fluids 13:561–564

36. Deinzer W, Stix M (1971) On the eigenvalues of Krause–Steenbeck's solar dynamo. Astron Astrophys 12:111–119

37. Roberts PH (1972) Kinematic dynamo models. Phil Trans R Soc Lond A272:663–703

38. Roberts PH, Stix M (1972) Alpha–effect dynamos by the Bullard–Gellman formalism. Astron Astrophys 18:453–466

39. Stix M (1973) Spherical $\alpha\omega$–dynamos by a variational method. Astron Astrophys 24:275–281

40. Rädler K-H (1975) Some new results on the generation of magnetic fields by dynamo action. Mem Soc R Sci Liege VIII:109–116

41. Stix M (1975) The galactic dynamo. Astron Astrophys 47:243–254

42. White MP (1978) Numerical models of the galactic dynamo. Astron Nachr 299:209–216

43. Krause F, Rädler K-H (1980) Mean–field magnetohydrodynamics and dynamo theory. Akademie, Berlin and Pergamon Press, Oxford

44. Rädler K-H (1995) Cosmic dynamos. Rev Mod Astron 8:295–321

45. Rädler K-H (2000) The generation of cosmic magnetic fields. In: Page D, Hirsch JG (eds) From the Sun to the Great Attractor (1999 Guanajuato Lectures in Astrophysics) Lecture Notes in Physics. Springer, Berlin, pp: 101–172

46. Ossendrijver M (2003) The solar dynamo. Astron Astrophys Rev 11:287–367

47. Roberts PH, Glatzmaier GA (2000) Geodynamo theory and simulations. Rev Mod Phys 72:1081–1123
48. Schrinner M, Rädler K-H, Schmitt D, Rheinhardt M, Christensen U (2005) Mean–field view on rotating magnetoconvection and a geodynamo model. Astron Nachr 326:245–249
49. Rädler K-H (1976) Mean–field magnetohydrodynamics as a basis of solar dynamo theory. In: Bumba V, Kleczek J (eds) Basic Mechanisms of Solar Activity. D Reidel Publishing Company Dordrecht: 323–344
50. Yoshizawa A, Yokoi N (1993) Turbulent magnetohydrodynamic dynamo for accretion disks using the cross-helicity effect. Astrophys J 407:540–548
51. Rädler K-H, Kleeorin N, Rogachevski I (2003) The mean electromotive force for MHD turbulence: The case of a weak mean magnetic field and slow rotation. Geophys Astrophys Fluid Dyn 97:249–274
52. Rädler K-H, Rheinhardt M (2006) Mean–field electrodynamics: critical analysis of various analytical approaches to the mean electromotive force. (In preparation)
53. Rädler K-H, Rheinhardt M, Apstein E, Fuchs H (2002) On the mean–field theory of the Karlsruhe dynamo experiment I Kinematic theory. Magnetohydrodynamics 38:41–71
54. Zheligovsky VA, Galloway DJ (1998) Dynamo action in Christopherson hexagonal flow. Geophys Astrophys Fluid Dyn 88:277–293
55. Roberts PH (1971) Dynamo theory. In: Reid WH (ed) Lectures on Applied Mathematics. American Mathematical Society, Providence, Cambridge, MA, 14:129–206
56. Brandenburg A, Subramanian K (2005) Astrophysical magnetic fields and nonlinear dynamo theory. Phys Rep 417:1-209

Astrophysical MHD: The Early Years

Leon Mestel

University of Sussex, Department of Physics and Astronomy, Brighton, East
Sussex BN1 9QH, United Kingdom (L.Mestel@sussex.ac.uk)

1 Cosmical magnetic fields: Earth, Sun, magnetic stars, interstellar medium; historical landmarks

(1) Halley, working before the discoveries of Oersted, Ampère, Faraday, and
Henry, pictured the Earth's interior with two massive blocks of permanently
magnetized material (lodestone): an outer shell and a concentric inner nucleus.
From the observed magnetic variations, he inferred the existence of a fluid
domain.

(2) Galileo and contemporaries observed dark patches on the Sun at low
latitudes – sunspots – from which they inferred the solar rotation.

(3) In the nineteenth century, Schwabe, Carrington, Spörer discovered the
11-year solar cycle and the associated latitude drift of sunspot pairs. The first
hint of a magnetic connection came with the correlation between the sunspot
cycle and geomagnetic storms.

(4) In 1908, Hale applied the recently discovered Zeeman effect to infer
sunspot fields of several kG. The observed field reversal in the leading spot
of a pair showed that the 11-year cycle is in fact one half of a basic 22-year
cycle. In 1913, Hale measured a weak general solar field.

(5) In 1959, Babcock discovered that the general solar field reverses along
with the sunspot cycle. Later work showed the existence of solar-type magnetic
activity in other "late-type" stars – the "Solar-Stellar Connection".

(6) From 1947, strongly magnetic "early-type" stars were discovered, forming a subclass of the "chemically peculiar" (CP) stars. The fields appear to
be stable, but show periodic variations, including polarity reversal in some.
The most plausible explanation is that we are witnessing the rotation of a
magnetic structure that is not symmetric about the rotation axis, e.g., the
"oblique rotator", with the magnetic field axis inclined to the rotation axis.

(7) From 1949 on, a galactic magnetic field has been inferred from several independent phenomena. The observed polarization of starlight is due to
selective absorption by magnetically aligned, non-spherical dust grains, yielding the optical electric vector parallel to the galactic **B**. Synchrotron radiation

S. Molokov et al. (eds.), Magnetohydrodynamics – Historical Evolution and Trends,
73–84. © 2007 *Springer.*

from relativistic electrons gyrating about $\mathbf{B}$ is recognized by its spectrum and by its polarization with $\mathbf{E}$ perpendicular to $\mathbf{B}$. Faraday rotation of the polarization of waves from radio galaxies is recognizable by its characteristic λ^2-dependence. And the Zeeman effect on e.g., the 21 cm line emitted by atomic hydrogen is detected, especially from clouds massive enough for gravitational amplification of the density and so also of the field strength.

2 Basic electrodynamics

1. Maxwell's equations, truncated – the displacement current is ignorable in high-conductivity media for low-frequency phenomena. In Gaussian units,

$$\nabla \times \mathbf{B} = \frac{4\pi}{c}\mathbf{j}, \qquad \nabla \times \mathbf{E} = -\frac{1}{c}\frac{\partial \mathbf{B}}{\partial t},$$

$$\nabla \cdot \mathbf{B} = 0, \qquad \rho_e = \frac{\nabla \cdot \mathbf{E}}{4\pi}. \tag{1}$$

2. Generalization of simplest form of Ohm's law $\mathbf{j} = \sigma\mathbf{E}$ to a moving conductor:

$$\mathbf{E}' \equiv \mathbf{E} + \frac{\mathbf{v} \times \mathbf{B}}{c} = \frac{\mathbf{j}}{\sigma}, \tag{2}$$

where $\mathbf{E}'$ is the electric field measured in the frame moving with the local bulk velocity $\mathbf{v}$ – the local 'rest-frame'. In non-relativistic problems, terms of order $(v/c)^2$ are ignorable. The theory is applicable to moving solids as well as to fluids, e.g., to a dynamo armature.

High-conductivity, large-length scales $\rightarrow$ large Magnetic Reynolds Number:

$$\mathbf{E} \simeq -\mathbf{v} \times \mathbf{B}/c \tag{3}$$

$\rightarrow$ 'freezing of the field into the moving fluid' found to be a good first approximation in many problems. Then in the local rest-frame $\mathbf{E}' \simeq 0$. From the Lorentz Transformation, $\mathbf{B}' = \mathbf{B}(1 + O(v/c)^2) = \mathbf{B}$, i.e., $\mathbf{B}$ is invariant in an essentially non-relativistic theory.

3. The 'two-fluid model' of a fully ionized gas, with electron/ion density/pressure $n_{e,i}, p_{e,i}$, yields a generalized Ohm's Law:

$$\mathbf{E}' = \frac{\mathbf{j}}{\sigma} + \frac{\mathbf{j} \times \mathbf{B}}{cn_e e}, \qquad \mathbf{E}' = \left(\mathbf{E} + \frac{\mathbf{v} \times \mathbf{B}}{c} + \frac{\nabla p_e}{n_e e}\right). \tag{4}$$

The conductivity $\sigma = n_e e^2 \tau/m_e$, with $-e$, m_e the electronic charge and mass, τ the time between successive scattering of electron by ions.

Note the new terms: (1) the Hall term $\mathbf{j} \times \mathbf{B}/cn_e e$; and (2) the 'battery term' $\nabla p_e/n_e e$, analogue of the Peltier effect.

$$\frac{\text{Hall term}}{\text{Ohmic term}} = \omega\tau, \qquad \omega = eB/mc = \text{electron gyration frequency.} \qquad (5)$$

When $\omega\tau \ll 1$, the relation between $\mathbf{j}$ and $\mathbf{E}'$ is isotropic. When $\omega\tau \gg 1$, the relation is anisotropic, sometimes described as a 'reduction of conductivity' for currents flowing across $\mathbf{B}$; but there is no significant increase in the dissipation, which remains j^2/σ, with σ the 'unreduced conductivity' given above. (Different results hold for a lightly ionized gas – see § 7 below).

One can write (4) as

$$\mathbf{E} + \frac{(\mathbf{v} + \mathbf{V}) \times \mathbf{B}}{c} = \frac{\mathbf{j}}{\sigma} - \frac{\nabla p_e}{n_e e}, \qquad (6)$$

where $\mathbf{V} = -\mathbf{j}/n_e e$ is the streaming velocity of the electrons relative to the ions. This yields

$$\frac{d}{dt}(\text{Magnetic Flux over Surface } S_e) = c \int_C \left(\frac{\nabla p_e}{n_e e} - \frac{\mathbf{j}}{\sigma} \right) \cdot d\mathbf{s}, \qquad (7)$$

where C is a circuit moving with the electron gas, bounding the surface S_e. In a non-turbulent medium, the Ohmic term $-\mathbf{j}/\sigma$ yields a slow destruction of flux. If n_e is not a function of p_e – i.e. ∇p_e not parallel to ∇n_e – then $\nabla p_e/n_e$ has a curl, its line-integral in (7) is non-zero, acting as the 'emf' of the 'Biermann battery', building up flux slowly against self-induction. In a medium of scale R, at rest, the characteristic time of growth or decay is the Cowling time $4\pi\sigma R^2/c^2$.

4. The equation of motion of the gas is the Navier–Stokes equation supplemented by the Lorentz force density $\rho_e\mathbf{E} + \mathbf{j} \times \mathbf{B}/c$. From (1), the magnetic force density is

$$|\mathbf{j} \times \mathbf{B}|/c \simeq B^2/4\pi D, \qquad (8)$$

where D is a characteristic scale of variation of $\mathbf{B}$ or $\mathbf{v}$. From (1) and (3), $|\rho_e| = |\nabla \cdot \mathbf{E}|/4\pi$, so $|\rho_e\mathbf{E}| \simeq (v/c)^2(B^2/4\pi D)$. Thus the electric part of the Lorentz force is negligible in a non-relativistic theory that ignores terms of order $(v/c)^2$, with the Lorentz γ factors all put equal to 1.

In 1942, Alfvén and Walén discovered the "Alfvén Wave". A uniform field $\mathbf{B}$ exerts a tension $(B^2/4\pi)A$ along a thin flux tube of cross-sectional area A. If the medium of volume density ρ is perfectly conducting, the same flux tube acquires an effective mass per unit length ρA. From elementary string theory, waves are propagated along the field with speed

$$v_A = \left(\frac{(B^2/4\pi)A}{\rho A} \right)^{1/2} = \frac{B}{(4\pi\rho)^{1/2}}. \qquad (9)$$

For more details, see, e.g., Mestel [1]. In the rest of this paper, a brief account is given – in approximate chronological order – of the early application of the theory to a number of cosmical problems.

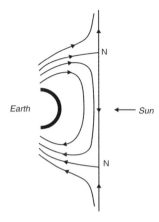

Fig. 1. The Chapman–Ferraro model of geomagnetic storm

3 Geomagnetic storms

The theory began with the highly idealized model of Chapman and Ferraro [2] – see Fig. 1. An ionized, non-magnetic plasma sheet, emitted from the active Sun, impinges on the Earth's field, supposed initially to be that of a vacuum dipole. Perfect conductivity is implicit, as the Earth's field does not penetrate into the sheet, but instead is squashed by it. The essentials of this pioneering work survive in the updated models, which include in the initial state a quasi-steady solar wind, confining the Earth's field in a magnetosphere in dynamic equilibrium, with the ram pressure balancing the pressure of the distorted field. Eruptions on the Sun cause further distortion, observed as a geomagnetic storm.

These models are essentially collective, with the Larmor gyration radii of the low-speed particles small compared with the macroscopic scales. There was a lot of controversy with Hannes Alfvén, whose approach differed, focusing rather on the motion of the highly energetic solar cosmic rays, which do not take cognizance of the magnetic field. Ian Axford (private communication) informs me that the updated models are a synthesis of these complementary approaches.

4 Magnetism and stellar rotation

The pioneering paper here was by Ferraro [3]: 'Non-uniform rotation of the Sun and its magnetic field'. His work is reproduced with some minor changes of notation. He worked with cylindrical polar coordinates, (ϖ, ϕ, z), with a prescribed time-independent poloidal field $\mathbf{B}_p$, symmetric about the rotation axis Oz, and maintained by toroidal (i.e., azimuthal) currents – see Fig. 2.

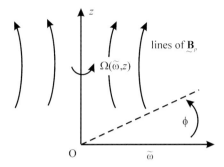

Fig. 2. Cylindrical polar coordinates (ϖ, ϕ, z). Rotation axis Oz. Poloidal magnetic field, symmetrical about Oz: $\mathbf{B_p} = [B_\varpi(\varpi, z), 0, B_z(\varpi, z)]$. Non-uniform rotation $\Omega(\varpi, z)\hat{\mathbf{z}}$ in general generates a toroidal field $\mathbf{B_t} = B_\phi\mathbf{t} = [0, B_\phi(\varpi, z), 0]$

The condition $\nabla \cdot \mathbf{B} = 0$ is satisfied by the introduction of the flux function $P(\varpi, z)$:

$$\mathbf{B}_p = -\frac{\nabla P \times \mathbf{t}}{\varpi}, \tag{10}$$

with $\mathbf{t}$ the unit toroidal vector. The poloidal field lines are given by $P = $ constant. Motions in meridian planes are assumed ignorable, but there is a rotational velocity

$$\mathbf{v} = v_\phi\mathbf{t} = \Omega\varpi\mathbf{t} \tag{11}$$

with the angular velocity an initially unprescribed function $\Omega = \Omega(\varpi, z)$.

Suppose the star has also a toroidal field $B_\phi\mathbf{t}$, maintained by poloidal currents, satisfying the simplest form of Ohm's Law. In a steady state,

$$\mathbf{j} = \sigma \left(\frac{v_\phi\mathbf{t} \times \mathbf{B_p}}{c} - \nabla V \right), \tag{12}$$

with V the electric potential. The curl of (12) has the ϕ-component

$$\nabla^2 B_\phi - \frac{B_\phi}{\varpi^2} - \frac{1}{\varpi}\nabla(\varpi B_\phi) \cdot \frac{\nabla\sigma}{\sigma} = -\frac{4\pi\sigma}{c^2}\frac{\partial(P, \Omega)}{\partial(\varpi, z)}. \tag{13}$$

Ferraro considers two possibilities.

(1) If $B_\phi = 0$, then $\mathbf{j}_p = 0$, and the vanishing of the Jacobian in (13) yields the law of *isorotation* – Ω constant on field lines:

$$\Omega = \Omega(P), \tag{14}$$

with an associated electrical polarization of the medium – $V = -\frac{1}{c}\int \Omega(P)\mathrm{d}P$.

(2) If $B_\phi \neq 0$, Ferraro solves for B_ϕ, and associated poloidal currents, for prescribed

$$\Omega = \Omega_0(r) + \Omega_1(r)\cos^2\theta + \dots\dots \qquad \sigma = \sigma_0(r). \tag{15}$$

Nowadays, there is a change of emphasis. If $\Omega = \Omega(P)$, each poloidal field line is rotated as a whole. If $\Omega \neq \Omega(P)$, but the medium is highly conducting, then we expect field freezing to hold to a high approximation. In (12), we put $1/\sigma = 0$, and replace $-\nabla V$ by a non-irrotational $\mathbf{E}$. An initially purely poloidal field is thus sheared so as to generate a toroidal component, by Faraday's Law:

$$\frac{\partial B_\phi}{\partial t} = (\varpi \mathbf{B_p}) \cdot \nabla \Omega. \tag{16}$$

Ferraro's state (2) is *kinematically* steady: slippage due to finite resistivity balances the effect of shearing. But in his words: 'It is highly improbable that these poloidal currents flow'. The consequent ϕ-component of the Lorentz force

$$\mathbf{j_p} \times \mathbf{B_p}/c = [\mathbf{B_p} \cdot \nabla(\varpi B_\phi)/(4\pi\varpi)]\mathbf{t} \tag{17}$$

"acts on the rotation to destroy shear".

Ferraro always kicked himself for not going on to discover the Alfvén wave. The back reaction of the growing Lorentz force (17), exerted by the field generated according to (16), would come into play well before the resistive terms become important. However, Ferraro's paper is seminal: strong magnetic coupling with rotation inside stars, in stellar coronae and winds, and in accretion discs, and even on the galactic scale, is built into contemporary astrophysics.

5 Sunspots; magneto-convection

This area can be traced back to a remarkable correspondence between Tom Cowling and Ludwig Biermann during the 1930s. A simple model for the lateral dynamic equilibrium of a spot is illustrated in Fig. 3.

Dynamical equilibrium requires

$$p_{\text{ext}} - p_{\text{int}} = \frac{B^2}{8\pi}; \tag{18}$$

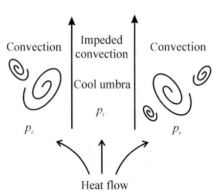

Fig. 3. Idealized sunspot model: umbra with vertical field lines

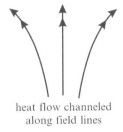

heat flow channeled
along field lines

Fig. 4. A more realistic sunspot model, with field lines splaying out

the combined magnetic and thermal pressures of the cool umbra are balanced
by the thermal pressure of the hotter surrounding penumbra. The long life of
a spot requires that the lateral heat flow into the umbra must be balanced
by a reduction in the vertical heat supply. Biermann suggested that this is
due to magnetic interference with the otherwise highly efficient turbulent heat
transport. Cowling noted that this will be important when the magnetic and
turbulent energy densities are comparable.

The magnetic field is thus identified as the primary cause of the sunspot
phenomenon. Detailed models require that there be not a suppression of the
turbulent heat flow, but a reduction by a factor that decreases monotonically
from unity as the ratio of magnetic to turbulent energy increases. More realistic field models (Schlüter and Temesvary) have the field lines splaying out, as
in Fig. 4. In studying heat flow, allowance must be made for the channelling
of the reduced turbulent heat flow along the field lines (Hoyle, Chitre).

6 Solar and stellar flares

Magnetohydrodynamic coupling of the high-density, sub-photospheric solar
convective envelope with the low-density solar atmosphere was suggested as
a plausible explanation of the general heating of the solar corona. To explain
flaring, one needs energy to be stored in a non-curl-free field, to be subsequently released rapidly, e.g., by a macro-instability. Sweet [4] proposed the
following model, to be developed subsequently by Parker and by Low and
Wolfson.

In Sweet's model, the initial state is as in Fig. 5a: above the y-axis – the
solar surface – there is a quadrupolar, curl-free field with an X-type neutral
point. Now suppose motions in the convective zone have shifted the foot-
points. A new *curl-free* field will be topologically similar to Fig. 5a, but with
different field-line connectivities – e.g., α linked with δ and β with γ. Such a
transition requires violation of flux-freezing. If strict flux-freezing holds, then
the new field will be as in Fig. 5b: the field is curl-free *nearly everywhere*,
but to keep the connectivities unchanged, there has to be a vertical pinched
current sheet. The field energy is greater than that in the initial curl-free

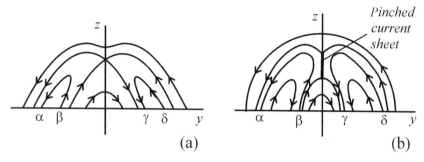

Fig. 5. (a) The initial state: a quadrupolar field with a neutral point; **(b)** field structure following motion of foot-points, with strict flux-freezing

field. The model can be generalized to more realistic geometry, yielding fields that are *force-free* nearly everywhere, but again with singular, magnetically pinched domains carrying current sheets.

There has developed a whole industry on the *reconnection*-problem – the processes by which deviation from flux-freezing allows for changes in field topology and associated release – slow or fast – of stored magnetic energy (Dungey, Sweet, Parker, Petschek, Priest, and collaborators).

7 Self-gravitating magnetic gas clouds

The estimated galactic field strengths suggested strongly that we should observe strongly magnetic gas clouds, with the Lorentz force comparable with self-gravitation. Chandrasekhar and Fermi [5] derived from the equations of motion the appropriate generalization of the virial theorem, used in basic stellar structure theory. For a gas cloud in equilibrium, this integral condition is

$$2T + (3\gamma - 1)U + \mathcal{M} + \mathcal{V} = \int \left(p + \frac{B^2}{8\pi} \right) (\mathbf{r} \cdot \mathbf{n}) \mathrm{d}S - \frac{1}{4\pi} \int (\mathbf{B} \cdot \mathbf{r})(\mathbf{B} \cdot \mathbf{n}) \mathrm{d}S,$$
(19)

where U is the thermal energy, with γ the ratio of principal specific heats (assumed constant), T the macroscopic kinetic energy (rotational, turbulent), $\mathcal{M}$ the magnetic energy, $\mathcal{V}$ the (negative) gravitational energy. The surface integral terms depend on the thermal pressure and the Maxwell stresses and are often (but not always) small enough to be dropped.

From (19), the 'virial limit' to the magnetic flux F in a gravitationally contracting cloud of mass M is given by dropping all the terms except $\mathcal{V}$ and $\mathcal{M}$, yielding $F_c \simeq kG^{1/2}M$, with k a constant of about 4. If the field remains frozen in, it will certainly inhibit or even prevent fragmentation into protostars, though it can be the dominant agent for the removal of excess angular momentum.

The strongest magnetic fields observed in upper main sequence stars are much below the virial limit. It was suggested that the required leakage of unwanted flux occurs in the early stages of star formation. In a lightly ionized cloud, it is a good approximation to picture the ions and electrons as a fully ionized "plasma", immersed in a much denser gas of neutral atoms or molecules. As the gravitationally distorted magnetic field tries to straighten itself, it drives the plasma plus the inductively coupled field through the neutral gas. This "ambipolar diffusion" occurs at the rate fixed by balance between the Lorentz force and the friction due to collisions between the ions and the neutral particles. The consequent reduction in the magnetic flux threading a cloud can be an important factor in the fragmentation of a cloud into protostars (Mestel and Spitzer [6]). Subsequent work on cosmical gas dynamics incorporates the differing electromagnetic properties of fully and partly ionized media.

Almost simultaneously, there appeared papers by Piddington [7] and Cowling [8], following earlier ones by Schlüter and Biermann [9] and Schlüter [10], discussing the general problem of dissipation of magnetic energy in partially ionized media. The crucial result is that whereas currents $\mathbf{j}_\parallel$ flowing along $\mathbf{B}$ suffer essentially the ordinary Ohmic resistivity, for currents $\mathbf{j}_\perp$ flowing perpendicular to $\mathbf{B}$, the effective resistivity is the very much larger quantity

$$\frac{F^3 Z \tau_i B^2}{n m_i c^2}, \tag{20}$$

where F is the fractional contribution of the non-ionized gas to the total density, n is the electron density, $Z = n/n_i$ with n_i the ion density, and τ_i is $m_n/(m_n + m_i)$ times the time interval between successive scatterings of an ion of mass m_i by a neutral particle of mass m_n.

8 Dynamo action

Biermann's slowly acting "battery" term $\mathbf{E_b} \equiv \nabla p_e / n_e e$ can generate significant fields; e.g., a toroidal field in a rotating star with a non-irrotational centrifugal field. The flux could be the seed from which a much larger flux could be generated by *dynamo action*, as suggested first by Larmor. Mass motions in the presence of an existing $\mathbf{B}$ generate currents

$$\mathbf{j} = \sigma[\mathbf{v} \times \mathbf{B}/c], \tag{21}$$

where the brackets "[....]" indicate the non-curl-free part. The "self-exciting dynamo" problem reduces to asking: can one find a velocity field $\mathbf{v}$ that yields from (21) the current field $\mathbf{j}$ which maintains $\mathbf{B}$ as given by Ampère's law?

Cowling [11] anticipated a "routine calculation" to find what fields can be maintained in this way. In a hypothetical steady state, Ohmic decay is offset by motional induction. If the velocities are increased, the field grows in strength until the back-reaction of the growing Lorentz force limits the velocities.

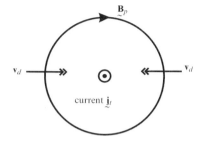

Fig. 6. Near an O-type neutral point of the poloidal field, the maintaining current flows into the paper. The Ohmic field yields loss of flux through diffusion into O with velocity $\mathbf{v_d}$

However: an axisymmetric poloidal field $\mathbf{B_p}$ has one or more O-type neutral points, with the field locally as in Fig. 6. By Ampère's law, the maintaining current $\mathbf{j_t}$ flows into the sheet. If $\mathbf{v_p} = 0$, then as noted by Sweet, the decay of the field can be pictured as the diffusion of the field lines into O with the velocity

$$\mathbf{v_d} = \frac{c(\mathbf{j_t} \times \mathbf{B_p})}{\sigma B_p^2} = \frac{(\nabla \times \mathbf{B_p}) \times \mathbf{B_p}}{(4\pi\sigma/c^2)B_p^2}, \qquad (22)$$

since (22) yields $\mathbf{E} + \mathbf{v_p} \times \mathbf{B_p}/c = \mathbf{E} - \mathbf{j_t}/\sigma = 0$. The spontaneous decay could be offset only if gas were to emerge from O.

This is the essence of Cowling's celebrated "anti-dynamo" theorem. The original theorem is extensible to non-axisymmetric fields that are topologically similar (Cowling, Bullard).

In his 1937 paper [3], recall that Ferraro stated that the initial axisymmetric (poloidal) field $\mathbf{B_p}$ is maintained by azimuthal currents, but by implication accepted Cowling's argument, applying the *steady state* Ohm's law just to the meridional currents maintaining his toroidal field $\mathbf{B_t} = B_\phi \mathbf{t}$. Non-uniform rotation generates $\mathbf{B_t}$ from $\mathbf{B_p}$, but under axisymmetric poloidal motions, the toroidal field component remains toroidal, so the cycle in not completed. In Elsasser's words, there is a 'topological asymmetry' between poloidal and toroidal fields.

An early dynamo existence theorem for an essentially non-axisymmetric system was given by Herzenberg [12]. Two spheres are rotating about mutually inclined axes within a bounded conducting medium. The toroidal field generated by the shear at the surface of one sphere serves as the poloidal field at the second; the rotation of the second sphere generates a toroidal field, which serves as poloidal field at the first.

In a seminal paper, Parker [13] produced a model with the essence of what is now called an "$\alpha\Omega$" dynamo in a late-type star. Again, a rotational shear generates $\mathbf{B_t}$ from $\mathbf{B_p}$, but the cycle is closed by an explicit appeal to non-axisymmetric and non-isotropic turbulent motions: small-scale vortices in a compressible, stratified medium, acting on $\mathbf{B_t}$, yield a new poloidal component. The process is modelled by writing

$$\mathbf{B_p} = \nabla \times (A\mathbf{t}), \qquad \frac{\partial A}{\partial t} = \alpha B_\phi + \text{turbulent resistive terms.} \qquad (23)$$

The first models yielded periodic solutions, with reversal of both poloidal and toroidal components, mimicking to some extent the observed solar behaviour. However, to yield something like a 22-year cycle period, one must appeal to a macro-resistivity.

Sweet [4] was one of the earliest to point out that the reduction of scale when the magnetic field is tangled by an inexorable turbulent field would yield a large effective resistivity. This is partially analogous to the turbulent viscosity introduced in early studies of stellar rotation, and indeed in models of the Kolmogoroff cascade. There is however the important difference that tangling of the field simultaneously increases the Lorentz forces opposing the tangling – a problem that is still with us.

The pioneering work by Parker and by Steenbeck and colleagues has been followed by a whole dynamo industry, including texts by Moffatt [14], Parker [15], Krause and Rädler [16], Zel'dovich et al. [17], and a host of papers on terrestrial, stellar and galactic dynamos.

9 MHD turbulence

If a magnetic field is supposed present, it is of interest and often of importance to study the effect on the field of a well-defined fluid motion. Batchelor [18] considered the effect of homogeneous, isotropic turbulence on an initial "seed" field, arguing for an analogy between $(\mathbf{B})$ and the vorticity $\boldsymbol{\omega} = \nabla \times \mathbf{v}$ in hydrodynamic turbulence. The word "dynamo" was not used in this paper, but in subsequent discussion he and others appeared to assume without further analysis that his model was of a self-exciting dynamo. Some confusion had arisen through differing usage. Some have called a "dynamo" any device for the continual conversion of kinetic energy into magnetic energy. In geophysics and astrophysics, one is usually interested in *self-exciting* AC or DC dynamos. A procedure which builds up and maintains a steady or fluctuating field, but only if the externally generated initial seed field persists, is best described as an "amplifier". And indeed a number of recent studies have shown that while tangling of an initial field by isotropic turbulence can yield a small-scale field, in approximate equipartition with the small-scale turbulence, to increase the net magnetic flux through the domain, the turbulence must have an essential anisotropy, e.g., a net magnetic helicity, as in the Parker model and its derivatives. The treatment in Moffatt [14] is pedagogically very helpful in showing how non-isotropic, unimpeded turbulence can yield both flux generation through the "α-effect" (23) and strongly enhanced Ohmic dissipation.

Macro-dissipation of unwanted, small-scale fields is essential for the successful generation of a large-scale field, e.g., a galactic field, detectable by Faraday rotation. And an unpalatable consequence of strict adherence to flux-freezing was noted early by Bondi and Gold [19]. If fluid motions are confined to a finite volume of *perfectly conducting* fluid, – e.g., a stellar convection zone – then outward diffusion of flux newly generated by internal dynamo action would be forbidden: the total flux, or equivalently the externally observed dipole moment would hardly change.

References

1. Mestel L (1999, 2003) Stellar Magnetism. Clarendon Press, Oxford
2. Chapman S, Ferraro VCA (1931) Terr Magn atmos Elect 36:77 and 171
3. Ferraro VCA (1937) Mon Not R Astr Soc 97:458
4. Sweet PA (1958) In: Electromagnetic Phenomena in Cosmical Physics. Cambridge University Press, Cambridge 123
5. Chandrasekhar S, Fermi E (1953) Ap J 118:113 and 116
6. Mestel L, Spitzer Jr L (1956) Mon Not R Astr Soc 116:583
7. Piddington JH (1955) Mon Not Roy Astr Soc 114:638 and 651
8. Cowling TG (1956) Mon Not Roy Astr Soc 116:114
9. Schlüter A, Biermann L (1950) Zeits f Naturforschung 5A:237
10. Schlüter A (1951) Zeits f Naturforschung 6A:73
11. Cowling TG (1934) Mon Not R Astr Soc 94:39
12. Herzenberg A (1958) Phil Trans R Soc A 250:543
13. Parker EN (1955) Ap J 122:293
14. Moffatt HK (1978) Magnetic Field Generation in Electrically Conducting Fluids. Cambridge University Press, Cambridge
15. Parker EN (1978) Cosmical magnetic fields. Oxford University Press, Oxford
16. Krause F, Rädler K-H (1980) Mean-field magnetohydrodynamics and dynamo theory. Pergamon, Oxford
17. Zel'dovich YaB, Ruzmaikin AA, Sokoloff DD (1983) Magnetic Fields in Astrophysics. Gordon and Breach, New York
18. Batchelor GK (1950) Proc R Soc A 201:405
19. Bondi H, Gold T (1950) Mon Not R Astr Soc 110:607

Turbulence and Magnetic Fields
in Astrophysical Plasmas

Alexander A Schekochihin[1] and Steven C Cowley[2]

[1] DAMTP, University of Cambridge, Cambridge CB3 0WA and Department of
Physics, Imperial College, London SW7 2BW, United Kingdom
(a.schekochihin@imperial.ac.uk)
[2] Department of Physics and Astronomy, UCLA, Los Angeles, CA 90095-1547,
USA and Department of Physics, Imperial College, London SW7 2BW, United
Kingdom (cowley@physics.ucla.edu)

1 Introduction

Magnetic fields permeate the Universe. They are found in planets, stars,
accretion discs, galaxies, clusters of galaxies, and the intergalactic medium.
While there is often a component of the field that is spatially coherent at the
scale of the astrophysical object, the field lines are tangled chaotically and
there are magnetic fluctuations at scales that range over orders of magnitude.
The cause of this disorder is the turbulent state of the plasma in these systems. This plasma is, as a rule, highly conducting, so the magnetic field lines
are entrained by (frozen into) the fluid motion. As the fields are stretched and
bent by the turbulence, they can resist deformation by exerting the Lorentz
force on the plasma. The turbulent advection of the magnetic field and the
field's back reaction together give rise to the statistically steady state of fully
developed MHD turbulence. In this state, energy and momentum injected
at large (object-size) scales are transfered to smaller scales and eventually
dissipated.

Despite over 50 years of research and many major advances, a satisfactory
theory of MHD turbulence remains elusive. Indeed, even the simplest (most
idealised) cases are still not fully understood. One would hope that there are
universal properties of MHD turbulence that hold in all applications – or
at least in a class of applications. Among the most important questions for
astrophysics that a successful theory of turbulence must answer are:

- How does the turbulence amplify, sustain and shape magnetic fields? What
  is the structure and spectrum of this field at large and small scales? The
  problem of turbulence in astrophysics is thus directly related to the fundamental problem of magnetogenesis.

S. Molokov et al. (eds.), Magnetohydrodynamics – Historical Evolution and Trends,
85–115. © 2007 *Springer.*

- How is energy cascaded and dissipated in plasma turbulence? In accretion discs and the solar corona, for example, one would like to know if the turbulence heats ions or electrons predominantly [1].
- How does the turbulent flow and magnetic field enhance or inhibit the transport of heat, (angular) momentum, and cosmic rays? Again in accretion discs, a key parameter is the effective turbulent viscosity that causes the transport of angular momentum and enables accretion [2]. In cluster physics, an understanding of how viscous heating and thermal conduction in a turbulent magnetised plasma balance the radiative cooling is necessary to explain the observed global temperature profiles [3].

In this chapter, we discuss the current understanding of the most basic properties of astrophysical MHD turbulence. We emphasise possible universal aspects of the theory. We shall touch primarily on two applications: turbulence in the solar wind and in clusters of galaxies. These are, in a certain (very approximate) sense, two "pure" cases of small-scale turbulence, where theoretical models of the two main regimes of MHD turbulence (discussed in § 2 and § 3) can be put to the test. They are also good examples of a complication that is more or less generic in astrophysical plasmas: the MHD description is, in fact, insufficient for astrophysical turbulence and plasma physics must make an entrance. Why this is so, will be explained in § 4.

The astrophysical plasma turbulence is even more of a *terra incognita* than the MHD turbulence, so we shall start with the equations of incompressible MHD – the simplest equations that describe (subsonic) turbulent dynamics in a conducting medium:

$$\frac{d\mathbf{u}}{dt} = -\boldsymbol{\nabla}p + \nu\Delta\mathbf{u} + \mathbf{B}\cdot\boldsymbol{\nabla}\mathbf{B} + \mathbf{f}, \quad \boldsymbol{\nabla}\cdot\mathbf{u} = 0, \tag{1}$$

$$\frac{d\mathbf{B}}{dt} = \mathbf{B}\cdot\boldsymbol{\nabla}\mathbf{u} + \eta\Delta\mathbf{B}, \tag{2}$$

where $\mathbf{u}$ is the velocity field, $d/dt = \partial_t + \mathbf{u}\cdot\boldsymbol{\nabla}$ the convective derivative, p the pressure (scaled by the constant density ρ and determined by the incompressibility constraint), $\mathbf{B}$ the magnetic field scaled by $(4\pi\rho)^{1/2}$, ν the kinematic viscosity, η the magnetic diffusivity, and $\mathbf{f}$ the body force that models large-scale energy input. The specific energy injection mechanisms vary: typically, in astrophysics, these are either background gradients (e.g., the temperature gradient in stellar convective zones, the Keplerian velocity shear in accretion discs), which mediate the conversion of gravitational energy into kinetic energy of fluid motion or direct sources of energy such as the supernovae in the interstellar medium or active galactic nuclei in galaxy clusters. What all these injection mechanisms have in common is that the scale at which they operate, hereafter denoted by L, is large, comparable with the size of the system. While the large-scale dynamics depend on the specific astrophysical situation, it is common to assume that, once the energy has cascaded down to scales substantially smaller than L, the non-linear dynamics are universal.

The universality of small scales is a cornerstone of all theories of turbulence. It goes back to Kolmogorov's 1941 dimensional theory (or K41 [4]; see [5], § 33 for a lucid and concise exposition). Here is an outline Kolmogorov's reasoning. Consider Eq. (1) without the magnetic term. Denote the typical fluctuating velocity difference across scale L by δu_L. The energy associated with these fluctuations is δu_L^2 and the characteristic time for this energy to cascade to smaller scales by non-linear coupling is $L/\delta u_L$. The total specific power (energy flux) going into the turbulent cascade is then $\epsilon = \langle \mathbf{u} \cdot \mathbf{f} \rangle \sim \delta u_L^3/L$. In a statistically stationary situation, all this power must be dissipated, so $\epsilon = \nu \langle |\nabla \mathbf{u}|^2 \rangle$. Since ϵ is a finite quantity completely defined by the large-scale energy-injection process, it cannot depend on ν. For very small ν, this implies that the velocity must develop very small scales so that $\nu \langle |\nabla \mathbf{u}|^2 \rangle$ has a constant limit as $\nu \to +0$. The only quantity with dimensions of length that one can construct out of ϵ and ν is $l_\nu \sim (\nu^3/\epsilon)^{1/4} \sim Re^{-3/4}L$, where $Re \sim \delta u_L L/\nu$ is the Reynolds number. In astrophysical applications, Re is usually large, so the viscous dissipation occurs at scales $l_\nu \ll L$. The energy injected at the large-scale L must be transfered to the small-scale l_ν across a range of scales (the inertial range). The hydrodynamic turbulence theory assumes that the physics in this range is universal, i.e., it depends neither on the energy-injection mechanism nor on the dissipation mechanism. Four further assumptions are made about the inertial range: homogeneity (no special points), scale invariance (no special scales), isotropy (no special directions), and locality of interactions (interactions between comparable scales dominate). Then, at each scale l such that $L \gg l \gg l_\nu$, the total power ϵ must arrive from larger scales and be passed on to smaller scales:

$$\epsilon \sim \delta u_l^2/\tau_l, \tag{3}$$

where δu_l is the velocity difference across scale l and τ_l the cascade time. Dimensionally, only one timescale can be constructed out of the local quantities δu_l and l: $\tau_l \sim l/\delta u_l$. Substituting this into Eq. (3) and solving for δu_l, we arrive at Kolmogorov's scaling: $\delta u_l \sim (\epsilon l)^{1/3}$, or, for the energy spectrum $E(k)$,

$$\delta u_l^2 \sim \int_{k=1/l}^{\infty} dk' E(k') \sim \epsilon^{2/3} k^{-2/3} \quad \Rightarrow \quad E(k) \sim \epsilon^{2/3} k^{-5/3}. \tag{4}$$

The history of the theory of MHD turbulence over the last half century has been that of a succession of attempts to adapt the K41-style thinking to fluids carrying magnetic fields. In the next two sections, we give an overview of these efforts and of the resulting gradual realisation that the key assumptions of the small-scale universality, isotropy and locality of interactions fail in various MHD contexts.

2 Alfvénic turbulence

Let us consider the case of a plasma threaded by a straight uniform magnetic field $\mathbf{B}_0$ of some external (i.e., large-scale) origin. Let us also consider weak forcing so that the fundamental turbulent excitations are small-amplitude wave-like disturbances propagating along the mean field. We will refer to such a limit as Alfvénic turbulence – it is manifestly anisotropic.

2.1 Iroshnikov–Kraichnan turbulence

If we split the magnetic field into the mean and fluctuating parts, $\mathbf{B} = \mathbf{B}_0 + \delta \mathbf{B}$, and introduce Elsasser [6] variables $\mathbf{z}^\pm = \mathbf{u} \pm \delta \mathbf{B}$, Eqs. (1) and (2) take a symmetric form:

$$\partial_t \mathbf{z}^\pm \mp v_A \nabla_\parallel \mathbf{z}^\pm + \mathbf{z}^\mp \cdot \nabla \mathbf{z}^\pm = -\nabla p + \frac{\nu + \eta}{2} \Delta \mathbf{z}^\pm + \frac{\nu - \eta}{2} \Delta \mathbf{z}^\mp + \mathbf{f}, \quad (5)$$

where $v_A = |\mathbf{B}_0|$ is the Alfvén speed and $\nabla_\parallel$ is the gradient in the direction of the mean field $\mathbf{B}_0$. The Elsasser equations have a simple *exact* solution: if $\mathbf{z}^+ = 0$ or $\mathbf{z}^- = 0$, the non-linear term vanishes and the other, non-zero, Elsasser field is simply a fluctuation of arbitrary shape and magnitude propagating along the mean field at the Alfvén speed v_A. Kraichnan [7] realised in 1965 that the Elsasser form of the MHD equations only allows nonlinear interactions between counterpropagating such fluctuations. The phenomenological theory that he and, independently, Iroshnikov [8], developed on the basis of this idea (the IK theory) can be summarised as follows.

Following the general philosophy of K41, assume that only fluctuations of comparable scales interact (**locality of interactions**) and consider these interactions in the inertial range, comprising scales l smaller than the forcing scale L and larger than the (still to be determined) dissipation scale. Let us think of the fluctuations propagating in either direction as trains of spatially localised Alfvén-wave[3] packets of parallel (to the mean field) extent $l_\parallel$ and perpendicular extent l (we shall not, for the time being, specify how $l_\parallel$ relates to l). Assume further that $\delta z_l^+ \sim \delta z_l^- \sim \delta u_l \sim \delta B_l$. We can again use Eq. (3) for the energy flux through scale l, but there is, unlike in the case of purely hydrodynamic turbulence, no longer a dimensional inevitability about the determination of the cascade time τ_l because two physical timescales are associated with each wave packet: the Alfvén time $\tau_A(l) \sim l_\parallel/v_A$ and the strain (or "eddy") time $\tau_s(l) \sim l/\delta u_l$. To state this complication in a somewhat more formal way, there are three dimensionless combinations in the problem of MHD turbulence: $\epsilon l/\delta u_l^3$, $\delta u_l/v_A$, and $l_\parallel/l$, so the dimensional analysis does not uniquely determine scalings and further physics input is needed.

---

[3] Waves in incompressible MHD can have either the Alfvén- or the slow-wave polarisation. Since both propagate at the Alfvén speed, we shall, for simplicity, refer to them as Alfvén waves. The differences between the Alfvén- and slow-wave cascades are explained in detail at the end of § 2.4.

Two counterpropagating wave packets take an Alfvén time to pass through each other. During this time, the amplitude of either packet is changed by

$$\Delta \delta u_l \sim \frac{\delta u_l^2}{l} \tau_A \sim \delta u_l \frac{\tau_A}{\tau_s}. \tag{6}$$

The IK theory now assumes weak interactions, $\Delta \delta u_l \ll \delta u_l \Leftrightarrow \tau_A \ll \tau_s$. The cascade time τ_l is estimated as the time it takes (after many interactions) to change δu_l by an amount comparable to itself. If the changes in amplitude accumulate like a random walk, we have

$$\sum^t \Delta \delta u_l \sim \delta u_l \frac{\tau_A}{\tau_s} \sqrt{\frac{t}{\tau_A}} \sim \delta u_l \quad \text{for} \quad t \sim \tau_l \quad \Rightarrow \quad \tau_l \sim \frac{\tau_s^2}{\tau_A} \sim \frac{l^2 v_A}{l_\parallel \delta u_l^2}. \tag{7}$$

Substituting the latter formula into Eq. (3), we get

$$\delta u_l \sim (\epsilon v_A)^{1/4} l_\parallel^{-1/4} l^{1/2}. \tag{8}$$

The final IK assumption, which at the time seemed reasonable in light of the success of the K41 theory, was that of **isotropy**, fixing the dimensionless ratio $l_\parallel/l \sim 1$, and, therefore, the scaling:

$$\delta u_l \sim (\epsilon v_A)^{1/4} l^{1/4} \quad \Rightarrow \quad E(k) \sim (\epsilon v_A)^{1/2} k^{-3/2}. \tag{9}$$

2.2 Turbulence in the solar wind

The solar wind, famously predicted by Parker [9], was the first astrophysical plasma in which direct measurements of turbulence became possible [10]. A host of subsequent observations (for a concise review, see [11]) revealed power-like spectra of velocity and magnetic fluctuations in what is believed to be the inertial range of scales extending roughly from 10^6 to 10^3 km. The mean magnetic field is $B_0 \sim 10$–10^2 μG, while the fluctuating part δB is a factor of a few smaller. The velocity dispersion is $\delta u \sim 10^2$ km/s, approximately in energy equipartition with δB. The **u** and **B** fluctuations are highly correlated at all scales and almost undoubtedly Alfvénic. It is, therefore, natural to think of the solar wind as a space laboratory conveniently at our disposal to test theories of Alfvénic turbulence in astrophysical conditions.

For nearly 30 years following Kraichnan's paper [7], the IK theory was accepted as the correct extension of K41 to MHD turbulence and, therefore, with minor modifications allowing for the observed imbalance between the energies of the $\mathbf{z}^+$ and $\mathbf{z}^-$ fluctuations [12], also to the turbulence in the solar wind. However, alarm bells were sounding already in 1970s and 1980s when measurements of the solar-wind turbulence revealed that it was strongly anisotropic with $l_\parallel > l_\perp$ [13] (see also [14]) and that its spectral index was closer to $-5/3$ than to $-3/2$ [15].[4] Numerical simulations have confirmed the anisotropy of MHD turbulence in the presence of a strong mean field [17–19].

---

[4] Another classic example of a $-5/3$ scaling in astrophysical turbulence is the spectrum of electron density fluctuations (thought to trace the velocity spectrum) in

2.3 Weak turbulence

The realisation that the isotropy assumption must be abandoned led to a reexamination of the Alfvén-wave interactions in MHD turbulence. If the assumption of weak interactions is kept, MHD turbulence can be regarded as an ensemble of waves, whose wavevectors $\mathbf{k}$ and frequencies $\omega^{\pm}(\mathbf{k}) = \pm k_{\|} v_{\mathrm{A}}$ have to satisfy resonance conditions in order for an interaction to occur. For three-wave interactions (1 and 2 counterpropagating, giving rise to 3),

$$\mathbf{k}_1 + \mathbf{k}_2 = \mathbf{k}_3 \qquad \Rightarrow \qquad k_{\|1} + k_{\|2} = k_{\|3}, \qquad (10)$$

$$\omega^{\pm}(\mathbf{k}_1) + \omega^{\mp}(\mathbf{k}_2) = \omega^{\pm}(\mathbf{k}_3) \quad \Rightarrow \quad k_{\|1} - k_{\|2} = k_{\|3}, \qquad (11)$$

whence $k_{\|2} = 0$ and $k_{\|3} = k_{\|1}$. Thus (i) interactions do not change $k_{\|}$; (ii) interactions are mediated by modes with $k_{\|} = 0$, which are quasi-2D fluctuations rather than waves [20–22].[5]

The first of these conclusions suggests a quick fix of the IK theory: take $l_{\|} \sim k_{\|0}^{-1} = \text{constant}$ (the wavenumber at which the waves are launched) and $l \sim l_{\perp}$ in Eq. (8) (no parallel cascade). Then the spectrum is [23]

$$E(k_{\perp}) \sim (\epsilon k_{\|0} v_{\mathrm{A}})^{1/2} k_{\perp}^{-2}. \qquad (12)$$

The same result can be obtained via a formal calculation based on the standard weak-turbulence theory [24, 25]. However, it is not uniformly valid at all $k_{\perp}$. Indeed, let us check if the assumption of weak interactions, $\tau_{\mathrm{A}} \ll \tau_{\mathrm{s}}$, is actually satisfied by the scaling relation (8) with $l_{\|} \sim k_{\|0}^{-1}$:

$$\frac{\tau_{\mathrm{A}}}{\tau_{\mathrm{s}}} \sim \frac{\epsilon^{1/4}}{\left(k_{\|0} v_{\mathrm{A}}\right)^{3/4} l_{\perp}^{1/2}} \ll 1 \quad \Leftrightarrow \quad l_{\perp} \gg l_* = \frac{\epsilon^{1/2}}{\left(k_{\|0} v_{\mathrm{A}}\right)^{3/2}} \sim \frac{\delta u_L^2}{v_{\mathrm{A}}^2} \frac{1}{k_{\|0}^2 L}, \quad (13)$$

where δu_L is the velocity at the outer scale (the rms velocity). Thus, if $Re = \delta u_L L / \nu$ and $R_m = \delta u_L L / \eta$ are large enough, the inertial range will always contain a scale l_* below which the interactions are no longer weak.[6]

---

the interstellar medium – the famous "power law in the sky", which appears to hold across 12 decades of scales [16].

[5] Goldreich and Sridhar [23] argued that the time it takes three waves to realise that one of them has zero frequency is infinite and, therefore, the weak-interaction approximation cannot be used. This difficulty can, in fact, be removed by noticing that the $k_{\|} = 0$ modes have a finite correlation time, but we do not have space to discuss this rather subtle issue here (two relevant references are [24, 25]).

[6] There is also an upper limit to the scales at which Eq. (12) is applicable. The boundary conditions at the ends of the "box" are unimportant only if the cascade time (7) is shorter than the time it takes an Alfvénic fluctuation to cross the box: $\tau_l \ll L_{\|} / v_{\mathrm{A}} \Leftrightarrow l_{\perp} \ll L_* = (\epsilon / k_{\|0} v_{\mathrm{A}}^3)^{1/2} L_{\|}$, where $L_{\|}$ is the length of the box along the mean field. Demanding that $L_* > L$, the perpendicular size of the box, we get a lower limit on the aspect ratio of the box: $L_{\|} / L > k_{\|0} L (v_{\mathrm{A}} / \delta u_L)^2$. If this

R.S. Iroshnikov (1937–1991)   R.H. Kraichnan

P. Goldreich   S. Sridhar

Fig. 1. IK and GS. (Photo of R.S. Iroshnikov courtesy Sternberg Astronomical Institute. Photo of R.H. Kraichnan courtesy of the Johns Hopkins University.)

2.4 Goldreich–Sridhar turbulence

In 1995, Goldreich and Sridhar [27] conjectured that the strong turbulence below the scale l_* should satisfy

$$\tau_A \sim \tau_s \quad \Leftrightarrow \quad l_\parallel/l_\perp \sim v_A/\delta u_l, \tag{14}$$

a property that has come to be known as **the critical balance**. Goldreich and Sridhar argued that when $\tau_A \ll \tau_s$, the weak turbulence theory "pushes"

---

is not satisfied, the physical (non-periodic) boundary conditions may impose a limit on the perpendicular field-line wander and thus effectively forbid the $k_\parallel = 0$ modes. It has been suggested [23] that a weak-turbulence theory based on 4-wave interactions [26] should then be used at $l_\perp > L_*$.

the spectrum towards the approximate equality (14). They also argued that when $\tau_A \gg \tau_s$, motions along the field lines are decorrelated and naturally develop the critical balance.

The critical balance fixes the relation between two of the three dimensionless combinations in MHD turbulence. Since now there is only one natural timescale associated with fluctuations at scale l, this timescale is now assumed to be the cascade time, $\tau_l \sim \tau_s$. This brings back Kolmogorov's spectrum (4) for the perpendicular cascade. The parallel cascade is now also present but is weaker: from Eq. (14),

$$l_\parallel \sim v_A \epsilon^{-1/3} l_\perp^{2/3} \sim k_{\parallel 0}^{-1} \left(l_\perp / l_*\right)^{2/3}. \tag{15}$$

The scalings (4) and (15) should hold at all scales $l_\perp \ll l_*$ and above the dissipation scale: either viscous $l_\nu \sim (\nu^3/\epsilon)^{1/4}$ or resistive $l_\eta \sim (\eta^3/\epsilon)^{1/4}$, whichever is larger. Comparing these scales with l_* [Eq. (13)], we note that the strong-turbulence range is non-empty only if $Re, R_m \gg \left(k_{\parallel 0} L\right)^3 \left(v_A/\delta u_L\right)^3$, a condition that is effortlessly satisfied in most astrophysical cases but should be kept in mind when numerical simulations are undertaken.

The Goldreich–Sridhar (GS) theory has now replaced the IK theory as the standard accepted description of MHD turbulence. The feeling that the GS theory is the right one, created by the solar wind [15] and ISM [16] observations that show a $k^{-5/3}$ spectrum, is, however, somewhat spoiled by the consistent failure of the numerical simulations to produce such a spectrum [17, 19]. Instead, a spectral index closer to IK's $-3/2$ is obtained (this seems to be the more pronounced the stronger the mean field), although the turbulence is definitely anisotropic and the GS relation (15) appears to be satisfied [17, 18]! This trouble has been blamed on intermittency [17], a perennial scapegoat of turbulence theory, but a non-speculative solution remains to be found.

The puzzling refusal of the numerical MHD turbulence to agree with either the GS theory or, indeed, with the solar-wind observations highlights the rather shaky quality of the existing physical understanding of what really happens in a turbulent magnetic fluid on the dynamical level. One conceptual difference between MHD and hydrodynamic turbulence is the possibility of long-time correlations. In the large-R_m limit, the magnetic field is determined by the displacement of the plasma, i.e., the time integral of the (Lagrangian) velocity. In a stable plasma, the field-line tension tries to return the field line to the unperturbed equilibrium position. Only "interchange" ($k_\parallel = 0$) motions of the *entire* field lines are not subject to this "spring-back" effect. Such motions are often ruled out by geometry or boundary conditions (cf. footnote 6). Thus, fluid elements in MHD cannot simply random walk as this would increase (without bound) the field-line tension. However, they may random walk for a substantial period before the tension returns them back to the equilibrium state. The role of such long-time correlations in MHD turbulence is unknown.

Reduced MHD, the decoupling of the Alfvén-wave cascade, and turbulence in the interstellar medium. We now give a rigorous demonstration of how the turbulent cascade associated with the Alfvén waves (or, more precisely, Alfvén-wave-polarised fluctuations) decouples from the cascades of the slow waves and entropy fluctuations. Let us start with the equations of compressible MHD:

$$\frac{d\rho}{dt} = -\rho \boldsymbol{\nabla} \cdot \mathbf{u}, \tag{16}$$

$$\rho \frac{d\mathbf{u}}{dt} = -\boldsymbol{\nabla}\left(p + \frac{B^2}{8\pi}\right) + \frac{\mathbf{B} \cdot \boldsymbol{\nabla}\mathbf{B}}{4\pi}, \tag{17}$$

$$\frac{ds}{dt} = 0, \quad s = \frac{p}{\rho^\gamma}, \quad \gamma = \frac{5}{3}, \tag{18}$$

$$\frac{d\mathbf{B}}{dt} = \mathbf{B} \cdot \boldsymbol{\nabla}\mathbf{u} - \mathbf{B}\boldsymbol{\nabla} \cdot \mathbf{u}. \tag{19}$$

Consider a uniform static equilibrium with a straight magnetic field, so $\rho = \rho_0 + \delta\rho$, $p = p_0 + \delta p$, $\mathbf{B} = \mathbf{B}_0 + \delta\mathbf{B}$. Based on observational and numerical evidence, it is safe to assume that the turbulence in such a system will be anisotropic with $k_\parallel \ll k_\perp$. Let us, therefore, introduce a small parameter $\epsilon \sim k_\parallel/k_\perp$ and carry out a systematic expansion of Eqs. (16)–(19) in ϵ. In this expansion, the fluctuations are treated as small, but not arbitrarily so: in order to estimate their size, we shall adopt the critical-balance conjecture (14), which is now treated not as a detailed scaling prescription but as an ordering assumption. This allows us to introduce the following ordering:

$$\frac{\delta\rho}{\rho_0} \sim \frac{u_\perp}{v_A} \sim \frac{u_\parallel}{v_A} \sim \frac{\delta p}{p_0} \sim \frac{\delta B_\perp}{B_0} \sim \frac{\delta B_\parallel}{B_0} \sim \frac{k_\parallel}{k_\perp} \sim \epsilon, \tag{20}$$

where we have also assumed that the velocity and magnetic-field fluctuations have the character of Alfvén and slow waves ($\delta\mathbf{B} \sim \mathbf{u}$) and that the relative amplitudes of the Alfvén-wave-polarised fluctuations ($u_\perp/v_A$, $\delta B_\perp/B_0$), slow-wave-polarised fluctuations ($u_\parallel/v_A$, $\delta B_\parallel/B_0$) and density fluctuations ($\delta\rho/\rho_0$) are all the same order.[7] We further assume that the characteristic frequency of the fluctuations is $\omega \sim k_\parallel v_A$, which means that the fast waves, for which $\omega \simeq k_\perp\sqrt{v_A^2 + c_s^2}$, where $c_s = \gamma p_0/\rho_0$ is the sound speed, are ordered out.

We start by observing that the Alfvén-wave-polarised fluctuations are 2D solenoidal: since $\boldsymbol{\nabla} \cdot \mathbf{u} = O(\epsilon^2)$ [from Eq. (16)] and $\boldsymbol{\nabla} \cdot \mathbf{B} = 0$, separating the $O(\epsilon)$ part of these divergences gives $\boldsymbol{\nabla}_\perp \cdot \mathbf{u}_\perp = \boldsymbol{\nabla}_\perp \cdot \delta\mathbf{B}_\perp = 0$. We may, therefore, express $\mathbf{u}_\perp$ and $\delta\mathbf{B}_\perp$ in terms of scalar stream (flux) functions:

$$\mathbf{u}_\perp = \hat{\mathbf{b}}_0 \times \boldsymbol{\nabla}_\perp\phi, \quad \frac{\delta\mathbf{B}_\perp}{\sqrt{4\pi\rho_0}} = \hat{\mathbf{b}}_0 \times \boldsymbol{\nabla}_\perp\psi, \tag{21}$$

where $\hat{\mathbf{b}}_0 = \mathbf{B}_0/B_0$. Evolution equations for ϕ and ψ are obtained by substituting the expressions (21) into the perpendicular parts of the induction Eq. (19) and the

---

[7] Strictly speaking, whether this is the case depends on the energy sources that drive the turbulence: as we are about to see, if no slow waves are launched, none will be present. However, it is safe to assume in astrophysical contexts that the large-scale energy input is random and, therefore, comparable power is injected in all types of fluctuations.

momentum Eq. (17) — of the latter the curl is taken to annihilate the pressure term. Keeping only the terms of the lowest order, $O(\epsilon^2)$, we get

$$\frac{\partial}{\partial t}\psi + \{\phi, \psi\} = v_A \nabla_\parallel \phi, \tag{22}$$

$$\frac{\partial}{\partial t}\nabla_\perp^2 \phi + \{\phi, \nabla_\perp^2 \phi\} = v_A \nabla_\parallel \nabla_\perp^2 \psi + \{\psi, \nabla_\perp^2 \psi\}, \tag{23}$$

where $\{\phi, \psi\} = \hat{\mathbf{b}}_0 \cdot (\boldsymbol{\nabla}_\perp \phi \times \boldsymbol{\nabla}_\perp \psi)$ and to lowest order,

$$\frac{d}{dt} = \frac{\partial}{\partial t} + \mathbf{u}_\perp \cdot \boldsymbol{\nabla}_\perp = \frac{\partial}{\partial t} + \{\phi, \cdots\}, \tag{24}$$

$$\frac{\mathbf{B}}{B_0} \cdot \boldsymbol{\nabla} = \nabla_\parallel + \frac{\delta \mathbf{B}_\perp}{B_0} \cdot \boldsymbol{\nabla}_\perp = \nabla_\parallel + \frac{1}{v_A}\{\psi, \cdots\}. \tag{25}$$

Eqs. (22) and (23) are known as the Reduced Magnetohydrodynamics (RMHD). They were first derived by Strauss [28] in the context of fusion plasmas. They form a closed set, meaning that the Alfvén-wave cascade decouples from the slow waves and density fluctuations.

In order to derive evolution equations for the latter, let us revisit the perpendicular part of the momentum equation and use Eq. (20) to order terms in it. In the lowest order, $O(\epsilon)$, we get the pressure balance

$$\boldsymbol{\nabla}_\perp \left(\delta p + \frac{B_0 \delta B_\parallel}{4\pi}\right) = 0 \quad \Rightarrow \quad \frac{\delta p}{p_0} = -\gamma \frac{v_A^2}{c_s^2}\frac{\delta B_\parallel}{B_0}. \tag{26}$$

Using Eq. (26) and the entropy Eq. (18), we get

$$\frac{d}{dt}\frac{\delta s}{s_0} = 0, \quad \frac{\delta s}{s_0} = \frac{\delta p}{p_0} - \gamma\frac{\delta\rho}{\rho_0} = -\gamma\left(\frac{\delta\rho}{\rho_0} + \frac{v_A^2}{c_s^2}\frac{\delta B_\parallel}{B_0}\right), \tag{27}$$

where $s_0 = p_0/\rho_0^\gamma$. On the other hand, from the continuity Eq. (16) and the parallel component of the induction Eq. (19),

$$\frac{d}{dt}\left(\frac{\delta\rho}{\rho_0} - \frac{\delta B_\parallel}{B_0}\right) + \frac{\mathbf{B}}{B_0}\cdot\boldsymbol{\nabla}u_\parallel = 0. \tag{28}$$

Combining Eqs. (27) and (28), we obtain

$$\frac{d}{dt}\frac{\delta\rho}{\rho_0} = -\frac{1}{1+c_s^2/v_A^2}\frac{\mathbf{B}}{B_0}\cdot\boldsymbol{\nabla}u_\parallel, \tag{29}$$

$$\frac{d\delta B_\parallel}{dt} = \frac{1}{1+v_A^2/c_s^2}\frac{\mathbf{B}}{B_0}\cdot\boldsymbol{\nabla}u_\parallel. \tag{30}$$

Finally, we take the parallel component of the momentum Eq. (17) and notice that, due to Eq. (26) and to the smallness of the parallel gradients, the pressure term is $O(\epsilon^3)$, while the inertial and tension terms are $O(\epsilon^2)$. Therefore,

$$\frac{du_\parallel}{dt} = v_A^2 \frac{\mathbf{B}}{B_0}\cdot\boldsymbol{\nabla}\frac{\delta B_\parallel}{B_0}. \tag{31}$$

Eqs. (30) and (31) describe the slow-wave-polarised fluctuations, while Eq. (27) describes the zero-frequency entropy mode. The non-linearity in these equations enters via the derivatives defined in Eqs. (24), (25) and is due solely to interactions with Alfvén waves.

Naturally, the reduced equations derived above can be cast in the Elsasser form. If we introduce Elsasser potentials $\zeta^\pm = \phi \pm \psi$, Eqs. (22) and (23) become

$$\frac{\partial}{\partial t}\nabla_\perp^2 \zeta^\pm \mp v_A \nabla_\parallel \nabla_\perp^2 \zeta^\pm = -\frac{1}{2}\left[\{\zeta^+,\nabla_\perp^2\zeta^-\} + \{\zeta^-,\nabla_\perp^2\zeta^+\} \mp \nabla_\perp^2\{\zeta^+,\zeta^-\}\right]. \tag{32}$$

This is the same as the perpendicular part of Eq. (5) with $\mathbf{z}_\perp^\pm = \hat{\mathbf{b}}_0 \times \boldsymbol{\nabla}_\perp \zeta^\pm$. The key property that only counterpropagating Alfvén waves interact is manifest here. For the slow-wave variables, we may introduce generalised Elsasser fields:

$$z_\parallel^\pm = u_\parallel \pm \frac{\delta B_\parallel}{\sqrt{4\pi\rho_0}}\left(1 + \frac{v_A^2}{c_s^2}\right)^{1/2}. \tag{33}$$

Straightforwardly, the evolution equation for these fields is

$$\frac{\partial z_\parallel^\pm}{\partial t} \mp \frac{v_A}{\sqrt{1 + v_A^2/c_s^2}}\nabla_\parallel z_\parallel^\pm = -\frac{1}{2}\left(1 \mp \frac{1}{\sqrt{1 + v_A^2/c_s^2}}\right)\{\zeta^+, z_\parallel^\pm\}$$
$$-\frac{1}{2}\left(1 \pm \frac{1}{\sqrt{1 + v_A^2/c_s^2}}\right)\{\zeta^-, z_\parallel^\pm\}. \tag{34}$$

This equation reduces to the parallel part of Eq. (5) in the limit $v_A \ll c_s$. This is known as the high-β limit, with the plasma beta defined by $\beta = 8\pi p_0/B_0^2 = (2/\gamma)c_s^2/v_A^2$. We see that only in this limit do the slow waves interact exclusively with the counterpropagating Alfvén waves. For general β, the phase speed of the slow waves is smaller than that of the Alfvén waves and, therefore, Alfvén waves can "catch up" and interact with the slow waves that travel in the same direction.

In astrophysical turbulence, β tends to be moderately high[8]: for example, in the interstellar medium, $\beta \sim 10$ by order of magnitude. In the high-β limit, which is equivalent to the incompressible approximation for the slow waves, density fluctuations are due solely to the entropy mode. They decouple from the slow-wave cascade and are passively mixed by the Alfvén-wave turbulence: $d\delta\rho/dt = 0$ [Eq. (27) or Eq. (29), $c_s \gg v_A$]. By a dimensional argument similar to K41, the spectrum of such a field is expected to follow the spectrum of the underlying turbulence [29]: in the GS theory, $k_\perp^{-5/3}$. It is precisely the electron-density spectrum (deduced from observations of the scintillation of radio sources due to the scattering of radio waves by the interstellar medium) that provides the evidence of the $k^{-5/3}$ scaling in the interstellar turbulence [16]. The explanation of this density spectrum in terms of passive mixing of the entropy mode, originally conjectured by Higdon [30], is developed on the basis of the GS theory in [31].

Thus, the anisotropy and critical balance taken as ordering assumptions lead to a neat decomposition of the MHD turbulent cascade into a decoupled Alfvén-wave cascade and cascades of slow waves and entropy fluctuations passively scattered/mixed

---

[8] The solar corona, where $\beta \sim 10^{-6}$, is one prominent exception.

by the Alfvén waves.[9] The validity of this decomposition and, especially, of the RMHD Eqs. (22) and (23) turns out to extend to collisionless scales where the MHD Eqs. (16)–(19) cannot be used: this will be briefly discussed in § 4.3.

3 Isotropic MHD turbulence

Let us now consider the case of isotropic MHD turbulence, i.e., a turbulent plasma where no mean field is imposed externally. Can the scaling theories reviewed above be adapted to this case? A popular view, due originally to Kraichnan [7], is that everything remains the same with the role of the mean field $\mathbf{B}_0$ now played by the magnetic fluctuations at the outer scale, δB_L, while at smaller scales, the turbulence is again a local (in scale) cascade of Alfvénic ($\delta u_l \sim \delta B_l$) fluctuations. This picture is only plausible if the magnetic energy is dominated by the outer-scale fluctuations, an assumption that does not appear to hold in the numerical simulations of forced isotropic MHD turbulence [32]. Instead, the magnetic energy is concentrated at small scales, where the magnetic fluctuations significantly exceed the velocity fluctuations, with no sign of the scale-by-scale equipartition implied for an Alfvénic cascade.[10] These features are especially pronounced when the magnetic Prandtl number $Pr_m = \nu/\eta = R_m/Re \gg 1$, i.e., when the magnetic cutoff scale lies below the viscous cutoff of the velocity fluctuations (Fig. 2). The numerically more accessible case of $Pr_m \gtrsim 1$, while non-asymptotic and, therefore, harder to interpret, retains most of the features of the large-Pr_m regime. A handy formula for Pr_m based on the Spitzer [34] values of ν and η for fully ionised plasmas is

$$Pr_m \sim 10^{-5} T^4/n, \tag{35}$$

where T is the temperature in Kelvin and n is the particle density in cm^{-3}. Eq. (35) tends to give very large values for hot diffuse astrophysical plasmas: e.g., 10^{11} for the warm interstellar medium, 10^{29} for galaxy clusters.

Let us examine the situation in more detail. In the absence of a mean field, all magnetic fields are generated and maintained by the turbulence itself,

---

[9] Eqs. (27), (32), and (34) imply that, at arbitrary β, there are five conserved quantities: $I_s = \langle|\delta s|^2\rangle$ (entropy fluctuations), $I_\perp^\pm = \langle|\nabla\zeta^\pm|^2\rangle$ (right/left-propagating Alfvén waves), $I_\parallel^\pm = \langle|z_\parallel^\pm|^2\rangle$ (right/left-propagating slow waves). $I_\perp^+$ and $I_\perp^-$ are always cascaded by interaction with each other, I_s is passively mixed by $I_\perp^+$ and $I_\perp^-$, $I_\parallel^\pm$ are passively scattered by $I_\perp^\mp$ and, unless $\beta \gg 1$, also by $I_\perp^\pm$.

[10] This is true for the case of *forced* turbulence. Simulations of the decaying case [33] present a rather different picture: there is still no scale-by-scale equipartition but the magnetic energy heavily dominates at the *large* scales – most likely due to a large-scale force-free component controlling the decay. The difference between the numerical results on the decaying and forced MHD turbulence points to another break down in universality in stark contrast with the basic similarity of the two regimes in the hydrodynamic case.

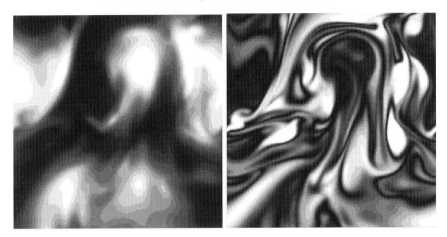

Fig. 2. Cross sections of the absolute values of **u** (*left panel*) and **B** (*right panel*) in the saturated state of a simulation with $Re \simeq 100$, $Pr_m = 10$ (run B of Ref. [32])

i.e., isotropic MHD turbulence is the saturated state of the turbulent (small-scale) dynamo. Therefore, we start by considering how a weak (dynamically unimportant) magnetic field is amplified by turbulence in a large-Pr_m MHD fluid and what kind of field can be produced this way.

3.1 Small-scale dynamo

Many specific deterministic flows have been studied numerically and analytically and shown to be dynamos [35]. While rigorously determining whether any given flow is a dynamo is virtually always a formidable mathematical challenge, the combination of numerical and analytical experience of the last 50 years suggests that smooth 3D flows with chaotic trajectories tend to have the dynamo property provided the magnetic Reynolds number exceeds a certain threshold, $R_m > R_{m,c} \sim 10^1$–10^2. In particular, the ability of Kolmogorov turbulence to amplify magnetic fields is a solid numerical fact first established by Meneguzzi et al. (1981) [36] and since then confirmed in many numerical studies with ever-increasing resolutions (most recently [32,37]). It was, in fact, Batchelor who realised already in 1950 [38] that the growth of magnetic fluctuations in a random flow should occur simply as a consequence of the random stretching of the field lines and that it should proceed at the rate of strain associated with the flow. In Kolmogorov turbulence, the largest rate of strain $\sim \delta u_l / l$ is associated with the smallest scale $l \sim l_\nu$ – the viscous scale, so it is the viscous-scale motions that dominantly amplify the field (at large Pr_m). Note that the velocity field at the viscous scale is random but smooth, so the small-scale dynamo in Kolmogorov turbulence belongs to the same class as fast dynamos in smooth single-scale flows [32,35].

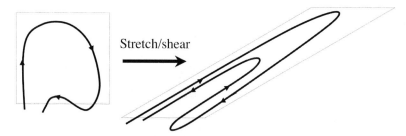

Fig. 3. Stretching/shearing a magnetic-field line

A repeated application of random stretching/shearing to a tangled magnetic field produces direction reversals at arbitrarily small scales, giving rise to a folded field structure (Fig. 3). It is an essential property of this structure that the strength of the field and the curvature of the field lines are anticorrelated: wherever the field is growing it is relatively straight (i.e., curved only on the scale of the flow), whereas in the bending regions, where the curvature is large, the field is weak. A quantitative theory of the folded structure can be constructed based on the joint statistics of the field strength $B = |\mathbf{B}|$ and curvature $\mathbf{K} = \hat{\mathbf{b}} \cdot \nabla \hat{\mathbf{b}}$, where $\hat{\mathbf{b}} = \mathbf{B}/B$ [39].[11] The curvature is a quantity easily measured in numerical simulations, which confirm the overall straightness of the field and the curvature–field-strength anticorrelation [32]. At the end of this section, we shall give a simple demonstration of the validity of the folded structure.

The scale of the direction reversals is limited from below only by Ohmic diffusion: for Kolmogorov turbulence, balancing the rate of strain at the viscous scale with diffusion and taking $Pr_m \gg 1$ gives the resistive cutoff l_η:

$$\delta u_{l_\nu}/l_\nu \sim \eta/l_\eta^2 \quad \Rightarrow \quad l_\eta \sim Pr_m^{-1/2} l_\nu. \tag{36}$$

If random stretching gives rise to magnetic fields with reversals at the resistive scale, why are these fields not eliminated by diffusion? In other words, how is the small-scale dynamo possible? It turns out that, in 3D, there are magnetic-field configurations that can be stretched without being destroyed by the concurrent refinement of the reversal scale and that add up to give rise to exponential growth of the magnetic energy. Below we give an analytical demonstration of this. It is a (somewhat modified) version of an ingenious argument originally proposed by Zeldovich et al. in 1984 [42]. A reader looking solely for a broad qualitative picture of isotropic MHD turbulence may skip to § 3.2.

---

[11] This is not the only existing way of diagnosing the field structure. Ott and co-workers studied field reversals by measuring magnetic-flux cancellations [40]. Chertkov et al. [41] considered two-point correlation functions of the magnetic field in a model of small-scale dynamo and found large-scale correlations along the field and short-scale correlations across.

Fig. 4. Ya. B. Zeldovich (1914–1987). (Photo courtesy of M. Ya. Ovchinnikova.)

Overcoming diffusion. Let us study magnetic fields with reversals at subviscous scales: at these scales, the velocity field is smooth and can, therefore, be expanded

$$u^i(t, \mathbf{x}) = u^i(t, \mathbf{0}) + \sigma^i_m(t)x^m + \dots, \tag{37}$$

where $\sigma^i_m(t)$ is the rate-of-strain tensor. The expansion is around some reference point $\mathbf{x} = \mathbf{0}$. We can always go to the reference frame that moves with the velocity at this point, so that $u^i(t, \mathbf{0}) = 0$. Let us seek the solution to Eq. (2) with velocity (37) as a sum of random plane waves with time-dependent wave vectors:

$$B^i(t, \mathbf{x}) = \int \frac{d^3k_0}{(2\pi)^3} \, \tilde{B}^i(t, \mathbf{k}_0)e^{i\tilde{\mathbf{k}}(t, \mathbf{k}_0)\cdot\mathbf{x}}, \tag{38}$$

where $\tilde{\mathbf{k}}(0, \mathbf{k}_0) = \mathbf{k}_0$, so $\tilde{B}^i(0, \mathbf{k}_0) = B^i_0(\mathbf{k}_0)$ is the Fourier transform of the initial field. Since Eq. (2) is linear, it is sufficient to ensure that each of the plane waves is individually a solution. This leads to two *ordinary* differential equations for every $\mathbf{k}_0$:

$$\partial_t \tilde{B}^i = \sigma^i_m \tilde{B}^m - \eta \tilde{k}^2 \tilde{B}^i, \qquad \partial_t \tilde{k}_l = -\sigma^i_l \tilde{k}_i, \tag{39}$$

subject to initial conditions $\tilde{B}^i(0, \mathbf{k}_0) = B^i_0(\mathbf{k}_0)$ and $\tilde{k}_l(0, \mathbf{k}_0) = k_{0l}$. The solution of these equations can be written explicitly in terms of the Lagragian transformation of variables $\mathbf{x}_0 \to \mathbf{x}(t, \mathbf{x}_0)$, where

$$\partial_t x^i(t, \mathbf{x}_0) = u^i(t, \mathbf{x}(t, \mathbf{x}_0)) = \sigma^i_m(t)x^m(t, \mathbf{x}_0), \qquad x^i(0, \mathbf{x}_0) = x^i_0. \tag{40}$$

Because of the linearity of the velocity field, the strain tensor $\partial x^i / \partial x^m_0$ and its inverse $\partial x^r_0 / \partial x^l$ are functions of time only. At $t = 0$, they are unit matrices. At $t > 0$, they satisfy

$$\partial_t \frac{\partial x^i}{\partial x^m_0} = \sigma^i_l \frac{\partial x^l}{\partial x^m_0}, \qquad \partial_t \frac{\partial x^r_0}{\partial x^l} = -\sigma^r_l \frac{\partial x^r}{\partial x^i_0}. \tag{41}$$

We can check by direct substitution that

$$\tilde{B}^i(t, \mathbf{k}_0) = \frac{\partial x^i}{\partial x_0^m} B_0^m(\mathbf{k}_0) \exp\left[-\eta \int_0^t dt' \tilde{k}^2(t')\right], \quad \tilde{k}_l(t, \mathbf{k}_0) = \frac{\partial x_0^r}{\partial x^l} k_{0r}. \quad (42)$$

These formulae express the evolution of one mode in the integral (38). Using the fact that $\det(\partial x_0^r / \partial x^l) = 1$ in an incompressible flow and that Eq. (42) therefore establishes a one-to-one correspondence $\mathbf{k} \leftrightarrow \mathbf{k}_0$, it is easy to prove that the volume-integrated magnetic energy is the sum of the energies of individual modes:

$$\langle B^2 \rangle(t) \equiv \int d^3x \, |\mathbf{B}(t, \mathbf{x})|^2 = \int \frac{d^3 k_0}{(2\pi)^3} \, |\tilde{\mathbf{B}}(t, \mathbf{k}_0)|^2. \quad (43)$$

From Eq. (42),

$$|\tilde{\mathbf{B}}(t, \mathbf{k}_0)|^2 = \mathbf{B}_0(\mathbf{k}_0) \cdot \hat{\mathsf{M}}(t) \cdot \mathbf{B}_0^*(\mathbf{k}_0) \exp\left[-2\eta \int_0^t dt' \, \mathbf{k}_0 \cdot \hat{\mathsf{M}}^{-1}(t') \cdot \mathbf{k}_0\right], \quad (44)$$

where the matrices $\hat{\mathsf{M}}$ and $\hat{\mathsf{M}}^{-1}$ have elements defined by

$$M_{mn}(t) = \frac{\partial x^i}{\partial x_0^m} \frac{\partial x^i}{\partial x_0^n} \quad \text{and} \quad M^{rs}(t) = \frac{\partial x_0^r}{\partial x^l} \frac{\partial x_0^s}{\partial x^l}, \quad (45)$$

respectively. They are the co- and contravariant metric tensors of the inverse Lagrangian transformation $\mathbf{x} \to \mathbf{x}_0$.

Let us consider the simplest possible case of a flow (37) with constant $\hat{\sigma} = \text{diag}\{\lambda_1, \lambda_2, \lambda_3\}$, where $\lambda_1 > \lambda_2 \geq 0 > \lambda_3$ and $\lambda_1 + \lambda_2 + \lambda_3 = 0$ by incompressibility. Then $\hat{\mathsf{M}} = \text{diag}\{e^{2\lambda_1 t}, e^{2\lambda_2 t}, e^{2\lambda_3 t}\}$ and Eq. (44) becomes, in the limit $t \to \infty$,

$$|\tilde{\mathbf{B}}(t, \mathbf{k}_0)|^2 \sim |B_0^1(\mathbf{k}_0)|^2 \exp\left[2\lambda_1 t - \eta\left(\frac{k_{01}^2}{\lambda_1} + \frac{k_{02}^2}{\lambda_2} + \frac{k_{03}^2}{|\lambda_3|} e^{2|\lambda_3|t}\right)\right], \quad (46)$$

where we have dropped terms that decay exponentially with time compared to those retained.[12] We see that for most $\mathbf{k}_0$, the corresponding modes decay superexponentially fast with time. The domain in the $\mathbf{k}_0$ space containing modes that are not exponentially small at any given time t is given by

$$\frac{k_{01}^2}{\lambda_1^2 t/\eta} + \frac{k_{02}^2}{\lambda_1 \lambda_2 t/\eta} + \frac{k_{03}^2}{\lambda_1|\lambda_3| t e^{2\lambda_3 t}/\eta} < \text{const}. \quad (47)$$

The volume of this domain at time t is $\sim \lambda_1^2 (\lambda_2 |\lambda_3|)^{1/2} (t/\eta)^{3/2} e^{\lambda_3 t}$. Within this volume, $|\tilde{\mathbf{B}}(t, \mathbf{k}_0)|^2 \sim |B_0^1(\mathbf{k}_0)|^2 e^{2\lambda_1 t}$. Using $\lambda_3 = -\lambda_1 - \lambda_2$ and Eq. (43), we get

$$\langle B^2 \rangle(t) \propto \exp\left[(\lambda_1 - \lambda_2)t\right]. \quad (48)$$

---

[12] If $\lambda_2 = 0$, k_{02}^2/λ_2 in Eq. (46) is replaced with $2k_{02}^2 t$. The case $\lambda_2 < 0$ is treated in a way similar to that described below and also leads to magnetic-energy growth. The difference is that for $\lambda_2 \geq 0$, the magnetic structures are flux sheets, or ribbons, while for $\lambda_2 < 0$, they are flux ropes.

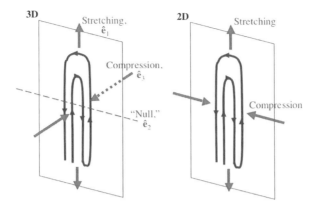

Fig. 5. Magnetic fields vs. the Lyapunov directions. (From [32].) Zeldovich et al. [42] did not give this exact interpretation of their calculation because the folded structure of the field was not yet clearly understood at the time

Let us discuss the physics behind the Zeldovich et al. calculation sketched above. When the magnetic field is stretched by the flow, it naturally aligns with the stretching Lyapunov direction: $\mathbf{B} \sim \hat{\mathbf{e}}_1 B_0^1 e^{\lambda_1 t}$. The wave vector $\mathbf{k}$ has a tendency to align with the compression direction: $\mathbf{k} \sim \hat{\mathbf{e}}_3 k_{03} e^{|\lambda_3|t}$, which makes most modes decay superexponentially. The only ones that survive are those whose $\mathbf{k}_0$'s were nearly perpendicular to $\hat{\mathbf{e}}_3$, with the permitted angular deviation from $90°$ decaying exponentially in time $\sim e^{-|\lambda_3|t}$. Since the magnetic field is solenoidal, $\mathbf{B}_0 \perp \mathbf{k}_0$, the modes that get stretched the most have $\mathbf{B}_0 \parallel \hat{\mathbf{e}}_1$ and $\mathbf{k}_0 \parallel \hat{\mathbf{e}}_2$ (Fig. 5). In contrast, in 2D, the field aligns with $\hat{\mathbf{e}}_1$ and must, therefore, reverse along $\hat{\mathbf{e}}_2$, which is always the compression direction (Fig. 5), so the stretching is always overwhelmed by the diffusion and no dynamo is possible (as should be the case according to the rigorous early result of Zeldovich [43]).

The above construction can be generalised to time-dependent and random velocity fields. The matrix $\hat{\mathsf{M}}$ is symmetric and can, therefore, be diagonalised by an appropriate rotation $\hat{\mathsf{R}}$ of the coordinate system: $\hat{\mathsf{M}} = \hat{\mathsf{R}}^T \cdot \hat{\mathsf{L}} \cdot \hat{\mathsf{R}}$, where, by definition, $\hat{\mathsf{L}} = \operatorname{diag}\left\{ e^{\zeta_1(t)}, e^{\zeta_2(t)}, e^{\zeta_3(t)} \right\}$. It is possible to prove that, as $t \to \infty$, $\hat{\mathsf{R}}(t) \to \{\hat{\mathbf{e}}_1, \hat{\mathbf{e}}_2, \hat{\mathbf{e}}_3\}$ and $\zeta_i(t)/2t \to \lambda_i$, where $\hat{\mathbf{e}}_i$ are constant orthogonal unit vectors, which make up the Lyapunov basis, and λ_i are the Lyapunov exponents of the flow [44]. The instantaneous values of $\zeta_i(t)/2t$ are called finite-time Lyapunov exponents. For a random flow, $\zeta_i(t)$ are random functions. Eq. (48) generalises to

$$\langle B^2 \rangle(t) \propto \overline{\exp\left[(\zeta_1 - \zeta_2)/2 \right]}, \tag{49}$$

where the overline means averaging over the distribution of ζ_i. The only random flow for which this distribution is known is a Gaussian white-in-time velocity first considered in the dynamo context in 1967 by Kazantsev [45].[13] The distribution of ζ_i for this flow is Gaussian in the long-time limit and Eq. (49) gives $\langle B^2 \rangle \propto e^{(5/4)\lambda_1 t}$,

---

[13] As the only analytically solvable model of random advection, Kazantsev's model has played a crucial role. Developed extensively in 1980s by Zeldovich and

where $\lambda_1 = \langle\zeta_1\rangle/2t$ [41]. For Kazantsev's velocity, it is also possible to calculate the magnetic-energy spectrum [45,49], which is the spectrum of the direction reversals. It has a peak at the resistive scale and a $k^{+3/2}$ power law stretching across the sub-viscous range, $l_\nu^{-1} \ll k \ll l_\eta^{-1}$. This scaling appears to be corroborated by numerical simulations [32,37].

Folded structure revisited. We shall now give a very simple demonstration that linear stretching does indeed produce folded fields with straight/curved field lines corresponding to larger/smaller field strength. Using Eq. (2), we can write evolution equations for the field strength $B = |\mathbf{B}|$, the field direction $\hat{\mathbf{b}} = \mathbf{B}/B$ and the field-line curvature $\mathbf{K} = \hat{\mathbf{b}} \cdot \nabla\hat{\mathbf{b}}$. Omitting the resistive terms,

$$\frac{dB}{dt} = (\hat{\mathbf{b}}\hat{\mathbf{b}} : \nabla\mathbf{u})B, \tag{50}$$

$$\frac{d\hat{\mathbf{b}}}{dt} = \hat{\mathbf{b}} \cdot (\nabla\mathbf{u}) \cdot (\hat{\mathbf{I}} - \hat{\mathbf{b}}\hat{\mathbf{b}}), \tag{51}$$

$$\frac{d\mathbf{K}}{dt} = \mathbf{K} \cdot (\nabla\mathbf{u}) \cdot (\hat{\mathbf{I}} - \hat{\mathbf{b}}\hat{\mathbf{b}}) - 2(\hat{\mathbf{b}}\hat{\mathbf{b}} : \nabla\mathbf{u})\mathbf{K} - [\hat{\mathbf{b}} \cdot (\nabla\mathbf{u}) \cdot \mathbf{K}]\hat{\mathbf{b}}$$
$$+ \hat{\mathbf{b}}\hat{\mathbf{b}} : (\nabla\nabla\mathbf{u}) \cdot (\hat{\mathbf{I}} - \hat{\mathbf{b}}\hat{\mathbf{b}}). \tag{52}$$

For simplicity, we again use the velocity field (37) with constant $\hat{\sigma} = \text{diag}\{\lambda_1, \lambda_2, \lambda_3\}$. Then the stable fixed point of Eq. (50) in the comoving frame is $\hat{\mathbf{b}} = \hat{\mathbf{e}}_1$ (magnetic field aligns with the principal stretching direction), whence $B \propto e^{\lambda_1 t}$. Since $\mathbf{K}\cdot\hat{\mathbf{b}} = 0$, we set $K_1 = 0$. Denoting $\Sigma = \hat{\mathbf{b}}\hat{\mathbf{b}} : \nabla\nabla\mathbf{u}$, we can now write Eq. (52) as

$$\frac{dK_2}{dt} = -(2\lambda_1 - \lambda_2)K_2 + \Sigma_2, \qquad \frac{dK_3}{dt} = -(3\lambda_1 + \lambda_2)K_3 + \Sigma_3. \tag{53}$$

Both components of $\mathbf{K}$ decay exponentially,[14] until they are comparable to the inverse scale of the velocity field (i.e., the terms containing $\nabla\nabla\mathbf{u}$ become important). The stationary solution is $K_2 = \Sigma_2/(2\lambda_1 - \lambda_2)$, $K_3 = \Sigma_3/(3\lambda_1 + \lambda_2)$.

  If the field is to reverse direction, it must turn somewhere (see Fig. 3). At such a turning point, the field must be perpendicular to the stretching direction. Setting $b_1 = 0$, we find two fixed points of Eq. (50) under this condition: $\hat{\mathbf{b}} = \hat{\mathbf{e}}_2$ and $\hat{\mathbf{b}} = \hat{\mathbf{e}}_3$. Only the former is stable, so the field at the turning point will tend to align with the "null" direction. Thus, stretching favours configurations with field reversals along the "null" direction, which are also those that survive diffusion (see Fig. 5). From Eq. (52) we find that at the turning point, $K_2 = 0$, $K_3 = \Sigma_3/(\lambda_1 + 3\lambda_2)$, while K_1 grows at the rate $\lambda_1 - 2\lambda_2$ (assumed positive). This growth continues until limited

---

co-workers [46], the model became a tool of choice in the theories of anomalous scaling and intermittency that flourished in 1990s [47] (in this context, it has been associated with the name of Kraichnan who, independently from Kazantsev, proposed to use it for the passive scalar problem [48]). It remains useful to this day as old theories are reevaluated and new questions demand analytical answers [39].

[14] K_3 decays faster than K_2. If the velocity is exactly linear ($\Sigma = 0$), $\mathbf{K}$ aligns with $\hat{\mathbf{e}}_2$ and decreases indefinitely, while the combination $BK^{1/(2-\lambda_2/\lambda_1)}$ stays constant. This rhymes with the result that can be proven for a linear Kazantsev velocity: at zero η, $B \propto e^{\zeta_1/2}$, while $BK^{1/2} \propto e^{\zeta_2/4}$ and $\lambda_2 = \langle\zeta_2\rangle/2t = 0$.

A. Schlüter   L. Biermann (1907–1986)

Fig. 6. Photo of A. Schlüter courtesy of MPI für Plasmaphysik; photo of L. Biermann courtesy of Max-Planck-Gesellschaft/AIP Emilio Segrè Visual Archives. Photo of G.K. Batchelor may be found in H.K. Moffatt's contribution to this volume

by diffusion at $K \sim 1/l_\eta$. The strength of the field in this curved region is $\propto e^{\lambda_2 t}$, so the fields are weaker than in the straight segments, where $B \propto e^{\lambda_1 t}$.

When the problem is solved for the Kazantsev velocity, the above solution generalises to a field of random curvatures anticorrelated with the magnetic-field strength and with a stationary PDF of K that has a peak at $K \sim$ flow scale^{-1} and a power tail $\sim K^{-13/7}$ describing the distribution of curvatures at the turning points [39]. Numerical simulations support these results [32].

3.2 Saturation of the dynamo

The small-scale dynamo gave us exponentially growing magnetic fields with energy concentrated at small (resistive) scales. How is the growth of magnetic energy saturated and what is the final state? Will magnetic energy stay at small scales or will it proceed to scale-by-scale equipartition via some form of inverse cascade? This basic dichotomy dates back to the 1950 papers by Batchelor [38] and Schlüter and Biermann [50]. Batchelor thought that magnetic field was basically analogous to the vorticity field $\boldsymbol{\omega} = \boldsymbol{\nabla} \times \mathbf{u}$ (which satisfies the same Eq. (2) except for the difference between η and ν) and would, therefore, saturate at a low energy, $\langle B^2 \rangle \sim Re^{-1/2} \langle u^2 \rangle$, with a spectrum peaked at the viscous scale. Schlüter and Biermann disagreed and argued that the saturated state would be a scale-by-scale balance between the Lorentz and inertial forces, with turbulent motions at each scale giving rise to magnetic fluctuations of matching energy at the same scale. Schlüter and Biermann's argument (and, implicitly, also Batchelor's) was based on the assumption of **locality of interaction** (in scale space) between the magnetic and velocity fields: locality both of the dynamo action and of the back reaction. As we saw in the previous section, this assumption is certainly incorrect for the dynamo: a linear velocity field, i.e., a velocity field of a formally infinitely large (in practice, viscous) scale can produce magnetic fields with reversals at the smallest scale allowed by diffusion. A key implication of the folded structure of these fields concerns the Lorentz force, the essential part of which, in the case of incompressible flow, is the curvature force $B^2\mathbf{K}$. Since it is a quantity that depends only on the parallel gradient of the magnetic field and does not know about direction reversals, it will possess a degree of velocity-scale spatial coherence necessary to oppose stretching. Thus, a field that is formally at the resistive scale will exert a back reaction at the scale of the velocity field. In other words, interactions are nonlocal: a random flow at a given scale l, having amplified the magnetic fields at the resistive scale $l_\eta \ll l$, will see these magnetic fields back react at the scale l. Given the nonlocality of back reaction, we can update Batchelor's and Schlüter and Biermann's scenarios for saturation in the following way [32, 51].

The magnetic energy is amplified by the viscous-scale motions until the field is strong enough to resist stretching, i.e., until $\mathbf{B}\cdot\boldsymbol{\nabla}\mathbf{B} \sim \mathbf{u}\cdot\boldsymbol{\nabla}\mathbf{u} \sim \delta u_{l_\nu}^2/l_\nu$. Since $\mathbf{B}\cdot\boldsymbol{\nabla}\mathbf{B} \sim B^2 K \sim B^2/l_\nu$ (folded field), this happens when

$$\langle B^2\rangle \sim \delta u_{l_\nu}^2 \sim Re^{-1/2}\langle u^2\rangle. \tag{54}$$

Let us suppose that the viscous motions are suppressed by the back reaction, at least in their ability to amplify the field. Then the motions at larger scales in the inertial range come into play: while their rates of strain and, therefore, the associated stretching rates are smaller than that of the viscous-scale motions, they are more energetic [see Eq. (4)], so the magnetic field is too weak to resist being stretched by them. As the field continues to grow, it will suppress the motions at ever larger scales. If we define a stretching scale $l_s(t)$ as the scale of the motions whose energy is $\delta u_{l_s} \sim \langle B^2\rangle(t)$, we can estimate

$$\frac{d}{dt}\langle B^2\rangle \sim \frac{\delta u_{l_s}}{l_s}\langle B^2\rangle \sim \frac{\delta u_{l_s}^3}{l_s} \sim \epsilon = \text{const} \quad \Rightarrow \quad \langle B^2\rangle(t) \sim \epsilon t. \tag{55}$$

Thus, exponential growth gives way to secular growth of the magnetic energy. This is accompanied by elongation of the folds (their length is always of the order of the stretching scale, $l_\parallel \sim l_s$), while the resistive (reversal) scale increases because the stretching rate goes down:

$$l_\parallel(t) \sim l_s(t) \sim \delta u_{l_s}^3/\epsilon \sim \sqrt{\epsilon}\, t^{3/2}, \quad l_\eta(t) \sim [\eta/(\delta u_{l_s}/l_s)]^{1/2} \sim \sqrt{\eta t}. \tag{56}$$

This secular stage can continue until the entire inertial range is suppressed, $l_s \sim L$, at which point saturation must occur. This happens after $t \sim \epsilon^{-1/3}L^{2/3} \sim L/\delta u_L$. Using Eqs. (55), (56), we have, in saturation,

$$\langle B^2\rangle \sim \langle u^2\rangle, \quad l_\parallel \sim L, \quad l_\eta \sim [\eta/(\delta u_L/L)]^{1/2} \sim R_m^{-1/2}L. \tag{57}$$

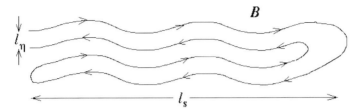

Fig. 7. Alfvén waves propagating along folded fields. (From [51].)

Comparing Eqs. (57) and (36), we see that the resistive scale has increased only by a factor of $Re^{1/4}$ over its value in the weak-field growth stage. Note that this imposes a very stringent requirement on any numerical experiment striving to distinguish between the viscous and resistive scales: $Pr_m \gg Re^{1/2} \gg 1$.

If, as in the above scenario, the magnetic field retains its folded structure in saturation, with direction reversals at the resistive scale, this explains qualitatively why the numerical simulations of the developed isotropic MHD turbulence with $Pr_m \geq 1$ [32] show the magnetic-energy pile-up at the small scales. What then is the saturated state of the turbulent velocity field? We assumed above that the inertial-range motions were "suppressed" – this applied to their ability to amplify magnetic field, but needed not imply a complete evacuation of the inertial range. Indeed, simulations at modest Pr_m show a powerlike velocity spectrum [32, 37]. The most obvious class of motions that can populate the inertial range without affecting the magnetic-field strength are a type of Alfvén waves that propagate not along a mean (or large-scale) magnetic field but along the folded structure (Fig. 7). Mathematically, the dispersion relation for such waves is derived via a linear theory carried out for the inertial-range perturbations ($L^{-1} \ll k \ll l_\nu^{-1}$) of the tensor $B_i B_j$ (cf. [52]). The unperturbed state is the average of this tensor over the sub-viscous scales: $\langle B_i B_j \rangle = \hat{b}_i \hat{b}_j \langle B^2 \rangle$, where $\langle B^2 \rangle$ is the total magnetic energy and the tensor $\hat{b}_i \hat{b}_j$ only varies at the outer scale L [Eq. (57)]. The resulting dispersion relation is $\omega = \pm |\mathbf{k} \cdot \hat{\mathbf{b}}| \langle B^2 \rangle^{1/2}$ [51]. The presence of these waves will not change the resistive-scale-dominated nature of the magnetic-energy spectrum, but should be manifest in the kinetic-energy spectrum. There is, at present, no theory of a cascade of such waves, although a line of argument similar to § 2 might work, since it does not depend on the field having a specific direction. A numerical detection of these waves is also a challenge for the future.

What we have proposed above can be thought of as a modernised version of the Schlüter and Biermann scenario, retaining the intermediate secular-growth stage and saturation with $\langle B^2 \rangle \sim \langle u^2 \rangle$, but not scale-by-scale equipartition. However, an alternative possibility, which is in a similar relationship to Batchelor's scenario, can also be envisioned. In Eqs. (55), (56), the scale l_η at which diffusion cuts off the small-scale magnetic fluctuations was assumed to be determined by the stretching rate $\delta u_{l_s}/l_s$. However, since the non-linear suppression of the viscous-scale eddies only needs to eliminate motions with $\hat{\mathbf{b}}\hat{\mathbf{b}} : \boldsymbol{\nabla}\mathbf{u} \neq 0$ [Eq. (50)], 2D "interchange" motions (velocity gradients $\perp \hat{\mathbf{b}}$) are, in principle, allowed to survive at the viscous scale. These could "two-dimensionally" mix the direction-reversing magnetic fields at the rate $\delta u_{l_\nu}/l_\nu$ – much faster than the unsuppressed larger-scale stretching can amplify the field, – with the consequence that the resistive scale is pinned at the value given by Eq. (36) and the field cannot grow above the Batchelor limit (54). The mixing efficiency of the suppressed motions is the key to choosing between the two saturation scenarios. Numerical simulations [32] corroborate the existence of an intermediate stage of slower-than-exponential growth accompanied by fold elongation and a modest increase of the resistive scale [Eq. (56)]. This tips the scales in favour of the first scenario, but, in view of limited resolutions, we hesitate to declare the matter definitively resolved.

3.3 Turbulence and magnetic fields in galaxy clusters

The intracluster medium (hereafter, ICM) is a hot ($T \sim 10^8$ K) diffuse ($n \sim 10^{-2}$–10^{-3} cm^{-3}) fully ionised plasma, which accounts for most of the luminous matter in the Universe (note that it is not entirely dissimilar from the ionised phases of the interstellar medium: e.g., the so-called hot ISM). It is a natural astrophysical environment to which the large-Pr_m isotropic regime of MHD turbulence appears to be applicable: indeed, Eq. (35) gives $Pr_m \sim 10^{29}$.

The ICM is believed to be in a state of turbulence driven by a variety of mechanisms: merger events, galactic and subcluster wakes, active galactic nuclei. One expects the outer scale $L \sim 10^2$–10^3 kpc and the velocity dispersions $\delta u_L \sim 10^2$–10^3 km/s (a fraction of the sound speed). Indirect observational evidence supporting the possibility of a turbulent ICM with roughly these parameters already exists (an apparently powerlike spectrum of pressure fluctuations found in the Coma cluster [53], broadened abundance profiles in Perseus believed to be caused by turbulent diffusion [54]), and direct detection may be achieved in the near future [55]. However, there is as yet no consensus on whether turbulence, at least in the usual hydrodynamic sense, is a generic feature of clusters [56]. The main difficulty is the very large values of the ICM viscosity obtained via the standard estimate $\nu \sim v_{\mathrm{th},i}\lambda_{\mathrm{mfp}}$, where $v_{\mathrm{th},i} \sim 10^3$ km/s is the ion thermal speed and $\lambda_{\mathrm{mfp}} \sim 1$–10 kpc is the ion mean free path. This gives $Re \sim 10^2$ if not less, which makes the existence of a well-developed inertial range problematic. Postponing the problem of viscosity until § 4, we observe that the small-scale dynamo does not, in fact, require a turbulent velocity field in the sense of a broad inertial range: in the weak-field regime discussed in § 3.1, the dynamo was controlled by the smooth single-scale random flow associated with the viscous-scale motions; in saturation, we argued in § 3.2 that the main effect was the direct nonlocal interaction between the outer-scale (random) motions and the magnetic field.

Given the available menu of large-scale stirring mechanisms in clusters, it is likely that, whatever the value of Re, the velocity field is random.[15]

The cluster turbulence is certainly magnetic. The presence of magnetic fields was first demonstrated for the Coma cluster, for which Willson detected in 1970 a diffuse synchrotron radio emission [57] and Kim et al. in 1990 were able to estimate directly the magnetic-field strength and scale using the Faraday rotation measure (RM) data [58]. Such observations of magnetic fields in clusters have now become a vibrant area of astronomy (reviewed most recently in [59]), usually reporting a field $B \sim 1$–10 μG at scales ~ 1–10 kpc [60].[16] All of this field is small-scale fluctuations: no appreciable mean component has been detected. The field is dynamically significant: the magnetic energy is less but not much less than the kinetic energy of the turbulent motions.

Do clusters fit the theoretical expectations reviewed above? The magnetic-field scale seen in clusters is usually 10–100 times smaller than the expected outer scale of turbulent motions and, indeed, is also smaller than the viscous scale based on $Re \sim 10^2$. However, it is certainly far above the resistive scale, which turns out be $l_\eta \sim 10^3$–10^4 km! Faced with these numbers, we must suspend the discussion of cluster physics and finally take account of the fact that astrophysical bodies are made of plasma, not of an MHD fluid.

4 Enter plasma physics

4.1 Braginskii viscosity

In all of the above, we have used the MHD Eqs. (1) and (2) to develop turbulence theories supposed to be relevant for astrophysical plasmas. Historically, such has been the approach followed in most of the astrophysical literature. The philosophy underpinning this approach is again that of universality: the "microphysics" at and below the dissipation scale are not expected to matter for the fluid-like dynamics at larger scales. However, in considering the MHD turbulence with large Pr_m, we saw that dissipation scales, determined by the values of the viscosity ν and magnetic diffusivity η, played a very prominent role: the growth of the small-scale magnetic fields was controlled by the turbulent rate of strain at the viscous scale and resulted in the magnetic energy piling up, in the form of direction-reversing folded fields, at the resistive scale – both in the growth and saturation stages of the dynamo. It is then natural to revisit the question of whether the Laplacian diffusion terms in Eqs. (1), (2) are a good description of the dissipation in astrophysical plasmas.

---

[15] Numerical simulations of the large-Pr_m regime at currently accessible resolutions also rely on a random forcing to produce "turbulence" with $Re \sim 1$–10^2 [32, 37].

[16] Because of the availability of the RM maps from extended radio sources in clusters, it is possible to go beyond field-strength and scale estimates and construct magnetic-energy spectra with spatial resolution of ~ 0.1 kpc [61].

The answer to this question is, of course, that they are not. A necessary assumption in the derivation of these terms is that the ion cyclotron frequency $\Omega_i = eB/m_ic$ exceeds the ion–ion collision frequency ν_{ii} or, equivalently, the ion gyroradius $\rho_i = v_{\mathrm{th},i}/\Omega_i$ exceeds the mean free path $\lambda_{\mathrm{mfp}} = v_{\mathrm{th},i}/\nu_{ii}$. This is patently not the case in many astrophysical plasmas: for example, in galaxy clusters, $\lambda_{\mathrm{mfp}} \sim 1$–$10$ kpc, while $\rho_i \sim 10^4$ km. In such a weakly collisional magnetised plasma, the momentum Eq. (1) assumes the following form, valid at spatial scales $\gg \rho_i$ and at timescales $\gg \Omega_i^{-1}$,

$$\frac{d\mathbf{u}}{dt} = -\nabla\left(p_\perp + \frac{B^2}{2}\right) + \nabla \cdot \left[\hat{\mathbf{b}}\hat{\mathbf{b}}(p_\perp - p_\parallel + B^2)\right] + \mathbf{f}, \qquad (58)$$

where $p_\perp$ and $p_\parallel$ are plasma pressures perpendicular and parallel to the local direction of the magnetic field, respectively, and we have used $\mathbf{B} \cdot \nabla\mathbf{B} = \nabla \cdot (\hat{\mathbf{b}}\hat{\mathbf{b}}B^2)$. The evolution of the magnetic field is controlled by the electrons – the field remains frozen into the flow and we may use Eq. (2) with $\eta = 0$.

If we are interested in subsonic motions, $\nabla(p_\perp + B^2/2)$ in Eq. (58) can be found from the incompressibility condition $\nabla \cdot \mathbf{u} = 0$ and the only quantity still to be determined is $p_\perp - p_\parallel$. The proper way to compute it is by a rather lengthy kinetic calculation due to Braginskii [62], which cannot be repeated here. The result of this calculation can, however, be obtained in the following heuristic way [63].

The fundamental property of charged particles moving in a magnetic field is the conservation of the first adiabatic invariant $\mu = m_iv_\perp^2/2B$.[17] When $\lambda_{\mathrm{mfp}} \gg \rho_i$, this conservation is only weakly broken by collisions. As long as μ is conserved, any change in B must be accompanied by a proportional change in $p_\perp$. Thus, the emergence of the pressure anisotropy is a natural consequence of the changes in the magnetic-field strength and vice versa: indeed, summing up the first adiabatic invariants of all particles, we get $p_\perp/B = \mathrm{constant}$. Then

$$\frac{1}{p_\perp}\frac{dp_\perp}{dt} = \frac{1}{B}\frac{dB}{dt} - \nu_{ii}\frac{p_\perp - p_\parallel}{p_\perp}, \qquad (59)$$

where the second term on the right-hand sight represents the collisional relaxation of the pressure anisotropy $p_\perp - p_\parallel$ at the rate $\nu_{ii} \sim v_{\mathrm{th},i}/\lambda_{\mathrm{mfp}}$.[18] Using Eq. (50) for B and balancing the terms in the rhs of Eq. (59), we get

$$p_\perp - p_\parallel = \nu_\parallel\frac{1}{B}\frac{dB}{dt} = \nu_\parallel\hat{\mathbf{b}}\hat{\mathbf{b}} : \nabla\mathbf{u}, \qquad (60)$$

where $\nu_\parallel \sim p/\nu_{ii} \sim v_{\mathrm{th},i}\lambda_{\mathrm{mfp}}$ is the "parallel viscosity". This equation turns out to be exact [62] up to numerical prefactors in the definition of $\nu_\parallel$.

---

[17] It may be helpful to the reader to think of this property as the conservation of the angular momentum of a gyrating particle: $m_iv_\perp\rho_i \propto m_iv_\perp^2/B = 2\mu$.

[18] This is only valid if the characteristic parallel scales $k_\parallel^{-1}$ of all fields are larger than λ_{mfp}. In the collisionless regime, $k_\parallel\lambda_{\mathrm{mfp}} \gg 1$, we may assume that the pressure anisotropy is relaxed in the time particles streaming along the field cover the distance $k_\parallel^{-1}$: this entails replacing ν_{ii} in Eq. (59) by $k_\parallel v_{\mathrm{th},i}$.

The energy conservation law based on Eqs. (58) and (2) is

$$\frac{d}{dt}\left(\frac{\langle u^2\rangle}{2} + \frac{\langle B^2\rangle}{2}\right) = \epsilon - \nu_{\|}\langle|\hat{\mathbf{b}}\hat{\mathbf{b}}:\boldsymbol{\nabla}\mathbf{u}|^2\rangle = \epsilon - \nu_{\|}\left\langle\left(\frac{1}{B}\frac{dB}{dt}\right)^2\right\rangle,\qquad (61)$$

where Ohmic diffusion has been omitted. Thus, the Braginskii viscosity only dissipates such velocity gradients that change the strength of the magnetic field. The motions that do not affect B are allowed to exist in the subviscous scale range. In the weak-field regime, these motions take the form of plasma instabilities. When the magnetic field is strong, a cascade of shear-Alfvén waves can be set up below the viscous scale. Let us elaborate.

4.2 Plasma instabilities

The simplest way to see that the pressure anisotropy in Eq. (58) leads to instabilities is as follows [63]. Imagine a "fluid" solution with $\mathbf{u}$, $p_\perp$, $p_\|$, $\mathbf{B}$ changing on viscous time and spatial scales, $t \sim |\boldsymbol{\nabla}\mathbf{u}|^{-1} \sim l_\nu/\delta u_{l_\nu}$ and $l \sim l_\nu$. Would such a solution be stable with respect to fast ($\omega \gg |\boldsymbol{\nabla}\mathbf{u}|^{-1}$) small-scale ($k \gg l_\nu^{-1}$) perturbations? Linearising Eq. (58) and denoting perturbations by δ, we get

$$-i\omega\delta\mathbf{u} = -i\mathbf{k}\left(\delta p_\perp + B\delta B\right) + \left(p_\perp - p_\| + B^2\right)\delta\mathbf{K}$$
$$+ i\hat{\mathbf{b}}\,k_\|\left[\delta p_\perp - \delta p_\| - \left(p_\perp - p_\| - B^2\right)\delta B/B\right],\qquad (62)$$

where the perturbation of the field curvature is $\delta\mathbf{K} = k_\|^2\delta\mathbf{u}_\perp/i\omega$ [see Eq. (52)]. We see that regardless of the origin of the pressure anisotropy, the shear-Alfvén-polarised perturbations ($\delta\mathbf{u} \propto \mathbf{k}\times\hat{\mathbf{b}}$) have the dispersion relation

$$\omega = \pm k_\|\left(p_\perp - p_\| + B^2\right)^{1/2}.\qquad (63)$$

When $p_\| - p_\perp > B^2$, ω is purely imaginary and we have what is known as the firehose instability [64–67]. The growth rate of the instability is $\propto k_\|$, which means that the fastest-growing perturbations will be at scales far below the viscous scale or, indeed, the mean free path. Therefore, adopting the Braginskii viscosity [Eq. (60)] exposes a fundamental problem with the use of the MHD approximation for fully ionised plasmas: the equations are ill posed wherever $p_\| - p_\perp > B^2$. To take into account the instability and its impact on the large-scale dynamics, the fluid equations must be abandoned and a kinetic description adopted. A linear kinetic calculation shows that the instability growth rate peaks at $k_\|\rho_i \sim 1$, so the fluctuations grow fastest at the ion gyroscale. While the firehose instability occurs in regions where the velocity field leads to a decrease in the magnetic-field strength [Eq. (60)], a kinetic calculation of the pressure perturbations in Eq. (62) shows that another instability, called the mirror mode [66], is triggered wherever the field increases ($p_\perp > p_\|$). Its growth rate is also $\propto k_\|$ and peaks at the ion gyroscale.

In weakly collisional astrophysical plasmas such as the ICM, the random motions produced by the large-scale stirring will stretch and fold magnetic fields, giving rise to regions both of increasing and decreasing field strength (§ 3.1). The instabilities should, therefore, be present in weak-field regions where $|p_\parallel - p_\perp| > B^2$ and, since their growth rates are much larger than the fluid rates of strain, their growth and saturation should have a profound effect on the structure of the turbulence. A quantitative theory of what exactly happens is not as yet available, but one might plausibly expect that the fluctuations excited by the instabilities will lead to some effective renormalisation of both the viscosity and the magnetic diffusivity. A successful theory of turbulence in clusters requires a quantitative calculation of this effective transport. In particular, this should resolve the uncertainties around the ICM viscosity and produce a prediction of the magnetic-field scale to be compared with the observed values reviewed in § 3.3.

In the solar wind, the plasma is magnetised ($\rho_i \sim 10^2$ km), while collisions are virtually absent: the mean free path exceeds the distance from the Sun (10^8 km). Ion pressure (temperature) anisotropies with respect to the field direction were directly measured in 1970s [68, 69]. As was first suggested by Parker [66], firehose and mirror instabilities (as well as several others) should play a major, although not entirely understood role [70]. A vast geophysical literature now exists on this subject, which cannot be reviewed here.

4.3 Kinetic turbulence

The instabilities are quenched when the magnetic field is sufficiently strong: B^2 overwhelms $p_\perp - p_\parallel$ in the second term on the right-hand side of Eq. (58). If we use the collisional estimate (60), this happens when

$$B^2 \gg \nu_\parallel \delta u_{l_\nu}/l_\nu \sim Re^{-1/2} \delta u_L^2. \tag{64}$$

The firehose-unstable perturbations become Alfvén waves in this limit. In the strong-field regime ($\delta B \ll B_0$), the appropriate mathematical description of the weakly collisional turbulence of Alfvén waves is the low-frequency limit of the plasma kinetic theory called the gyrokinetics [71, 72].[19] It is obtained under an ordering scheme that stipulates

$$k_\perp \rho_i \sim 1, \qquad \omega/\Omega_i \sim k_\parallel/k_\perp \sim \delta u/v_A \sim \delta B/B_0 \ll 1. \tag{65}$$

The second relation in Eq. (65) coincides with the GS critical-balance conjecture (14) if the latter is treated as an ordering assumption. The gyrokinetics can be cast as a systematic expansion of the full kinetic description of the

---

[19] While originally developed and widely used for fusion plasmas, this "kinetic-fluid" description has only recently started to be applied to astrophysical problems such as the relative heating of ions and electrons by Alfvénic turbulence in advection-dominated accretion flows [1].

plasma in the small parameter $\epsilon \sim k_\parallel / k_\perp$ – the direct generalisation of the similar expansion of MHD equations given at the end of § 2.4. It turns out that the decoupling of the Alfvén-wave cascade that we demonstrated there is also a property of the gyrokinetics and that this cascade is correctly described by the RMHD Eq. (22) and (23) all the way down to the ion gyroscale, $k_\perp \rho_i \sim 1$ [72]. Broad fluctuation spectra observed in the solar wind [11] and in the ISM [16] are likely to be manifestations of just such a cascade. The slow waves and the entropy mode are passively mixed by the Alfvén-wave cascade, but Eqs. (29)–(31) have to be replaced by a kinetic equation.

When magnetic fields are not stronger than the turbulent motions – as is the case for clusters, where the magnetic energy is, in fact, quite close to the threshold (64) – the situation is more complicated and much more obscure because small-scale dynamo (§ 3.1), back reaction (§ 3.2), plasma instabilities (in weak-field regions such as, for example, the bending regions of the folded fields, § 3.1), and Alfvén waves (possibly of the kind discussed in § 3.2) all enter into the mix and remain to be sorted out.

5 Conclusion

We conclude here, in the hope that we have provided the reader with a fair overview of the state of affairs to which the MHD turbulence theory has arrived after its first 50 years. Perhaps, despite much insight gained along the way, not very far. It is clear that a simple extension of Kolmogorov's theory has so far proven unattainable. Two of the key assumptions of that theory – **isotropy** and **locality of interactions** – are manifestly incorrect for MHD. Indeed, even the applicability of the fundamental principle of small-scale universality is suspect. Although a fair amount is known about Alfvénic turbulence, progress in answering many important astrophysical questions (see the Introduction) has been elusive because there is little knowledge of the general spectral and structural properties of the fully developed turbulence in an MHD fluid and, more generally, in magnetised weakly collisional or collisionless plasmas. Thus, while many unanswered questions demand further effort on MHD turbulence, there is also an imperative, mandated by astrophysical applications, to go beyond the fluid description.

Acknowledgements. We would like to thank Russell Kulsrud who is originally responsible for our interest in these matters. We are grateful to N. A. Lipunova, M. Bernard, R. A. Sunyaev, R. Buchstab, and H. Lindsay for helping us obtain the photos used in this chapter. Our work was supported by the UKAFF Fellowship, PPARC Advanced Fellowship, King's College, Cambridge (A.A.S.) and in part by the US DOE Center for Multiscale Plasma Dynamics (S.C.C.).

References

1. Quataert E, Gruzinov A (1999) Turbulence and particle heating in advection-dominated accretion flows. Astrophys J 520:248–255
2. Shakura NI, Sunyaev RA (1973) Black holes in binary systems. Observational appearance. Astron Astrophys 24:337–355
3. Dennis TJ, Chandran BDG (2005) Turbulent heating of galaxy-cluster plasmas. Astrophys J 622:205–216
4. Kolmogorov AN (1941) The local structure of turbulence in incompressible viscous fluid at very large Reynolds numbers. Dokl Akad Nauk SSSR 30:299–303 [English translation: Proc R Soc Lond A 434:9–13 (1991)]
5. Landau LD, Lifshitz EM (1987) Fluid Mechanics. Butterworth-Heinemann, Oxford
6. Elsasser WM (1950) The hydromagnetic equations. Phys Rev 79:183
7. Kraichnan RH (1965) Inertial-range spectrum of hydromagnetic turbulence. Phys Fluids 8:1385–1387
8. Iroshnikov RS (1963) Turbulence of a conducting fluid in a strong magnetic field. Astron Zh 40:742–750 [English translation: Sov Astron 7:566–571 (1964) Note that Iroshnikov's initials are given incorrectly as PS in the translation]
9. Parker EN (1958) Dynamics of the interplanetary gas and magnetic fields. Astrophys J 128:664–676
10. Coleman PJ (1968) Turbulence, viscosity, and dissipation in the solar-wind plasma. Astrophys J 153:371–388
11. Goldstein ML, Roberts DA (1999) Magnetohydrodynamic turbulence in the solar wind. Phys Plasmas 6:4154–4160
12. Dobrowolny M, Mangeney A, Veltri P (1980) Fully developed anisotropic hydromagnetic turbulence in the interplanetary space. Phys Rev Lett 45:144–147
13. Belcher JW, Davis L (1971) Large-amplitude Alfvén waves in interplanetary medium, 2. J Geophys Res 76:3534–3563
14. Matthaeus WH, Goldstein ML, Roberts DA (1990) Evidence for the presence of quasi-two-dimensional nearly incompressible fluctuations in the solar wind. J Geophys Res 95:20673–20683
15. Matthaeus WH, Goldstein ML (1982) Measurement of the rugged invariants of magnetohydrodynamic turbulence in the solar wind. J Geophys Res 87:6011–1028
16. Armstrong JW, Rickett BJ, Spangler SR (1995) Electron density power spectrum in the local interstellar medium. Astrophys J 443:209–221
17. Maron J, Goldreich P (2001) Simulations of incompressible magnetohydrodynamic turbulence. Astrophys J 554:1175–1196
18. Cho J, Lazarian A, Vishniac ET (2002) Simulations of magnetohydrodynamic turbulence in a strongly magnetized medium. Astrophys J 564:291–301
19. Müller W-C, Biskamp D, Grappin R (2003) Statistical anisotropy of magntohydrodynamic turbulence. Phys Rev E 67:066302
20. Montgomery D, Turner L (1981) Anisotropic magnetohydrodynamic turbulence in a strong external magnetic field. Phys Fluids 24:825–831
21. Shebalin JV, Matthaeus WH, Montgomery D (1983) Anisotropy in MHD turbulence due to a mean magnetic field. J Plasma Phys 29:525–547
22. Ng CS, Bhattacharjee A (1996) Interaction of shear-Alfvén wave packets: implication for weak magnetohydrodynamic turbulence in astrophysical plasmas. Astrophys J 465:845–854

23. Goldreich P, Sridhar S (1997) Magnetohydrodynamic turbulence revisited. Astrophys J 485:680–688
24. Galtier S, Nazarenko SV, Newell AC, Pouquet A (2000) A weak turbulence theory for incompressible magnetohydrodynamics. J Plasma Phys 63:447–488
25. Lithwick Y, Goldreich P (2003) Imbalanced weak magnetohydrodynamic turbulence. Astrophys J 582:1220–1240
26. Sridhar S, Goldreich P (1994) Toward a theory of interstellar turbulence. I. Weak Alfvénic turbulence. Astrophys J 432:612–621
27. Goldreich P, Sridhar S (1995) Toward a theory of interstellar turbulence. II. Strong Alfvénic turbulence. Astrophys J 438:763–775
28. Strauss HR (1976) Nonlinear, three-dimensional magnetohydrodynamics of non-circular tokamaks. Phys Fluids 19:134–140
29. Obukhov AM (1949) Structure of a temperature field in a turbulent flow. Izv Akad Nauk SSSR Ser Geogr Geofiz 13:58–69
30. Higdon JC (1984) Density fluctuations in the interstellar medium: evidence for anisotropic magnetogasdynamic turbulence. I. Model and astrophysical sites. Astrophys J 285:109–123
31. Lithwick Y, Goldreich P (2001) Compressible magnetohydrodynamic turbulence in interstellar plasmas. Astrophys J 562:279–296
32. Schekochihin AA, Cowley CS, Taylor SF, Maron JL, McWilliams JC (2004) Simulations of the small-scale turbulent dynamo. Astrophys J 612:276–307
33. Biskamp D, Müller W-C (2000) Scaling properties of three-dimensional isotropic magnetohydrodynamic turbulence. Phys Plasmas 7:4889–4900
34. Spitzer L (1962) Physics of fully ionized gases. Wiley, New York
35. Childress S, Gilbert AD (1995) Stretch, Twist, Fold: The Fast Dynamo. Springer, Berlin
36. Meneguzzi M, Frisch U, Pouquet A (1981) Helical and nonhelical turbulent dynamos. Phys Rev Lett 47:1060–1064
37. Haugen NEL, Brandenburg A, Dobler W (2004) Simulations of nonhelical hydromagnetic turbulence. Phys Rev E 70:016308
38. Batchelor GK (1950) On the spontaneous magnetic field in a conducting liquid in turbulent motion. Proc R Soc Lond A 201:405–416
39. Schekochihin A, Cowley S, Maron J, Malyshkin L (2002) Structure of small-scale magnetic fields in the kinematic dynamo theory. Phys Rev E 65:016305
40. Ott E (1998) Chaotic flows and kinematic magnetic dynamos: A tutorial review. Phys Plasmas 5:1636–1646
41. Chertkov M, Falkovich G, Kolokolov I, Vergassola M (1999) Small-scale turbulent dynamo. Phys Rev Lett 83:4065–4068
42. Zeldovich YaB, Ruzmaikin AA, Molchanov SA, Sokoloff DD (1984) Kinematic dynamo problem in a linear velocity field. J Fluid Mech 144:1–11
43. Zeldovich YaB (1956) The magnetic field in the two-dimensional motion of a conducting turbulent liquid. Zh Exp Teor Fiz 31:154–155 [English translation: Sov Phys JETP 4:460–462 (1957)]
44. Goldhirsch I, Sulem P-L, Orszag SA (1987) Stability and Lyapunov stability of dynamical systems: a differential approach and a numerical method. Physica D 27:311–337
45. Kazantsev AP (1967) Enhancement of a magnetic field by a conducting fluid. Zh Eksp Teor Fiz 53:1806–1813 [English translation: Sov Phys JETP 26:1031–1034 (1968)]

46. Zeldovich YaB, Ruzmaikin AA, Sokoloff DD (1990) The Almighty Chance. World Scientific, Singapore
47. Falkovich G, Gawędzki K, Vergassola M (2001) Particles and fields in fluid turbulence. Rev Mod Phys 73:913–975
48. Kraichnan RH (1968) Small-scale structure of a scalar field convected by turbulence. Phys Fluids 11:945–953
49. Kulsrud RM, Anderson SW (1992) The spectrum of random magnetic fields in the mean field dynamo theory of the galactic magnetic field. Astrophys J 396:606–630
50. Schlüter A, Biermann L (1950) Interstellare Magnetfelder. Z Naturforsch 5a:237–251
51. Schekochihin AA, Cowley CS, Hammett GW, Maron JL, McWilliams JC (2002) A model of nonlinear evolution and saturation of the turbulent MHD dynamo. New J Phys 4:84
52. Gruzinov AV, Diamond PH (1996) Nonlinear mean field electrodynamics of turbulent dynamos. Phys Plasmas 3:1853–1857
53. Schuecker P, Finoguenov A, Miniati F, Böhringer H, Briel UG (2004) Probing turbulence in the Coma galaxy cluster. Astron Astrophys 426:387–397
54. Rebusco P, Churazov E, Böhringer H, Forman W (2005) Impact of stochastic gas motions on galaxy cluster abundance profiles. Mon Not R Astr Soc 359:1041–1048
55. Inogamov NA, Sunyaev RA (2003) Turbulence in clusters of galaxies and X-ray-line profiles. Astron Lett 29:791–824
56. Fabian AC, Sanders JS, Crawford CS, Conselice CJ, Gallagher JS, Wyse RFG (2003) The relationship between optical Hα filaments and the X-ray emission in the core of the Perseus cluster. Mon Not R Astr Soc 344:L48–L52
57. Willson MAG (1970) Radio observations of the cluster of galaxies in Coma Berenices — The 5C4 survey. Mon Not R Astr Soc 151:1–44
58. Kim K-T, Kronberg PP, Dewdney PE, Landecker TL (1990) The halo and magnetic field of the Coma cluster of galaxies. Astrophys J 355:29-37
59. Govoni F, Feretti L (2004) Magnetic fields in clusters of galaxies. Int J Mod Phys D 13:1549–1594
60. Clarke TE, Kronberg PP, Böhringer H (2001) A new radio-X-ray probe of galaxy cluster magnetic fields. Astrophys J 547:L111–L114
61. Vogt C, Enßlin TA (2005) A Bayesian view on Fraday rotation maps — Seeing the magnetic power spectra in galaxy clusters. Astron Astrophys 434:67–76
62. Braginskii SI (1965) Transport processes in a plasma. Rev Plasma Phys 1:205–310
63. Schekochihin AA, Cowley CS, Kulsrud RM, Hammett GW, Sharma P (2005) Plasma instabilities and magnetic-field growth in clusters of galaxies. Astrophys J 629:139–142
64. Rosenbluth MN (1956) Stability of the pinch. Los Alamos Scientific Laboratory Report, LA-2030
65. Chandrasekhar S, Kaufman AN, Watson KM (1958) The stability of the pinch. Proc R Soc Lond A 245:435–455
66. Parker EN (1958) Dynamical instability in an anisotropic ionized gas of low density. Phys Rev 109:1874–1876
67. Vedenov AA, Sagdeev RZ (1958) Some properties of the plasma with anisotropic distribution of the velocities of ions in the magnetic field. Sov Phys Dokl 3:278

68. Feldman WC, Asbridge JR, Bame SJ, Montgomery MD (1974) Interpenetrating solar wind streams. Rev Geophys Space Phys 12:715–723
69. Marsch E, Ao X-Z, Tu C-Y (2004) On the temperature anisotropy of the core part of the proton velocity distribution function in the solar wind. J Geophys Res 109:A04102
70. Gary SP (1993) Theory of space plasma microinstabilities. Cambridge University Press, Cambridge
71. Frieman EA, Chen L (1982) Nonlinear gyrokinetic equations for low-frequency electromagnetic waves in general plasma equilibria. Phys Fluids 25:502–508
72. Schekochihin AA, Cowley CS, Dorland WD, Hammett GW, Howes GG, Quataert E, Tatsuno T (2007) Kinetic and fluid turbulent cascades in magnetized weakly collisional astrophysical plasmas. Astrophys J (submitted; e-print arXiv:0704.0044)

Transient Pinched Plasmas and Strong Hydromagnetic Waves

John E Allen

University College, Oxford OX1 4BH, United Kingdom
(john.allen@eng.ox.ac.uk)

Summary. Some researches are described concerning early work on the transient self-magnetic "pinch effect" in plasmas at both high and low pressures. Theoretical work motivated by interest in shock waves led to the discovery in 1958 of solitary waves in plasma physics, they were later to become known as solitons.

1 Magnetohydrodynamics and plasma physics

I am not at all sure whether I am well qualified to be an invited contributor to this volume on "Magnetohydrodynamics: Historical Evolution and Trends"; this is because most of my work has been in Plasma Physics rather than the magnetohydrodynamics of liquids. The two subjects are, of course, closely related, mainly because both depend on Maxwell's equations; MHD can be defined as the branch of continuum mechanics that deals with the motion of an electrically conducting fluid in the presence of a magnetic field. The usual approach employs the simplest (isotropic) form of Ohm's Law in which the current is in the direction of the electric field. This is valid because the Hall parameter $(\omega\tau)$ is much less than unity in a liquid, where ω is the electron gyrofrequency and τ is the mean free time between collisions. The displacement current is neglected. Neither of these assumptions is generally true in a plasma. In MHD the model of an infinitely conducting fluid is sometimes employed, this leads to the concept of "frozen-in" magnetic lines of force as pioneered by Alfvén. The concept is analogous to Kelvin's theorem in fluid mechanics, which applies in the case of negligible viscosity.

Research workers in Plasma Physics employ three different models to describe a plasma; mixed models are also used. The simplest model is to consider the plasma as an electrically conducting fluid and again the perfect conductor is sometimes invoked. More accurately it can be considered as two (or more) fluids coexisting in space. Quite often the ion fluid is considered to be cold whereas the electron fluid is always considered to be hot. The terms "electron temperature" and "ion temperature" are employed, although

S. Molokov et al. (eds.), Magnetohydrodynamics – Historical Evolution and Trends,
117–127. © 2007 *Springer.*

the system is very rarely in a state of thermal equilibrium for which the term "temperature" is normally defined. More detailed study of plasmas is based on kinetic theory; new phenomena, such as wave damping in the absence of collisions are then encountered (Landau damping), which lie beyond the reach of fluid mechanics and our fluid models have to be abandoned.

In this chapter I shall describe some of the experimental and theoretical work on plasmas that I was involved in during the 1950s and 1960s. The first two plasma models briefly described above were employed in the theoretical work, although the "two-fluid" case was restricted to cold fluids.

2 Research at the University of Liverpool

Plasma Physics is, of course, related to research on controlled thermonuclear reactions (fusion). In the United Kingdom research on fusion began, not in Government Laboratories, but in the Universities. Shortly after World War II research groups were engaged in this activity at Liverpool (J.D.Craggs), Imperial College (G.P.Thomson), and Oxford (P.C.Thonemann). The experiments at that time were based on the self-magnetic "pinch -effect", i.e., the tendency of a plasma column carrying a high current to be constricted by the inwardly directed $\mathbf{j} \times \mathbf{B}$ force. It is not generally known that one of the first attempts to produce a thermonuclear reaction in the laboratory was carried out in the late 1940s at Liverpool University when Reynolds and Craggs employed a high current spark discharge, of about 300 kA in deuterium at atmospheric pressure. The experiment gave a negative result in that no neutrons were detected, but Professor Skinner later persuaded the authors to publish their work [1].

Fig. 1. The high current generator viewed from above. The arrangement of connections can be clearly seen [2]. To fix the scale, the top of each capacitor measured approximately 50 cm × 30 cm

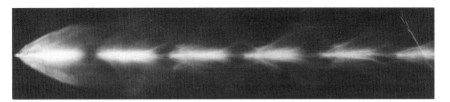

Fig. 2. Symmetrical 188 kA spark in hydrogen; mirror speed 238 rev/s [3]

As a Ph.D. student at Liverpool University I later (1949) inherited the capacitor bank [2] shown in Fig. 1; it consisted of 114 0·5 µF 25kV capacitors (then known as condensers) stacked in 19 columns. After altering the central spark chamber assembly, I carried out research on high current spark channels in various gases [3]. The highest current in these experiments was 265 kA and it reached its maximum value after 7.7 µs. Fig. 2 shows a streak photograph of a 188kA spark in hydrogen, the radius of the core at the current maximum was 0.55 cm. This value, together with measurements of the voltage gradient in the channel, gave an electron temperature of 94,000 K, assuming that the conductivity depended on collisions between electrons and ions. This temperature is much lower than that required for a thermonuclear reaction (in deuterium). An estimate of the pressure at the centre of the discharge, based on a steady-state calculation, gave a value of 370 atmospheres.

I shall now digress and refer to the first paper that I have found referring to the "pinch effect". In 1907 Northrup [4] carried out experiments in which he passed current through liquid mercury and observed the pressure difference existing between the axis and the outer edge of the column. This effect resulted from the $\mathbf{j} \times \mathbf{B}$ force associated with the self-magnetic field. The apparatus employed by Northrup is shown in Fig. 3; it is seen that he summed a number of such pressure differences in order to obtain a measurable effect. The mercury was contained in a tube so that the instabilities later encountered by workers in fusion were not able to develop. His friend, Carl Hering, suggested that the phenomenon should be called the "pinch effect".

3 Research at the Atomic Energy Research Establishment (Harwell)

On leaving the University in 1952 my first job was at the Atomic Research Establishment at Harwell (instead of doing military service!). Peter Thonemann had just moved to Harwell from Oxford to continue his research on Controlled Thermonuclear Reactions (fusion). One of my tasks, with Peter Reynolds, was to repeat and extend the experiments carried out on the transient pinch discharge, at low pressures, by Cousins and Ware at Imperial College [5]. In this case the $\mathbf{j} \times \mathbf{B}$ force associated with the self-magnetic field causes the discharge to contract, rather than oppose its expansion as in the Liverpool work. Figure 4 illustrates a simple theory of the phenomenon [6].

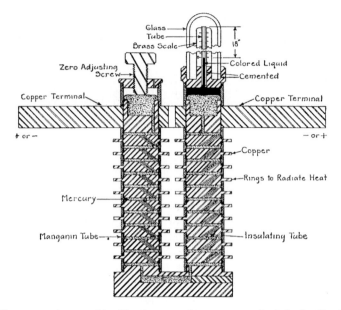

Fig. 3. The apparatus used by Northrup to demonstrate the "pinch effect" in liquid mercury in 1907. Summing the pressure differences developed in a number (22) of mercury conductors enabled him to measure the effect [4]

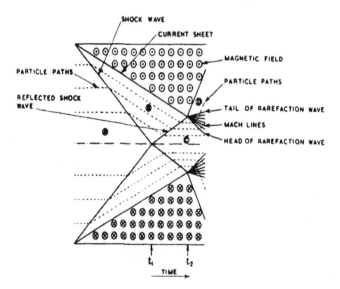

Fig. 4. Distance–time diagram illustrating the sequence of events [6]

A shock wave travels ahead of the "magnetic piston". On arrival at the axis the shock wave is reflected and the magnetic pressure is then no longer sufficiently great to contain the plasma column, which thereby expands. The

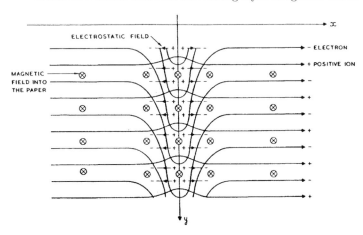

Fig. 5. Schematic diagram showing the trajectories of the electrons and ions, viewed in a coordinate system moving with the wave. An induced electric field exists in the y-direction [8]

experimental results that we obtained [7] were remarkably similar to those shown in the diagram, given that the electrical conductivity was finite rather than infinite as in the model.

Returning to the question of shock waves, the question then arose as to whether one could have a shock wave in the absence of collisions. John Adlam and I tackled this problem by considering a wave propagating into an undisturbed plasma across a magnetic field; the particle trajectories are illustrated in Fig. 5, where a coordinate system moving with the wave was chosen. An induced electric field in the y-direction exists in the moving frame, and an electrostatic field in the x-direction maintains quasi-neutrality for the case in question. Such quasi-neutrality exists when the electron plasma frequency is much greater than the electron cyclotron frequency. Not surprisingly, since no dissipative mechanism was involved, we did not find a shock wave but instead we found a solitary wave. The equations could be solved analytically and the results are displayed in Fig. 6. The uppermost curve shows the magnetic field strength and the lowest curve illustrates the electron density. It can be noted that electrons are not "tied to the magnetic lines of force" as often stated in the literature. The quantity α is the Alfvén Mach number and the analytical solution is valid for $1 < \alpha < 2$. We can note here that the "Alfvén velocity" is a characteristic velocity in both Plasma Physics and Magnetohydrodynamics, although the physics is quite different in the two cases; Alfvén considered wave propagation in an electrically conducting fluid. In a second paper Adlam and I [9] considered the experimental arrangement shown in Fig. 7, where a plasma containing a magnetic field is compressed by a stronger field. Such an arrangement became known as a "theta-pinch", although another name for it was the orthogonal pinch, because the magnetic lines of force are straight

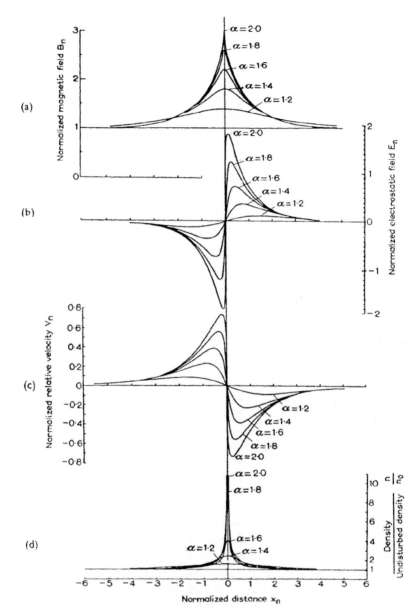

Fig. 6. The variation throughout the wave of (a) the magnetic field, (b) the electrostatic field, (c) the relative velocity between electrons and ions, and (d) the ion density [8]

and the lines of current flow are circles. The resulting magnetic force is again inwards. The numerical work that we carried out in this case showed the generation of a series of "solitary waves", as depicted in Fig. 8. Later Zabusky

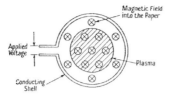

Fig. 7. Schematic diagram of an experiment for the compression of a plasma by a magnetic field which is parallel to the internal magnetic field [9]

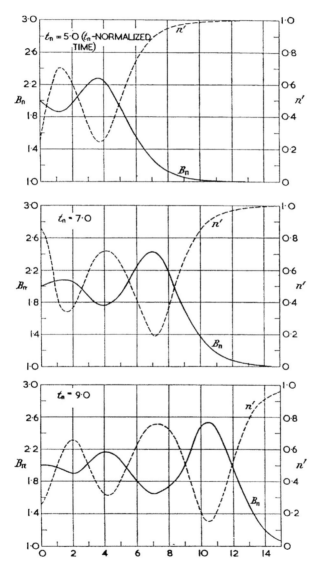

Fig. 8. Magnetic field strength and ion density, as functions of h_n (a Lagrangian coordinate) computed for various times after the start of the compression [9]

and Kruskal studied the same plasma wave numerically and showed that two such waves retained their identity after colliding. In view of this "particle-like" behaviour Zabusky and Kruskal introduced the term "soliton" to describe the solitary wave [10]. It is interesting to note that this phenomenon was known early in the nineteenth century, for the case of shallow water waves [11].

My work with John Adlam seems to have been largely overlooked until very recently [12], presumably because it pre-dated the term "soliton".

4 Research at Rome and Frascati

My next post was in Rome where Professor Amaldi had invited me to help to set up a Plasma Group in collaboration with Bruno Brunelli. The group started work at the Università di Roma (La Sapienza) and then moved to Frascati in the nearby Alban hills. Part of the programme, that dealing with "fusion research", was to study the theta pinch, similar to that illustrated in Fig. 7. We were not especially concerned with the details of the electrical breakdown, i.e., the transition from an insulating gas to a conducting one, but in designing the experiment it was necessary to consider the distribution of the electric field existing before breakdown. This is illustrated in Fig. 9 for the usual theta-pinch geometry [13]; it can be noted that the field distribution bears little resemblance to the concentric circles that were assumed by other workers at the time. Our theta-pinch, which was designed to study the rapid compression of a plasma by a magnetic field [14], was more complex and six feed-points were employed, as shown in Fig. 10. The apparatus was

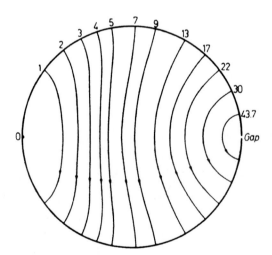

Fig. 9. The lines of force of the electric field inside a single turn coil. The numbers are the values of the flux function in arbitrary units [13]

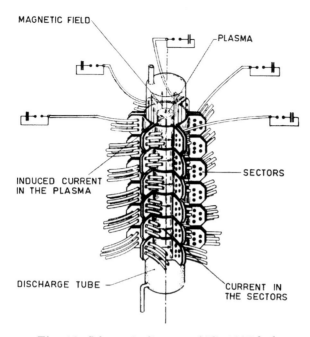

Fig. 10. Schematic diagram of "Cariddi" [14]

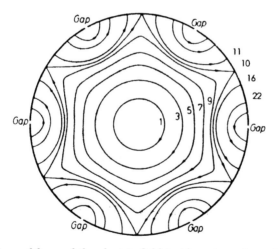

Fig. 11. The lines of force of the electric field inside a six sector coil. The numbers are the values of the flux function in arbitrary units [13]

named "Cariddi", which is Italian for Charybdis [15], since the name Scylla had already been employed in some other theta-pinch research. Figure 11 shows the electric field before breakdown; in the absence of screening, the

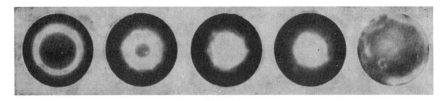

Fig. 12. Image converter photographs (exposure time 90 ns) of a hydrogen discharge taken at 50 μHg (6.7 Pa). From left to right, the times are: 2.9 μs, 3.15 μs, 3.3 μs, 3.55 μs, and 5.2 μs, respectively [14]

electric field configuration consists of concentric circles near the centre, but strong electrostatic fields appear at each feed-point gap [13].

Figure 12 shows a series of image converter photographs of a hydrogen discharge taken at 50 μHg (6.7 Pa). The centre photograph refers to the maximum compression, examination of the original film shows that less light was emitted from the centre. The first two photographs show that the luminous region had an irregular outer surface and the final photograph (5.2 μs) shows a complicated structure which corresponds to a new plasma just leaving the walls. In general electrical breakdown took place during the first half cycle. Magnetic probe measurements showed that the imploding plasma subsequently compressed the internal magnetic field.

These measurements simply indicate one avenue of research that was being pursued at the time (1962). Since then, research on magnetically confined plasmas for fusion has been largely carried out with tokamak machines, with their more complex configurations of magnetic field. Time will show whether this approach is the right one, or indeed whether power generation using controlled thermonuclear reactions (fusion) is possible. The present writer hopes so.

5 Epilogue

After 6 years in Rome I returned to England and, after a brief sojourn at Cambridge, I took up a post at the University of Oxford. Since then I have mainly worked on low-temperature plasmas.[1] In this work magnetic forces are usually negligible, although a changing magnetic field is sometimes employed to produce an electric field and the magnetic field plays a rôle via the Poynting vector in the vast majority of cases.

---

[1] I have done just a little work on liquid MHD, some of it with Sergei Molokov, one of the editors of the present volume, and the references are appended below [16–19].

References

1. Reynolds P, Craggs JD (1952) An attempt to produce a thermo-nuclear reaction in deuterium by means of a high current spark discharge. Phil Mag 43:258
2. Allen JE (1953) High current discharges. Ph.D. thesis, University of Liverpool, UK
3. Allen JE, Craggs JD (1954) High current spark channels. Brit J Appl Phys 5:446
4. Northrup EW (1907) Some newly observed manifestations in the interior of an electric conductor. Phys Rev 24:474
5. Cousins SW, Ware AA (1951) Pinch effect oscillations in a high current toroidal ring discharge. Proc Phys Soc B 64:159
6. Allen JE (1957) An elementary theory of the transient pinched discharge. Proc Phys Soc B 70:24
7. Allen JE, Reynolds P (1957) Experiments with a ring discharge of short duration. In: Proceedings of the Third International Conference on Ionization Phenomena in Gases 33
8. Adlam JH, Allen JE (1958) The structure of strong collision-free hydromagnetic waves. Phil Mag 3:448
9. Adlam JH, Allen JE (1960) Collision-free hydromagnetic disturbance of large amplitude in a plasma. Proc Phys Soc 75:640
10. Zabusky NJ, Kruskal MD (1965) Interactions of "solitons" in a collisionless plasma and the recurrence of initial states. Phys Rev Lett 15:240
11. Robison J, Scott Russell J (1837) Report of the Committee on Waves. In: 7th Meeting of the British Association for the Advancement of Science, Liverpool, UK, 417–496 and 5 plates
12. Verheest F, Cattaert T (2005) Oblique propagation of large amplitude electromagnetic solitons in pair plasmas. Phys Plasmas 12:032304
13. Allen JE, Segre SE (1961) The electric field in singleturn and multi sector coils. Nuovo Cimento 21:980
14. Allen JE, Bartoli C, Brunelli B, Nation JA, Rumi B, Toschi R (1962) Observations on an orthogonal pinch discharge (Cariddi). Nuclear Fusion Supplement Part 2:621
15. Alighieri Dante. La Divina Commedia, Inferno, Canto VII, 22
16. Allen JE, Auer PL, Endean VG (1976) On the law of isorotation and laboratory experiments. Plasma Phys 18:143
17. Allen JE (1986) A note on the magnetic Reynolds number. J Phys D: Appl Phys 19:L133
18. Molokov SY, Allen JE (1992) On the theory of the Heiser and Shercliff experiment. Part 1 MHD flow in an open channel in a strong magnetic field. J Phys D: Appl Phys 25:393–400
19. Molokov SY, Allen JE (1992) On the theory of the Heiser and Shercliff experiment. Part 2 MHD flow between two cylinders in a strong radial magnetic field. J Phys D: Appl Phys 25:933–937

Part II

High Magnetic Fields

Early Years of MHD at Cambridge University Engineering Department

Martin Cowley

Trinity College, Cambridge, CB2 1TQ, United Kingdom
(mdc1000@eng.cam.ac.uk)

1 Early interest in liquid metals

How the study of magnetohydrodynamics came to the Engineering Department of Cambridge University was never fully recorded. What is known is
that an interest in the heat-transfer properties of liquid metals began with
the research undertaken by L.M. Trefethen, who entered the Department as a
research student in 1946. Trefethen's earlier education had been in the United
States, culminating in a master's degree from the Massachusettes Institute of
Technology. His subject of research at Cambridge was approved initially as
"Gas turbines – turbine blade cooling".

At some stage a research contract was negotiated with the Ministry of
Supply, acting for the Atomic Energy Research Establishment at Harwell.
How the switch in general area of interest from gas turbines to nuclear energy
came about is not known, but both areas were of course potential users of
cooling systems capable of carrying high heat loads. A key contact in the
negotiations with the AERE was Dr Stefan Bauer of the Chemical Engineering
Section there. He was an enthusiastic believer in the future of nuclear energy
for power production, including the use of fast reactors, and for the latter he
realized that the only practical media for heat removal would be liquid metals
or molten salts.

Trefethen's research developed into a study of heat transfer to and from
the turbulent flow of liquid metals in pipes and annuli. In setting up a flow
circuit to conduct tests with mercury as the circulating fluid, the decision
was taken to drive the mercury and measure its flow rate by electromagnetic
means and, so, magnetohydrodynamics came to the Department.

In 1947 a new student was recruited to the research effort in liquid-metal
heat transfer. He was W. Murgatroyd, a Cambridge graduate in engineering.
His subject of research was approved as "Heat transfer from liquid metals" and
he collaborated in the design and commissioning of Trefethen's liquid-metal
flow circuit, developing in the process an interest in electromagnetic pumps.

S. Molokov et al. (eds.), Magnetohydrodynamics – Historical Evolution and Trends,
131–154. © 2007 Springer.

The pump on Trefethen's rig was a simple linear DC one, but Murgatroyd began to consider other possibilities, such as a centrifugal type in which a liquid-metal disk was spun by electromagnetic forces. He was naturally concerned with losses in electromagnetic pumps and this led him to investigate "magnetic viscosity", taking measurements of the friction factor for flow in insulated rectangular channels of high sectional aspect ratio (15:1). The transverse magnetic field was normal to the longer sides, which were 6.8 mm apart.

It is for his results on transition to turbulence [1] that Murgatroyd's name remains well known in the MHD community. The apparatus was such that he could conduct experiments at considerably higher values of Reynolds number Re than had been achieved earlier in Denmark by Hartmann and Lazarus [2]. He was able to show that, given high enough values of Hartmann number Ha such that the Hartmann layers on the long sides of the channel were much thinner than their distance apart, non-dimensional quantities such as friction factor would depend on the ratio Re/Ha. Departure from the predictions of analysis assuming laminar-flow Hartmann layers took place experimentally when $Re/Ha \approx 225$,[1] which has been normally used since then as the criterion for the onset of turbulence.

At a much later date (1998/99) interest in the stability of Hartmann layers was revived in the Department of Engineering by a young member of the academic staff, T. Alboussière, in collaboration with a postdoctoral research fellow, R.J. Lingwood. Over the years various researchers elsewhere had taken up the subject, both analytically and experimentally. A classic result had been that of Lock [3], who had calculated the threshold for growth of infinitesimal disturbances and had obtained 50,000 approximately for the critical value of Re/Ha. The discrepancy between Lock's analytical value and Murgatroyd's experimental result was embarrassing, but the latter was used by Lykoudis [4] and confirmed by Branover [5]. Lingwood and Alboussière [6] made the important point that what was being found in the experiments was the condition for laminarization as a turbulent flow entered a region of magnetic field, and by means of an energy analysis they calculated the condition for *decay* of finite perturbations to be $Re/Ha < 26.5$. They also calculated the growth condition more precisely as $Re/Ha > 48250$. It is clear that the stability problem is not yet resolved.

In another paper the same authors [7] developed a mixing-length model for fully developed turbulent Hartmann layers, giving friction factor as a function of Re/Ha. Their results agreed well with experimental data for channel flows when the layers on either side were thin enough not to overlap.

Before Murgatroyd completed his Ph.D. a new recruit to the liquid-metal group arrived in 1951. J.A. Shercliff had been awarded the Rex Moir prize in 1947 for being the most distinguished undergraduate of his year in the

---

[1] Murgatroyd presents the figure as 900, but this is a consequence of his using a length scale in Re four times that in Ha.

Department of Engineering. After graduating he had taken a master's degree at Harvard and completed a graduate apprenticeship with the aircraft manufacturers A.V. Roe & Co. Ltd. His subject of research was approved as "Problems in magnetohydrodynamics", although he seems to have taken a little while to settle on a particular topic. Nevertheless Shercliff interacted well with Murgatroyd and was soon involved in research on channel flows and electromagnetic flowmeters. In his first year as a research student he was already submitting a paper [8] which was to become a classic one in the literature of laminar flow of electrically conducting fluids through insulated channels with an imposed transverse magnetic field. From analysis of the problem when the cross-section of the channel is rectangular he deduced the important result for high Hartmann number that not only would there be layers of thickness order $1/Ha$ times channel dimension on walls with a normal magnetic-field component, but also layers of thickness order $\sqrt{1/Ha}$ on walls with no normal magnetic-field component. This was effectively the first discussion of "parallel layers" in MHD.

A part of the same paper was concerned with insulated channels no longer rectangular in section, but having symmetry about a plane normal to the magnetic field. Shercliff showed how the properties of Hartmann layers lead to current return in the layers being the dominant influence on the velocity distribution in an inviscid core between the layers. Later Moreau of the MHD group at Grenoble would coin the phrase "active Hartmann layers" to describe such influence. The predictions of Shercliff's analysis compared well with the results of experiments conducted by Hartmann and Lazarus [2] in channels of square and circular section. In the case of the former experimental results were closely approximated by asymptotic analysis for high Ha, even though values of Ha were only moderate, whereas this seemed not to be so in the case of the circular channels. Later (1962) Shercliff was introduced to work by Gold [9] which led him to realize that an effect had been missed in his asymptotic analysis, namely that in a circular channel core velocity varies in the direction mutually perpendicular to the magnetic field and the axis of the channel. In consequence there are order $1/Ha$ viscous forces in the core and these forces have an overall effect of the same magnitude as the order $1/Ha$ thickness of the Hartmann layers. When viscous forces in the core were included in the analysis, much improved agreement was obtained between theory and experiment.

Shercliff continued his Ph.D. work with further study of problems arising in the context of flow measurement by electromagnetic means. It had been known for some time that a potential difference across a circular channel with uniform transverse magnetic field would be directly proportional to the total flow rate and independent of the detailed velocity distribution provided that the distribution is axially symmetric. Furthermore there was a general belief that the calibration of electromagnetic flowmeters would be nearly independent of the velocity distribution even if the flow were not axially symmetric. As Shercliff showed [10, 11], this was just not true and much of his work was devoted to quantifying effects which might influence the calibration, such as the distortion to the velocity distribution caused by entry to a region of magnetic field [12]. The work culminated in his monograph of 1962 on the theory of electromagnetic flow measurement [13].

Shercliff was appointed to a University Demonstratorship (a title which is no longer used, but was roughly equivalent to an Assistant Lectureship), followed by appointment to a University Lectureship in 1957. Being established on the academic staff of the Department, he was in a position to expand research activity in the field of magnetohydrodynamics.

2 Ionized gases

In the spring of 1956 Soviet Academician Kurchatov was invited to give a lecture at the Atomic Energy Research Establishment, Harwell. He let it be known then that the Russians were working on the control of thermonuclear fusion of hydrogen isotopes for the purpose of peaceful power production. In fact the British and the Americans had also been working on controlling fusion, but the research had been classified. The Kurchatov lecture marked perhaps the most significant event in the general opening up of fusion research in the three countries, and a major exchange of information occurred at the United Nations conference on the peaceful uses of atomic energy held at Geneva in 1958. The principal effort of all three countries was concentrated on finding the means to contain the hydrogen isotopes at high enough temperature for the fusion reaction to produce net energy. Since the gas would be ionized and electrically conducting, the route everyone favoured at that time was the formation of a "magnetic bottle". A detailed history of the early work on fusion research in the UK is to be found in a report by Hendry and Lawson [14].

Within the Atomic Energy Authority it was proposed that Universities be encouraged to contribute to the research effort and Shercliff seized the opportunity to involve Cambridge in work on ionized gases and their interaction with magnetic fields. A research contract was negotiated with the UK Atomic Energy Authority to finance the construction of a combustion-driven shock tube, capable of propagating shock waves into argon at Mach numbers (shock speed divided by the speed of sound in the argon ahead of the shock) up to 17. Argon flow behind the shock would typically have a temperature of $13,700\,\mathrm{K}$ and an ionization level of 17%.

The Cambridge effort was not a venture into the totally unknown. There was a small shock-tube group at Harwell under W. Miller, who acted as monitor for the research contract with Cambridge. The Harwell shock tube differed from that proposed at Cambridge only in size, being just under 35 mm internal diameter as compared to 127 mm at Cambridge. The main Harwell experiments concerned the interaction of the plasma flow behind a shock wave with the magnetic field due to a coil mounted coaxially with the shock tube. Typical conditions in those experiments behind a Mach 17 shock wave traveling into argon at a pressure of 10 mm Hg yielded a magnetic Reynolds number $R_m \approx 1.5$ based on tube diameter d and flow velocity v, so that only a modest distortion of the magnetic field might be expected. On the other hand, for 3 T, the maximum magnetic field available on the coil axis, and with an initial pressure in the argon of 10 mm Hg again, the interaction parameter $N = \sigma B^2 d / \rho v$ took the more promising value of about 4. Here σ, B and ρ are the scales for electrical conductivity, magnetic field strength and density. Certainly, when the Harwell experiments were reported by Dolder and Hide [15] the patterns of luminosity appearing on photographs of the radiating ionized gas provided strong evidence of the flow being affected.

In the Harwell experiments axial symmetry, a poloidal magnetic field and only weak variation of the field in time implied that the electric field could be taken as zero and current density given by $\mathbf{j} = \sigma \mathbf{v} \times \mathbf{B}$. However, the configuration meant that $\mathbf{v}$ and $\mathbf{B}$ were not close to being perpendicular to each other over much of the magnetic-field region. It was therefore thought that the first task with a larger shock tube at Cambridge would be to conduct similar experiments to those at Harwell except that the flow would pass axially through an annulus and interact with a mainly radial magnetic field between pole pieces. Current density would be primarily azimuthal, giving an axial retarding force, and the flow situation, if it settled to a steady state, would nearly correspond to the 1D Fanno-line process of ordinary gasdynamics.

The Cambridge shock-tube group under the leadership of Shercliff came into being at the start of the academical year 1956/57, two research students having been recruited. The first was E.J. Morgan a Shell Scholar from Australia who had graduated from the University of Sidney with a first-class degree in electrical engineering and physics. His subject of research was approved as "Ionized gas flows". I was the second recruit, returning to Cambridge after a year spent in the gas-turbine design office of Brown Boveri, Switzerland. My subject of research was approved as "Shock-tube flows". I found that the prospect of being connected to the glamorous field of plasma physics was exciting, but the task of designing, constructing, instrumenting, and commissioning the shock tube was somewhat daunting.

In the end it was three full years before the shock tube was ready to be used as a regular experimental tool (see Fig. 1), Morgan having taken a major share of the development work. A particular aspect of this had been the design and testing of ionization gauges, which would have to respond reliably to the passage of a shock wave. The time of travel between two such gauges was the major diagnostic, used for determining shock speed and hence determining conditions in the plasma flow. The problem is that too sensitive a gauge and it is likely to be triggered by precursor ionization arising from radiation being absorbed ahead of the shock wave. With an insufficiently sensitive gauge and a time interval between arrival of the shock wave and the onset of substantial ionization, a delay in triggering the gauge is likely to occur. Morgan decided to build on his experience with development of the gauges and to use them

COMBUSTION
CHAMBER

PUMPING AND LOW
PRESSURE MEASUREMENT.

VELOCITY MEASURING
SECTION.

DUMP CHAMBER.

Fig. 1. The shock tube in 1960. A test section is positioned between the velocity measuring section and the dump chamber. The latter replaced the blank end of the shock tube when avoidance of high pressures due to shock reflection was desired

for measuring ionization rates. He was particularly interested in determining the influence of adding small amounts of gaseous impurity on the rates.

After completing his work on ionization rates, little time was left to Morgan for the experiment which had been originally planned – the study of a nearly 1D flow through a transverse magnetic field. One expected outcome can be illustrated by means of the hydraulic analogy to unsteady one-dimensional gas flow. In Fig. 2a hydraulic jump, analogous to a shock wave (S_1), has been generated to the right of the picture by lifting a sluice gate at O and it is propagating leftward into still water, which has been dyed black. The horizontal strips in the figure are pictures taken at successive instants in time, creating a distance-time diagram for phenomena in the water channel. The jump interacts with a gauze obstacle (online GG') equivalent to a region of electromagnetic forces opposing the flow. In the picture shown the consequence is a transmitted jump (S_2) and a reflected jump (S_3). However, if the gauze were to provide only a weak blockage effect and S_1 were strong enough to

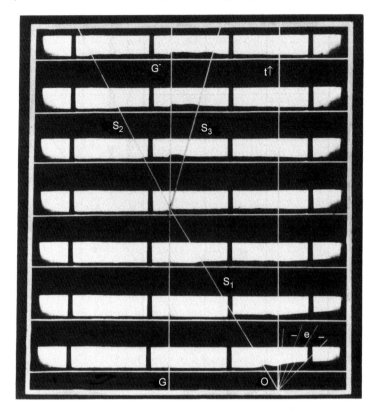

Fig. 2. The hydraulic analogy to shock-tube flow. The thin horizontal lines mark the floor of the channel and the black region immediately above is the water (dyed)

have sufficiently super-critical flow behind it (supersonic flow), there would be no upstream propagation, i.e., S_3 could not exist. In the shock tube, the dividing condition between there being a reflected shock and there not being one is that the shock is stationary at the entrance to the magnetic field region. Fair agreement was shown between theory and observation of this behaviour in a few runs, but, by the time the experiments on the blockage effect got under way, experiments elsewhere (e.g., R.M. Patrick and T.R. Brogan, working under A.R. Kantrowitz at Avco in the late 1950s [16]) meant that the Cambridge work was virtually superseded.

The value of the magnetic Reynolds number based on diameter of the Cambridge shock tube was approximately 6 and it seemed likely that magnetic fields applied in shock-tube experiments would suffer considerable distortion due to the flow of the electrically conducting medium. With this in mind I began a general study of kinematic problems, where the flow field is taken as known [17]. Solution of the particular problem of magnetic-field distortion in a configuration similar to that of the Harwell shock tube [15] was then compared with measurements of flow influence in the Cambridge apparatus [18] with good agreement, which is perhaps surprising in view of the finite length of the plasma (about two shock-tube diameters instead of being infinitely long) and the temperature variation along the plasma due to cooling by radiation. However, it did confirm that the shock tube was capable of supplying a useful sample of plasma.

3 Magnetogasdynamic shock waves

Although the shock tube saw two more research students through to the Ph.D. degree after Morgan and me, its main significance, sad to say, was to act as an encouragement to thinking about gasdynamics and magnetogasdynamics rather than providing direct experimental evidence of phenomena. Shercliff [19] initiated theoretical work with a generalized treatment of steady 1D processes in ordinary gasdynamics. Typical of such processes is the Fanno line, which could arise in the context of the action of electromagnetic forces without exchange of electromagnetic energy (i.e., $\mathbf{E} = 0$ and $\mathbf{j} \times \mathbf{B} = \sigma(\mathbf{v} \times \mathbf{B}) \times \mathbf{B}$), and the Rayleigh line which could be generated by energy exchange $\mathbf{j} \cdot \mathbf{E}$ whilst $\mathbf{B} = 0$. Shercliff showed how qualitative differences between supersonic and subsonic flow could be deduced from general requirements of thermodynamic stability, together with the assumption of $\partial^2 p / \partial v^2 > 0$ on isentropics, v here being specific volume, a condition formulated by Weyl [20] in the context of shock waves. At a qualitative level there is no need to invoke particular gas laws with the consequent complexity which that would imply for ionized gases.

Linking magnetogasdynamic behaviour to ordinary gasdynamic processes was an important step forward in understanding, which applied primarily to flows with magnetic Reynolds number R_m tending to zero. However, it was clearly going to be difficult to treat problems analytically with both effective electromagnetic forces and R_m of order unity. This provided justification for gaining a better understanding of flows at the other extreme, i.e., $R_m \rightarrow \infty$, while for much of the work still keeping an interest in cases where the electric field is short-circuited ($\mathbf{E} = 0$). Ohm's law under steady conditions then requires that $\mathbf{v} \times \mathbf{B}$ be zero, i.e., $\mathbf{v}$ parallel to $\mathbf{B}$, and continuity of both magnetic flux and flow implies that $B/\rho v$ be constant along streamlines. Analysis under these conditions turned out to be easier than first thought, when it was realized that the governing equations could be recast in a form analogous to those of ordinary gasdynamics. The quantity corresponding to velocity in the analogous flow is $\mathbf{v}^* = \mathbf{v}(1 - b^2/v^2)$, where $\mathbf{v}$ is the actual velocity and b the local Alfvén wave speed, but there were some peculiarities, such as the possibility of the analogue to Mach number becoming imaginary and the density negative [21]!

Meanwhile Shercliff was extending his 1D analysis to perfectly conducting ($R_m \rightarrow \infty$) flows, covering cases where $\mathbf{v}$ and $\mathbf{B}$ are parallel [22] and cases where $\mathbf{B}$ is perpendicular both to $\mathbf{v}$ and the direction of variation [23].

This was also a period when in the world at large there was an upsurge of interest in perfectly conducting magnetogasdynamic flows. The analogy which I had spotted was also discovered by Imai [24], while various authors worked directly on flows with $\mathbf{v}$ and $\mathbf{B}$ parallel, e.g., Sears [25]. Of particular interest was the topic of magnetogasdynamic shock waves in such flows and in his paper Shercliff [22] gave what is probably the clearest account of the classification of the states which could occur on either side of oblique-field shock waves and their ordering $1\rightarrow2\rightarrow3\rightarrow4$ in accordance with transitions permitted by the second law of thermodynamics. Here oblique refers to the magnetic field direction being at an angle to the shock normal. If u is the velocity component normal to the shock wave, b the Alfvén wave speed, c_f the fast magnetoacoustic speed and c_s the slow, all in the normal direction, the states are defined by $u_1 > c_{f1} > b_1$, $c_{f2} > u_2 > b_2$, $b_3 > u_3 > c_{s3}$, $b_4 > c_{s4} > u_4$. It turned out that slightly earlier and independently Germain [26], who was working on the structure of magnetogasdynamics shock waves, adopted and is credited with, the same classification of states.

A contribution to shock-wave theory which was due mainly to Russian workers, e.g., Akhiezer, Liubarski, and Polovin [27] concerned the stability of transitions when perturbed by weak magnetoacoustic or Alfvén waves. Determination of whether a particular transition is stable (or has the ability to evolve) depends on counting the number of waves which may emanate from it as a consequence of the perturbation. What is needed for the problem to be well posed and to satisfy boundary conditions at the transition is that there can be three magnetoacoustic waves (contact surface not counted) emanating from the transition if the perturbation is due to a magnetoacoustic wave or that there can be two Alfvén waves emanating if the perturbation is due to a transverse Alfvén wave. Note that the waves may include ones which are travelling upstream relative to the flow, but are actually being swept downstream. Counting the waves shows that fast shocks $1\rightarrow2$ and slow shocks $3\rightarrow4$ can emit just the right number of waves to react to magnetoacoustic or Alfvén waves. Intermediate shock waves $1,2\rightarrow3,4$ are not able to emit enough Alfvén waves to react in an orderly way to perturbations carried by Alfvén waves and $1\rightarrow4$ shocks are not able to react to magnetoacoustic waves. What happens in these cases was elucidated by Todd (see below). For all other transitions between shock states too many waves can be emitted and this can mean that shock waves will decay spontaneously. A comprehensive monograph on shock waves, including their structure, was written by Anderson [28], a graduate student at MIT who interacted with Shercliff when the latter was spending a year (1960–61) there on sabbatical leave. What was also emerging was that, although not all transitions which satisfy the second law of thermodynamics may evolve in an orderly fashion when perturbed, there seems to be a match between the results of the stability analysis and those of investigations into shock structure when ohmic diffusivity is the largest of the diffusivities. Too many waves in the stability analysis seemed to match there being no steady structure solution. Too few waves matched the structure problem becoming non-unique.

On his return to Cambridge UK from MIT Shercliff recruited L.Todd who was just starting research and was in the Department of Applied Mathematics and Theoretical Physics. Shercliff was appointed supervisor and Todd's subject of research was approved as "Magnetogasdynamic shock-wave stability". His first project was an analytic and computational demonstration of how a normal 1→4 shock wave evolves when perturbed by an infinitesimal Alfvén wave. In the unperturbed state, such a shock wave could be an ordinary gasdynamic one with both velocity and magnetic field normal to the shock plane, so that there is no electromagnetic interaction. It must therefore be the case that a structure for the shock wave exists involving viscous and thermal diffusivities. Now consider small transverse perturbations to the velocity and magnetic fields acting under the influence of large ohmic diffusivity. The gasdynamic shock wave becomes a sub-shock in a wide region where the combination of Ohm's law and transverse momentum requires that

$$\frac{\partial^2 B_t}{\partial n^2} = \mu \sigma u \left(1 - b^2/u^2\right) \frac{\partial B_t}{\partial n},$$

where n and t denote the normal coordinate and transverse direction and b is again the Alfvén wave speed based on normal component of magnetic field. The assumption of weak transverse components implies that $\mu \sigma u (1 - b^2/u^2)$ may be taken as approximately constant on the upstream side of the sub-shock and a different constant on the downstream side. It follows that, as $n \to -\infty$, B_t may decay to zero from some value β at the sub-shock provided $u > b$ upstream. As $n \to \infty$, B_t may decay from β to zero provided $u < b$. Thus an ordinary gasdynamic shock wave which can be classed as intermediate 1→4 can absorb arbitrary amounts of transverse magnetic flux (but in actuality non-linear effects limit). The distribution of B_t was computed by Todd [29] for that case and also for $u > b$ both upstream and downstream. His calculations showed the two emitted waves of the latter case forming. Todd went on to extend his analysis to generally oriented intermediate shock waves [30] followed by an exhaustive analysis of switch-on and switch-off shocks [31].

Before leaving the topic of magnetogasdynamic shock waves two proposals for experiments using the shock tube deserve a brief mention. The idea behind the proposals was that for very strong magnetic fields achieving a highly conducting flow pattern depends on electromagnetic forces being strong enough to make the *gas flow adjust* so that $\mathbf{E} + \mathbf{v} \times \mathbf{B} \approx 0$, whilst the *magnetic field is unaffected*. What then sets the length scale over which conditions may depart from perfectly conducting is the length which makes the interaction parameter of order unity instead of the magnetic Reynolds number. The interaction parameter N and the magnetic Reynolds number are related by $N = R_m (b^2/v^2)$, so that modest R_m with large value of N implies $b >> v$ and any attempt to generate a magnetogasdynamic shock in the shock tube would only be feasible for a slow shock. Although an attempt was made to study the diversion of ionized-gas flow over a truncated cone mounted coaxially with the shock tube in the presence of an axial field, it proved too difficult to separate effects at the cone from disturbances where the upstream axial flow interacts with the radial field components at entry to the magnetic-field region. Some background to the experiment is provided by analysis of plane flow over a wedge [32].

The second proposal was for an experiment on shocks which are ionizing a cold gas as well as being affected by the presence of a magnetic field. Since the directions of magnetic field and flow were different from the cone experiment, it was believed that the problems at entry to the field region might be avoided more readily. However, before the experiment was attempted, the decision was taken to wind up the shock-tube work and all that was left was some analysis of the ionizing shock waves [33]. At a much later date the analytical techniques proved useful in the elucidation of behaviour in a magnetogasdynamic thruster model [34].

4 Magnetohydrodynamic waves

Returning to an earlier stage in this history, 1958 saw an expansion of the group, whose interests now began to extend beyond shock-tube plasmas. Two new research students were recruited. The first was G.A. Jameson, a Cambridge graduate, and the second J.A. Decker, who had come from the United States with a degree from MIT. Shercliff had attempted in 1953 a demonstration of Alfvén waves in an electrically conducting liquid. He failed to achieve a convincing demonstration, but there was a clear indication that a mild improvement in parameters would lead to success. So Shercliff suggested to Jameson that he might like to try the experiment on a larger scale and with sodium as liquid medium rather than sodium–potassium eutectic and it turned out that Jameson was indeed interested.

The sodium was contained in a sealed, stainless-steel torus, which was mounted with its axis vertical in a nearly uniform vertical magnetic field B_0 as shown in Fig. 3a. The cross-section of the toroidal passage was rectangular, being 237 mm in the radial direction by 195 mm in the vertical. The strength of magnetic field available was limited by the need to prevent over-heating of the field windings, but the magnet could run at 1 T for short periods and 0.7 T indefinitely. The aim of the set-up was to generate vertically running Alfvén waves by means of an exciting coil wrapped around the toroidal passage in such a way that velocity and magnetic-field perturbations would be directed azimuthally. What was happening in the sodium would then be inferred from search coils, one wrapped around the toroidal passage in the same way as the exciting winding and another small coil mounted in a sealed casing to measure $\partial B/\partial t$ at the centre of the flow passage. The final feature of the apparatus was a heater winding to maintain the sodium in the liquid state, the temperature being held at 120°C. The construction and filling of the torus was undertaken by the Atomic Energy Authority and, as Shercliff delighted in saying, it could have had strawberry jam inside it for all we knew at Cambridge.

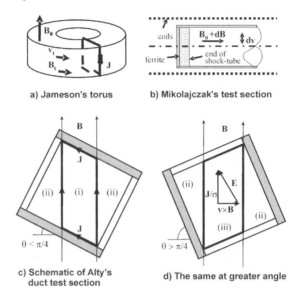

a) Jameson's torus

b) Mikolajczak's test section

c) Schematic of Alty's
duct test section

d) The same at greater angle

Fig. 3. The geometry of various experiments. In (a) only one turn of the perturbing coil is shown. In (c) the thick-lined loop indicates the position of current sheets, two in the Hartmann layers on the insulating walls and two in free shear layers. In (d) a similar loop indicates the position of passive Hartmann layers and free layers carrying a change in the tangential component of the electric field [41]

Consider the simpler geometry of two *infinite* parallel planes with magnetic field perpendicular to them. The dispersion relationship for plane Alfvén waves traveling in the field direction and carrying perturbations in velocity v_t and field B_t components which are perpendicular to the main field is

$$\left(i\omega + \lambda k^2\right)\left(i\omega + \nu k^2\right) = -b^2 k^2.$$

Here ω and k are frequency and wave number respectively, λ and ν the ohmic and viscous diffusivities and b the Alfvén wave speed. In the limit of both diffusivities tending to zero by comparison with ω/k^2, the relationship reduces to that for ideal Alfvén waves, $k = \pm\omega/b$. Taking account of the ohmic diffusivity, while still treating the fluid as inviscid (since $\nu << \lambda$ in practice), the relationship yields a decaying wave, $k \approx \pm(\omega/b)(1 - i\omega\lambda/2b^2)$. The parameter controlling the damping of the wave $\omega\lambda/2b^2$ may be written approximately as $k\lambda/2b$ provided the damping is weak and in this form it is recognizable as π times the inverse of a magnetic Reynolds number based on wavelength and wave speed, i.e., the inverse of a Lundquist number. In an ideal wave the magnitude of the velocity perturbation is related to the magnitude of the magnetic-field perturbation by $|v_t| = |B_t|/\sqrt{\mu\rho}$. This is not what is needed at the horizontal boundaries in Jameson's apparatus where $|v_t| = 0$ and $|B_t| \neq 0$.

The second solution for k^2 in the dispersion relation must also be invoked in order to satisfy the boundary conditions. However, the second solution turns out to be not so obvious as the wave-like one, but inspection of the dispersion relation shows that an approximate solution is $k^2 = -b^2/\nu\lambda$, provided that ω is much less than νk^2 and λk^2. Clearly k is imaginary and not dependent on ω. The second solution can therefore represent an exponentially decaying layer with thickness of order $|k^{-1}| = \sqrt{\nu\lambda/b^2} = \sqrt{\rho\nu/\sigma B_0^2}$, i.e., a Hartmann layer. The change of perturbation velocity Δv_t across the layer is related to the change of perturbation magnetic field ΔB_t by the standard result $\Delta v_t = (\Delta B_t/\mu)\sqrt{\rho\nu\sigma}$. It is then easily shown that ΔB_t across the layer is of order $\sqrt{\nu/\lambda}$ times the amplitude of the magnetic field perturbation being propagated by Alfvén waves if the velocity condition is to be met. Since this ratio of diffusivities is small, wave motion in the sodium is strongly coupled to perturbations introduced by the exciting winding of Jameson's apparatus, whereas, if the ratio had been large, it would have been more appropriate to generate the waves by a mechanical oscillation, a point discussed in more detail in Shercliff's textbook [35]. This book was being written soon after Jameson completed his Ph.D. work.

Jameson [36] looked for resonances in the torus and the results provided a convincing demonstration of wave-like behaviour. Signals from the central search coil peaked at a magnitude 9.6 times the magnitude for a free-space field there. Comparison with theory was good, in spite of uncertainty about the exact influence of the stainless-steel walls and their contact resistance with the sodium. Jameson also ran a series of experiments with a step input of exciting current, although the results were not made public until Shercliff quoted them in a paper (1976) [37].

Meanwhile Decker was building apparatus to study magnetoacoustic waves in a low-pressure caesium plasma. The first discharge tube in which the plasma was contained was 640 mm long and 47 mm in diameter (a second tube was subsequently made to similar measurements). The tube was placed on the axis of a solenoidal winding, capable of giving a longitudinal magnetic field of 0.5 T. Oscillations observed in the tube were attributed to longitudinal ion-wave resonance, but the experiment owed more to plasma physics than to MHD (further details of the work were published by Decker (1964) [38]).

The next recruit to the shock-tube group was P.L. Read (in October 1959), who already had some experience of liquid-metal work. As a final-year undergraduate he had undertaken a project with Shercliff related to the use of liquid-metal "brushes" on commutators of electric motors. However, his first task was to assist Morgan with the experiments on 1D flow through a transverse magnetic field and it may well have been partly that experience which led him to appreciate the need for more diagnostic facilities on the shock tube, although another factor was the early proposal that he investigate boundary layers in ionized gases. Read's main Ph.D. project became the development of apparatus to measure the small-angle scattering of an electron beam when it passes through a shock-tube plasma. The underlying idea was to use such measurement to deduce the density of both the charged and neutral particles present. However, like Decker's, Read's work turned out not to have direct connection with magnetohydrodynamics.

Experimental work with a strong MHD flavour began again in October 1961 when A.A. Mikolajczak joined the group and took on a project aimed at forming magnetoacoustic waves in the shock tube. As illustrated by the water-channel analogy (see Fig. 2), a sample of stationary ionized gas is created when a shock wave reflects at the end of the tube, if it is a blank end. In Mikolajczak's experiment a solenoidal winding around the tube, as shown in Fig. 3b, provided a longitudinal magnetic field of up to 1.4 T there, with current supplied from a capacitor bank over a sufficiently long time for the field to be regarded as steady. An additional winding superimposed a longitudinal perturbation field. Search coils were mounted so as to detect radial propagation of the field.

The essence of Mikolajczak's experiment is most easily understood in terms of an analogy with ordinary gasdynamics discovered by Grad [39] for the case of plane 2D flow with magnetic field everywhere perpendicular to the flow plane. The equation of motion is then the same as that of ordinary gasdynamics except that pressure is replaced by $p^* = p + B^2/2\mu$. If the gas has infinite conductivity, the flux-freezing theorem implies that B/ρ is maintained constant following a fluid particle. With B/ρ constant *throughout*, p^* becomes a function directly of the thermodynamic state and the speed of ideal magnetoacoustic waves for the geometry of the analogy will be given by $\sqrt{dp^*/d\rho}$, which evaluates under the dubious assumption of isentropic conditions as $c_f = \sqrt{a^2 + b^2}$. Here a is the ordinary speed of sound and b is the Alfvén-wave speed again (but note that the waves are fast magnetoacoustic carrying radial variation of radial velocity).

The problem of how the waves should be excited bears a resemblance to the similar problem in Jameson's experiment. The wave phenomenon imposes a relation between velocity and perturbation magnetic field, which is not consistent with a boundary condition of zero normal velocity at the tube wall. What is needed is a boundary layer to match wave phenomena to the boundary conditions. Treating the region near the wall as one with flow nearly uniform in direction and as inviscid, the dispersion relation there is approximately

$$\left(\lambda k^2 - i\omega\right)\left(a^2 k^2 - i\omega^2\right) = i\omega b^2 k^2.$$

Suppose that $k^2\lambda/\omega \to 0$. The dispersion relation then reduces to $k^2 = \omega^2/c_f^2$, representing the wave propagation proposed above. The criterion justifying this infinite-conductivity result is found by eliminating k^2, so that $\lambda\omega/c_f^2 \to 0$, i.e., the inverse of the magnetic Reynolds number based on frequency and wave speed tends to zero. With large, but finite magnetic Reynolds number, the dispersion relation would represent damped magnetoacoustic waves. However the full dispersion relation is quadratic in k^2 and the second root has been lost. Suppose the lost root does refer to a thin layer at the wall (large k) and take $\omega^2/a^2 k^2 \to 0$ to yield $k^2 = (i\omega/\lambda)(c_f^2/a^2)$, meaning that the length scale of the solution is the normal skin depth $\sqrt{\lambda/\omega}$ modified by the factor (a/c_f). The modification arises because compressibility in the skin causes movement and in contrast to the effect in solids $\mathbf{v} \times \mathbf{B}$ is non-zero (without which waves would not be generated). Another interesting feature of the skin-effect limit is that the neglect of the ω^2 term implies negligible inertia, the force balance in the skin being between pressure and magnetic pressure.

Using the boundary condition that the two solutions combine at the surrounding cylindrical wall in order to give zero normal velocity there, it is easily shown that the perturbation magnitude of the magnetic field is made up of contributions with ratio $\sqrt{\lambda\omega/a^2}(b/c_f)^2$ between the skin-effect and wave-like solutions. The implication is that, if the value of the magnetic Reynolds number is very high, most of the perturbations would be associated with the skin. On the other hand too low a magnetic Reynolds number, and the waves will be heavily damped. Typical conditions in Mikolajczak's experiments gave a value order one for $\sqrt{\lambda\omega/a^2}(b/c_f)^2$, so that it was more a case of modified skin effect than a real demonstration of magnetoacoustic waves. Nevertheless calculations taking account of the axial symmetry, but not of the finite length of plasma sample or the decay of ionization due to radiation agreed reasonably well with measurements.

5 Return to liquid metals

October 1962 heralded a brief golden age for what had now become the magnetohydrodynamics, rather than shock-tube, group. Three new research students joined, C.J.N. Alty, J.A. Baylis, and C.J. Stephenson, all of whom were Cambridge graduates. In the outside world magnetohydrodynamics was becoming more popular as a subject of research, although there were some unfortunate attempts to liven up moribund topics in fluid mechanics by adding an arbitrary magnetic field. This was not always the case and other papers were truly enlightening. I can remember the excitement in the group over the paper by Hasimoto [40] on flow parallel to the axis of an infinite cylinder in the presence of a transverse magnetic field. This was the first intimation that *free* shear layers parallel to the magnetic field have an important role to play in external flows.

Shercliff realized that Hasimoto's work also had implications for flow in square ducts (and other cross sections) when two opposite walls are insulating and the other two highly conducting, while a uniform magnetic field is oriented transversely to the flow but *at an angle to the walls*. He suggested to Alty that it would be interesting to undertake an experiment on such a configuration and Alty set about building a mercury flow circuit in which the test section was a square channel of side 12.7 mm machined out of a high-conductivity copper rod and of length 254 mm. The bottom of the channel was insulated by a strip of sellotape and an insulated cover formed the opposing wall. The test section was mounted in the field of a magnet capable of providing up to 1.27 T. The flow was driven through the test section by a longitudinal pressure gradient acting against electromagnetic forces induced by current crossing between the channel sides with exposed copper. The channel construction provided a short-circuit path between those sides.

Being a linear problem, use of super-position is helpful when determining how the electrical current flows. It is easier to visualize what is happening with a flow being driven electrically along the channel, there being no longitudinal pressure gradient, i.e., the electrodes are now at appropriately different potentials. The difference between electrical drive and pressure-gradient drive is just the addition of a uniform current flow with the consequent $\mathbf{j} \times \mathbf{B}$ balanced by a uniform pressure gradient. The key then to predicting asymptotic behaviour in the channel at high Hartmann number is the well-known fact that in the inviscid (core) regions current density does not vary along field lines. With electrical drive $\mathbf{j} \times \mathbf{B}$ must in fact be zero in such regions, so that the direction of $\mathbf{j}$ can only be parallel to $\mathbf{B}$. Three cases may be envisaged for a field line crossing the channel at an angle in an inviscid region: (i) the field line intersects two insulating walls, (ii) it intersects one insulating wall and one conducting wall, (iii) it intersects two conducting walls.

In case (i) current entering a Hartmann layer on an insulating wall from a core region would induce vorticity normal to the wall, but at the other Hartmann layer vorticity is of opposite sign, so that there is a contradiction and it is to be concluded that there is no current in the region. In a similar fashion for case (ii) it is found that current is zero and also that the stationary conducting wall holds the velocity at zero. Case (iii) has a voltage-gradient component along the field line, but the overall voltage gradient must be normal to the conducting wall. Since $\mathbf{j}$ is parallel to $\mathbf{B}$, a vector diagram of $\mathbf{j}/\sigma$, $\mathbf{E}$ and $\mathbf{v} \times \mathbf{B}$ shows that Ohm's law requires $\mathbf{v} \times \mathbf{B}$ to be non-zero.

Denote by θ the angle of tilt of the channel away from the position where the magnetic field is normal to the insulating walls. For $0 < \theta < \pi/4$, the situation is made up of two case (ii) regions separated from a case (i) region by parallel shear layers (see Fig. 3c). The Hartmann layers on the insulating walls where region (i) field lines meet the walls are active, so that the velocity in the region is given by the standard Hartmann layer result $J/\sqrt{\rho\nu\sigma}$, where J is half the current supplied by the electrical drive per unit length of channel.

When the tilt is such that $\pi/4 < \theta < \pi/2$ (see Fig. 3d), there are again two case (ii) regions, but they are now separated from a case (iii) region by free layers of the type described by Moffatt [41], which carry a change of electric field tangential to the layer or, in other words, varying potential difference across it. That potential difference is balanced by the emf which the flow in the layer induces and a jet is formed there, which contributes a total flow of the same order as that in the case (iii) region.

Although the only measurements taken were the pressure difference down the channel, the total flow rate and potential distributions at the insulated walls, the results are remarkably close to the asymptotic theory for high Hartmann number [42]. In addition, before the results were published, Hunt [43] and Hunt and Stewartson [44] had developed analytical solutions for channel flows corresponding to $\theta = \pi/2$ and $\theta = 0$ and Alty was able to make comparisons between their solutions and his experiments. At a much later date I incurred a debt to this work of Shercliff and Alty for its influence on my choosing a tilted container for a model in which to investigate principles of buoyancy-driven MHD flow [45].

A particular preoccupation in the outside world during the 1960s was production of electrical power directly from flow of combustion gases in a MHD duct. The aim was to gain a thermodynamic advantage by deriving this power at a much higher temperature than would be possible in a gas turbine, say, because with the latter there is a need limit the temperature of highly stressed blading. To turn the combustion gases into a moderately good conductor, it was proposed that they be seeded with small amounts of potassium or caesium and one of the difficulties of the whole scheme was going to be the recovery of the seed material before the gases were exhausted to the atmosphere. (Shercliff's joke was that MHD power generation with potassium seeding should be linked to tomato-growing schemes down wind of the stack.) The group at Cambridge were skeptical about the viability of MHD power generation, but gaining experience with seeded gas was given consideration. The plan was to construct a torus of similar dimensions to the one used by Jameson and mount it in his magnet (Jameson had completed his Ph.D. work by this stage). Radial current would be passed from an outer conducting wall to an inner one so as to heat a gas-potassium mixture within the torus and interaction of the current with the applied axial magnetic field would drive the gas azimuthally. It was realized that the Hartmann number would not be large, so that resistance to motion would probably involve secondary-flow effects. Although the project did not come to fruition, it left an interest in secondary flows, which was picked up by Baylis.

Baylis' subject of research was approved all inclusively as "Magnetohydrodynamics" (the same had been true of Alty) but his thesis title "Fully developed secondary flow in magnetohydrodynamics" is a more reasonable indication of his work. He used the solenoid which had been part of Decker's apparatus and mounted mercury-filled toruses of rectangular cross section and of various aspect ratios, coaxially with the solenoid. Flow was driven azimuthally by radial current.

At Shercliff's suggestion Baylis' first experiment was conducted in a torus formed by concentric copper cylinders with a narrow gap between them. The criterion for stability of a pressure-driven Couette flow in the gap has been given by Chandrasekhar [46], although full axial symmetry with such a drive is clearly not possible. On the other hand, if conditions are such that $j_r = C/r$, where C is a constant, drive by electromagnetic force is precisely equivalent to a continuous pressure gradient around the gap. Detection of when the flow becomes unstable, i.e., forms Taylor vortices, is achieved from overall voltage-current characteristics, which are linear between zero current and the value at instability. The break-away point from the original line of the characteristic occurs where the flow resistance begins to increase more strongly as current increases, and it is easily located.

The gap between the cylinders in Baylis' experiment [47] was just over 1 mm, the radius R_1 of the inner cylinder was 26 mm and the length 50 mm. Values of the magnetic field strength ranged from 0.1 T to 0.4 T, corresponding to Hartmann numbers based on gap width from 3 to slightly more than 12. Experimental results for the critical Dean number, $Re_{\text{crit}}\sqrt{d/R_1}$, compared well with the predictions of Chandrasekhar [46], e.g., 35.94 as $Ha \to 0$ (the pressure-driven case), 80 for $Ha = 10$. The results tended to be slightly above the predicted values and this was considered most likely due to uncertainty over the gap width when amalgam layers were building up on the copper.

For the next series of experiments Baylis constructed a toroidal channel of square cross section. Although the initial aim had been to investigate secondary-flow effects and their interaction with the MHD [48], it turned out that the apparatus provided an excellent test bed [49] for the asymptotic ($Ha \to \infty$) theory of channel flow due to Hunt and Stewartson [44], mentioned above in the context of Alty's work.

Baylis' final series of experiments was conducted in a disk-like torus. It had been noticed that the axially symmetric boundary-layer equations for flow over stationary disks appeared to admit a family of similarity solutions with an azimuthal primary flow and a radial/axial secondary flow. The similarity solutions would only be valid for $Ha \to 0$, but they were expected to be qualitatively like that of Bödewadt [50], which is the member of the family having solid-body rotation outside the boundary layer. Other members of the family may be characterized by the index m in $v_\theta \propto r^m$, where v_θ is the azimuthal velocity. For driven flow between two stationary disks continuity is satisfied when the radial inflow in the boundary layers balances radially outward core flow with velocity $v_r \propto r^{(1+m)/2}$. The Coriolis term in the equation of motion for the core must then have a dependence on radius to the power $(1 + 3m)/2$. To be consistent with a $\mathbf{j} \times \mathbf{B}$ force varying inversely with radius, m should be -1. For the Bödewadt solution velocities oscillate as the edge of the layer is approached. Unfortunately the oscillations become more violent as the power is decreased, until the solution collapses just at $m = -1$. Nevertheless, Baylis' thesis contains a wide range of experimental data for secondary-flow situations. Calculations of inflow in the family of boundary layers at various powers were confirmed shortly afterwards by the work of a group in California [51].

A more successful foray into secondary flow in boundary layers on non-conducting disks was made by Stephenson, who started experiments on mercury flow between one stationary and one rotating disk [52]. The apparatus was designed to fit into the gap in Jameson's magnet giving a uniform magnetic field parallel to the axis of the disks and this limited the diameter of the disks to 254 mm. The gap between them was set at either 12.7 mm or 25.4 and the stationary disk was fitted with a shroud at its edge. Platinum voltage probes were mounted in a radial line at the surface of the stationary disk so that the average azimuthal velocity could be deduced from the radial gradient of voltage.

The addition of a uniform axial magnetic field does not affect the fact that similarity solutions exist for flows with radial dependence of azimuthal velocity to the power 1, such as flow over a spinning disk, and this is true for the full Navier–Stokes equations, not merely the boundary-layer equations. There is, however, the constraint on the electromagnetic aspects that the magnetic Reynolds number be much less than unity. The ordinary hydrodynamic problem has received much attention over the years, because it represents an exact solution to the Navier–Stokes equations, although there has been controversy over whether in practice it works for flow between disks of *finite* diameter. In the MHD case the layers on the disks can range from pure Hartmann type to pure secondary flow. Matching the radial current flow inwards on the stationary disk to the flow outwards on the spinning disk when electromagnetic effects are dominant leads to the angular velocity of the fluid half way between the disks being half the angular velocity of the spinning disk. Matching the fluid inflow to the outflow when secondary-flow effects are dominant leads to the angular velocity between secondary flow layers being 0.31 times that of the spinning disk. Stephenson's measurements of voltage confirmed these predictions and reproduced the expected variation in angular velocity ratio between the two extremes of high Hartmann number and high secondary-flow effect. The ranges covered by Stephenson were $100 < \Omega d^2/\nu < 800$, where Ω is the angular velocity of the spinning disk and $0 < Ha^2 < 800$.

Two more Cambridge graduates were recruited in October 1963, R.C. Baker and J.C.R. Hunt. One of the proposals for MHD generators had been that a striated fluid might be made to flow through the duct. The striations would arise from the injection of a conducting fluid and they would act as "pistons" against which the non-conducting medium would do work, but Lemaire [53] had pointed out that the interface between conductor and non-conductor would be subject to Rayleigh–Taylor instability. Shercliff suggested to Baker that he investigate the stability of mercury partly filling a trough and set in a horizontal magnetic field. A horizontal electric current would interact with the magnetic field so as to give an upward vertical force and it was expected that with unstable conditions 2D waves would grow exponentially with time constant τ given by

$$\tau^{-2} = \frac{2}{3\sqrt{3T}} \frac{(|\mathbf{j} \times \mathbf{B}| - g\rho)^{\frac{3}{2}}}{\rho},$$

where T is the surface tension. The two-dimensionality refers to the wave crests running parallel to the current and it was observed that just before complete breakdown of the surface a crest would rise slightly, stretching from electrode to electrode. Although there was considerable scatter in a plot of τ^{-1} against $|\mathbf{j} \times \mathbf{B}|$, the results were nevertheless convincing [54]. Later, when Shercliff had left Cambridge for Warwick University, he set a research student there, I.R. Robinson, onto a more comprehensive investigation of surface waves [55].

After completion of the stability experiment, Baker decided to work on the sensitivity and optimization of the type of electromagnetic flowmeter which had been patented in 1917 by Smith and Slepian for application as a ship's log. The meter is also known as a wall velometer and works by having a magnetic field emanating from the wall, field coils lying behind the surface. The voltage difference is measured between two flush-mounted electrodes. The response of the meter is likely to be influenced most by the flow nearest the wall and the influence of flow further away becomes progressively weaker with distance from the wall. Such a sensitivity distribution is not what is wanted in the face of the boundary layer on the wall surface, and, in marine applications, of variability caused by marine growth.

Shercliff [13] had derived expressions for induced voltage on the electrodes in two cases, a uniform velocity and uniform velocity gradient. Baker made a remarkable advance by showing how the response would depend on velocity V and its distribution when it varied as a function of distance x from the wall. Baker's formula for the overall potential field is

$$\int_0^\infty V\left(\xi\right) B_y\left(2\xi + x, y, z\right) d\xi + \int_x^\infty V\left(\xi\right) B_y\left(2\xi - x, y, z\right) d\xi,$$

where the x-axis is directed normal to the wall and the z-axis is directed parallel to the flow. I always found the appearance of the factor 2 multiplying ξ intriguing, but there is a physical explanation and it appears in the published derivation [56]. Further generalizations of flowmeter theory were to be made by M.K. Bevir at Warwick University later [57]. Baker used a water channel to test the response of a wall velometer with magnetic field designed in the light of his formula. Subsequently he conducted practical tests with a wall velometer mounted in the flat bottom of a punt!

Although Hunt was registered as a Ph.D. student of Cambridge University, he spent only the first year of research (subject "Magnetohydrodynamics") at Cambridge. He was being supervised by Shercliff and it was at the end of Hunt's first year that Shercliff left to become head of the engineering department at the new university of Warwick. Hunt followed, being given leave to work away from Cambridge.

Hunt was supported as a research student by the Central Electricity Generating Board. As this was a time of interest in MHD power generation, it was natural that he should become involved with channel-flow problems, albeit at higher Hartmann number than seemed likely in the power generation context. During that first year of research, while still at Cambridge, he found that he could extend the analysis used by Shercliff for rectangular-section channels with insulating walls all round to the problem of a channel with conducting walls perpendicular to a uniform magnetic field and insulating walls parallel to the field. Alternatively the analysis would work for perfectly conducting walls perpendicular to the field and imperfectly conducting walls parallel [43]. What at that stage eluded Hunt was a solution of the practically important case for pumps and generators of insulating walls perpendicular to the magnetic field and conducting walls parallel. However that was to wait only a short time until it yielded to a collaboration between Hunt and Stewartson [44]. As is well known in MHD circles Hunt then went on to a highly productive period with other collaborations in a wide range of duct flows and electrically driven flows.

6 The end of an era

The departure of Shercliff in October 1964 marked the end of the early years of MHD in the Engineering Department. Alty and Baylis stayed on for a time to complete their Ph.D. work. Baker remained and was awarded a Research Fellowship at St John's College. During the academic year 1966/1967 I took Sabbatical leave and spent it at MIT, where I picked up an interest in electric arcs, initially in the context of heaters for high-stagnation-enthalpy flows, but developing into the context of high-current switch gear. Baker joined in and experimental work on arcs in gas flow was initiated.

Hunt returned to Cambridge, having been awarded a Research Fellowship at Trinity College and reinvigorated activity in the field of liquid-metal magnetohydrodynamics, a well-known student of his being R.J. Holroyd, who undertook experiments on flow in a duct through a strong, but spatially varying magnetic field. The practical interest in this was the power requirements for circulating lithium in fusion-reactor blankets. Another student of Hunt was P.A. Davidson, who worked on electromagnetic stirring of liquid metals in continuous casting processes. After a period away from Cambridge Davidson returned to a University Lectureship in the Engineering Department in 1994 and resurrected the MHD group up with the aid of T. Alboussière.

A tragedy for MHD and cause of sadness to those who knew Arthur Shercliff personally was that, having returned to Cambridge in 1980 and begun to foster his latest enthusiasm, thermoelectric MHD, he died from cancer in 1983 at the early age of 56. We can still obtain some flavour of his personality because we are fortunate in having a film record [58] of him demonstrating some classic MHD experiments at the time when he was a young firebrand at Cambridge (see Fig. 4).

Fig. 4. Arthur Shercliff illustrating the action of a magnetic field on vorticity at low-magnetic Reynolds number by spinning a brass loop in a field. (Still taken from the educational film "Magnetohydrodynamics" [58].)

References

1. Murgatroyd W (1953) Experiments in magnetohydrodynamic channel flow. Phil Mag 44:1348–1354
2. Hartmann J, Lazarus F (1937) Experimental investigations on the flow of mercury in a homogeneous magnetic field. K Dan Vidensk Selsk Mat–Fys Medd 15(7):1–45
3. Lock RC (1955) The stability of the flow of an electrically conducting fluid between parallel planes under a transverse magnetic field. Proc R Soc Lond Ser A 233:105–125
4. Lykoudis PS (1960) Transition from laminar to turbulent flow in magneto-fluid mechanic channels. Rev Mod Phys 32:796–798
5. Branover GG (1967) Resistance of magnetohydrodynamic channels. Magnetohydrodynamics 3:1–11
6. Lingwood RJ, Alboussière T (1999) On the stability of the Hartmann layer. Phys Fluids 11:2058–2068
7. Alboussière T, Lingwood RJ (2000) A model for the turbulent Hartmann layer. Phys Fluids, 12:1535–1543
8. Shercliff JA (1953) Steady motion of conducting fluids in pipes under transverse magnetic fields. Proc Cam Phil Soc 49:136–144
9. Gold RR (1962) Magnetohydrodynamic pipe flow, Pt I. J Fluid Mech 13:505–512

10. Shercliff JA (1954) Relation between the velocity profile and the sensitivity of electromagnetic flowmeters. J Appl Phys 25:817–818
11. Shercliff JA (1955) Experiments on the dependence of sensitivity on velocity profile in electromagnetic flowmeters. J Sci Instrum 32:441–412
12. Shercliff JA (1956) Edge effects in electromagnetic flowmeters. J Nucl Energy 3:305–311
13. Shercliff JA. (1962) The theory of electromagnetic flow measurement. Cambridge University Press, Cambridge
14. Hendry J, Lawson JD (1993) Fusion research in the UK 1945–1960. AEA Technology Report AHO 1
15. Dolder K, Hide R (1960) Experiments on the passage of a shock wave through a magnetic field. Rev Mod Phys 32:770–779
16. Patrick RM, Brogan TR (1959) One-dimensional flow of an ionized gas through a magnetic field. J Fluid Mech 5:289–309
17. Cowley MD (1961) On some kinematic problems in magnetohydrodynamics. Quart J Mech Appl Math 14:319–333
18. Cowley MD (1961) The distortion of a magnetic field by flow in a shock tube. J Fluid Mech 11:567–576
19. Shercliff JA (1958) Some generalizations in steady one-dimensional gas dynamics. J Fluid Mech 3:645–657
20. Weyl H (1949) Shock waves in arbitrary fluids. Comm Pure Appl Math 2:103–122
21. Cowley MD (1960) A magnetogasdynamic analogy. ARS J 30:271–272
22. Shercliff JA (1960) One-dimensional magnetogasdynamics in oblique fields. J Fluid Mech 9:481–505
23. Shercliff JA (1960) Some generalizations in one-dimensional magnetogasdynamics. Rev Mod Phys 32:980–986
24. Imai I (1960) On flows of conducting fluids past bodies. Rev Mod Phys 32:992–999
25. Sears WR (1959) Magnetohydrodynamic effects in aerodynamic flows. ARS J 20:397–406
26. Germain P (1960) Shock waves and shock-wave structure in magneto-fluid dynamics. Rev Mod Phys 32:951–958
27. Akhiezer AI, Liubarski GI, Polovin RV (1959) The stability of shock waves in magnetohydrodynamics. Soviet Phys JETP 8:507
28. Anderson JE (1963) Magnetohydrodynamic shock waves. MIT press, Cambridge, MA
29. Todd L (1964) Evolution of the trans-Alfvénic normal shock in a gas of finite electrical conductivity. J Fluid Mech 18:321–336
30. Todd L (1965) The evolution of trans-Alfvénic shocks in gases of finite electrical conductivity. J Fluid Mech 21:193–209
31. Todd L (1966) The evolution of switch-on and switch-off shocks in a gas of finite electrical conductivity. J Fluid Mech 24:597–608
32. Cowley MD (1963) On the plane flow of gas with finite electrical conductivity in a strong magnetic field. J Fluid Mech 15:577–596
33. Cowley MD (1967) Gas-ionizing shocks in a magnetic field. J Plasma Phys 1:37–54
34. Cowley MD, Horlock JH (1994) On one-dimensional flow of a conducting gas between electrodes - with application to MHD thrusters. J Fluid Mech 266:147–173

35. Shercliff JA (1965) A textbook of magnetohydrodynamics. Pergamon Press, Oxford

36. Jameson A (1964) A demonstration of Alfvén waves, Part I. Generation of standing waves. J Fluid Mech 19:513–527

37. Shercliff JA (1976) Technological Alfvén waves. Proc Instn Elect Engrs, 123:1035–1042

38. Decker JA (1964) Ion wave resonance and plasma instability in a caesium discharge. J Appl Phys 35:497–501

39. Grad H (1960) Reducible problems in magneto-fluid dynamic steady flows. Rev Mod Phys 32:830–847

40. Hasimoto H (1960) Steady longitudinal motion of a cylinder in a conducting fluid. J Fluid Mech 8:61–81

41. Moffatt HK (1964) Electrically driven steady flows in magneto-hydrodynamics. In: Gortler H (ed) Proc. XIth Int Cong Appl Mech Munich, Springer-Berlin, pp: 946–953

42. Alty CJN (1971) Magnetohydrodynamic duct flow in a uniform transverse magnetic field of arbitrary orientation. J Fluid Mech 48:429–461

43. Hunt JCR (1965) Magnetohydrodynamic flow in rectangular ducts. J Fluid Mech 21:577–590

44. Hunt JCR, Stewartson K (1965) Magnetohydrodynamic flow in rectangular ducts. Part 2. J Fluid Mech 23:563–581

45. Cowley MD (1996) Natural convection in rectangular enclosures of arbitrary orientation with magnetic field vertical. Magnetohydrodynamics 32:390–398

46. Chandrasekhar S. (1961) Hydrodynamic and hydromagnetic stability. Clarendon Press, Oxford

47. Baylis JA (1964) Detection of the onset of instability in cylindrical magnetohydrodynamic flow. Nature 204:563

48. Baylis JA (1971) Experiments on laminar flow in curved channels of square section. J Fluid Mech 48:417–422

49. Baylis JA, Hunt JCR (1971) MHD flow in an annular channel; theory and experiment. J Fluid Mech 48:423–428

50. Bödewadt UT (1940) Die Drehstromung überfestem Grunde. Z Ang Math Mech 20:241

51. King WS, Lewellen WS (1964) Boundary-layer similarity solutions for rotating flows with and without magnetic interaction. Phys Fluids 7:1674–1680

52. Stephenson CJ (1969) Magnetohydrodynamic flow between rotating co-axial disks. J Fluid Mech 38:335–352

53. Lemaire A (1962) Centre d'etudes Nucleaires de Saclay. Rapp. IFP/7713 CEA/PA.IGn/RT.150

54. Baker RC (1965) Maximum growth rate of Rayleigh-Taylor instabilities due to an electromagnetic force. Nature 207:65–66

55. Robinson IS (1975) A novel form of the MHD Rayleigh-Taylor instability. J Fluid Mech 72:135–143

56. Baker RC (1968) On the potential distribution resulting from flow across a magnetic field projecting from a plane wall. J Fluid Mech 33:73–86

57. Bevir MK, Shercliff JA (1968) Theory of electromagnetic flowmeters with non-uniform fields. 12th IUTAM Congress of Applied Mechanics, Stanford, Cambridge

58. Shercliff JA (1965) Magnetohydrodynamics (a 30 min. educational film), Educational Services Inc. for the National Committee on Fluid Mechanic Films, USA

Julius Hartmann and His Followers: A Review on the Properties of the Hartmann Layer

René Moreau[1] and Sergei Molokov[2]

[1] Laboratoire EPM, ENSHMG, BP 95, 38402 St Martin d'Hères Cedex, France
(Rene.Moreau@hmg.inpg.fr)
[2] AMRC, Department of Mathematical Sciences, Coventry University, Priory
Street, Coventry CV1 5FB, United Kingdom (s.molokov@coventry.ac.uk)

1 Historical introduction

Julius Hartmann, born in 1881 (November 11th) and deceased in 1951 (November 6th), was a leading Professor at the Technical University of Denmark, in Copenhagen, where he founded the *Laboratoriet for teknisk fysik*, which was the basis for today's Department of Applied Physics. In this laboratory, he worked on different technical processes, inventing in particular the device now called the electromagnetic conduction pump to drive the flow of electrically conducting liquids, such as molten metals. He may be the first scientist using mercury in a hydraulic circuit, and applying a magnetic field and a DC current in two orthogonal directions, both perpendicular to the duct axis, to generate an electromagnetic force capable to drive a fluid flow against friction. In the archives of the Copenhagen Technical University, there are still reminiscences of this pump, whose construction dates back to probably 1915–1917. But it is now extremely difficult to get published papers related to this pioneering work (Moerch [1]).

Julius Hartmann is well known within the MHD community for his discovery in 1937 [2] of the now well-known distributions of the velocity and current density in the fully established flow of an electrically conducting fluid between two parallel solid walls, both being electrically insulating, in the presence of a uniform magnetic field applied from outside in the direction perpendicular to the walls. Those distributions exhibit exponential functions, often expressed in terms of $\cosh(zB\sqrt{\sigma/\rho\nu})$, because of the duct symmetry (here B denotes the magnetic field intensity, σ the fluid electrical conductivity, ρ its density, ν its kinematic viscosity, and z stands for the coordinate in the magnetic field direction). Those expressions show that, when the magnetic field is high enough, there exist a core flow with a uniform velocity profile, between two boundary layers (now called the *Hartmann layers*), where all the velocity variations takes place. The layer's thickness is then $\delta = \frac{1}{B}\sqrt{\frac{\rho\nu}{\sigma}}$, and

S. Molokov et al. (eds.), Magnetohydrodynamics – Historical Evolution and Trends,
155–170. © 2007 *Springer.*

the ratio between the duct's half-width h and δ is now called the *Hartmann number*,

$$Ha = \frac{h}{\delta} = Bh\sqrt{\frac{\sigma}{\rho\nu}}.$$

Hartmann also understood that the induced current loop is confined within the cross section. This implied that the total current is zero, and that the total current passing in the core is exactly opposed to that within the layers, however thin they are. His results demonstrate that these layers, having a thickness inversely proportional to the applied magnetic field, may be as thin as one desires, depending on the strength of the magnetic field.

Hartmann's aim in his 1937 theoretical paper was, indeed, to understand the origin of the rather poor efficiency of the conduction pump. In this paper, he first derives a simple analytical solution of the basic equations of motion, in which the Lorentz force $\mathbf{j} \times \mathbf{B}$ is introduced. Then he obtains an expression for the local shear stress at the wall (now walls transverse to the field are called the *Hartmann walls*), and for the pressure gradient necessary to maintain a given flow rate against this friction. Then, he discusses the origin of the strong head losses that they had measured with Lazarus [3]. As a consequence, his paper is not limited to the theoretical investigation of the fully established regime (now called the *Hartmann flow*), since he qualitatively describes the mechanisms of the important pressure variations, which appear both at the entrance to and at the exit from the region of a uniform magnetic field.

In this review, we discuss the Hartmann layer properties in a more general context, since we escape from the fully established regime and we consider core flows carrying inertia and vorticity (possibly turbulence). In § 2, we start with the Hartmann layers at plane solid walls, focusing on the particular case of an electrically insulating wall for the sake of simplicity. In § 3, we discuss particular properties of the Hartmann layer when it develops in the vicinity of a free surface, attempting to underline the differences between the two cases. Finally, in § 4, we try to conclude this paper with remarks on the quite singular character of the Hartmann layer. Each paragraph aims at illustrating the remarkably large number of important and specific properties of that layer. For each of them, it has been our intention to select and to quote the first paper in which it is theoretically established or experimentally demonstrated, but not to present an exhaustive list of the published papers where it is discussed or applied to investigate a particular flow.

2 The Hartmann layer at a plane solid wall

Let us consider a fluid flow in the vicinity of an insulating plane wall at $z = 0$, whose velocity scale is U_0, in the presence of a uniform magnetic field $\mathbf{B}$, which is transverse to two parallel electrically insulating walls. Outside the Hartmann layer, where viscosity cannot play any role, we assume that inertia

Fig. 1. Julius Hartmann

is negligible in comparison with the Lorentz force (later on the validity of this assumption will be specified). Then, combining the expressions for the electric potential φ_0 derived from the equation of conservation of the electric charge and Ohm's law, one gets:[1]

$$\frac{\partial^2 \varphi_0}{\partial z^2} + \Delta_\perp \varphi_0 = \mathbf{B} \cdot \boldsymbol{\omega}_0, \tag{1}$$

where $\boldsymbol{\omega}_0 = \nabla_\perp \times \mathbf{u}_{\perp 0}$ stands for vorticity in the core, $\nabla_\perp = (\partial_x, \partial_y, 0)$, and $\Delta_\perp = \nabla_\perp^2$.

Using Navier–Stokes equation yields:

$$\Delta_\perp \varphi_0 = \mathbf{B} \cdot \boldsymbol{\omega}_0. \tag{2}$$

This follows from the strong property that, in the core flow, most quantities, such as $\mathbf{u}_{\perp 0}$ and φ_0, are asymptotically independent of the z-coordinate. This result, derived by Ludford [4] (see also [5,6]), may be seen as analogous to the Proudman–Taylor theorem for rotating fluids. It implies that any structure in the core must have the form of a column parallel to the magnetic field, or that all planes perpendicular to the magnetic field are strongly correlated.

---

[1] In the following, the zero index systematically refers to quantities in the core and the $()_\perp$ index for vector quantities refers to their components in the plane (x, y).

But, within the Hartmann layer, the equation for the potential is:

$$\frac{\partial^2 \varphi}{\partial z^2} + \Delta_\perp \varphi = \mathbf{B} \cdot (\nabla_\perp \times \mathbf{u}_\perp). \tag{3}$$

Then, assuming that $\Delta_\perp \varphi$ and $\Delta_\perp \varphi_0$ are close to each other and combining Eqs. (2) and (3) yields new expression

$$\frac{\partial^2 \varphi}{\partial z^2} = \mathbf{B} \cdot [\nabla_\perp \times (\mathbf{u}_\perp - \mathbf{u}_{\perp 0})], \tag{4}$$

which, after two integrations, confirms that the variation of the electric potential across the Hartmann layer is of the order of δ^2 and becomes negligible when δ is very small. This allows to simplify the equation of motion within the Hartmann layer as follows:

$$\frac{\rho \nu}{\sigma B^2} \frac{\partial^2 \mathbf{u}_\perp}{\partial z^2} - \mathbf{u}_\perp = -\mathbf{u}_{\perp 0}, \tag{5}$$

and to get the exponential velocity distribution

$$\mathbf{u}_\perp = \mathbf{u}_{\perp 0} \left[1 - \exp(-Haz/h)\right] = \mathbf{u}_{\perp 0} \left[1 - \exp(-z/\delta)\right]. \tag{6}$$

This expression is a minor generalisation to the case of non-established flows, where $\mathbf{u}_{\perp 0}$ is non-uniform, of the equivalent relations obtained by Hartmann [2] and Shercliff [5], for fully established flows. It introduces the Hartmann number defined above, but one could notice that the thickness δ is independent of the duct half-width h.

Since the electric potential and the velocity are known within the Hartmann layer, the expression for the current density immediately follows from Ohm's law. It also has an exponential form similar to Eq. (6), but its main feature is the integral quantity

$$\mathbf{J}_\perp = \int_H (\mathbf{j}_\perp - \mathbf{j}_{\perp 0}) dz = \sqrt{\sigma \rho \nu} (\mathbf{u}_{\perp 0} \times \mathbf{e_B}), \tag{7}$$

where integration is performed over the Hartmann layer thickness.

A somewhat analogous relation deserves to be mentioned. It is the expression

$$(j_z)_0 = \sqrt{\sigma \rho \nu} \omega_0 \tag{8}$$

valid at the edge of the Hartmann layer. Equation (7) means that one can consider the layer as an electric current sheet carrying a current $\mathbf{J}_\perp$ proportional and perpendicular to the local core velocity $\mathbf{u}_{\perp 0}$. And expression (8) explains how this sheet may be fed from the electric current passing in the neighbouring core. These expressions, discovered almost at the same time by Hunt and Ludford [7] and by Kulikovskii [8], are of particular interest for non-established flows where $\mathbf{u}_{\perp 0}$ is non-uniform (see Hunt and Shercliff [9]). For instance, they imply that each quasi-two-dimensional (quasi-2D) vortical structure present or moving in the core flow carries an electric circuit, which must loop through the Hartmann layer. They are also of prime importance to match local solutions in the core, in the Shercliff layers[2] present along the walls parallel to the magnetic field, or in free shear layers [6, 9–11]). Remarkably, the local value of the magnetic field does not appear in Eq. (7) (the only information on **B** which is relevant is its orientation $\mathbf{e_B}$); this implies that relation (7) is also valid in the presence of a non-uniform magnetic field, as at the entry into or at the exit from the region where the magnetic field is uniform.

From the above ideas follows that the current density within the core flow is Ha times smaller than in the Hartmann layer, where it is of the same order of magnitude as the other two terms of Ohm's law: σBU_0. This has two strong implications. First, it implies that the relevant non-dimensional number to estimate the ratio between the Lorentz force and inertia in the core is not the interaction parameter $N = \frac{Ha^2}{Re}$, but rather the number $\frac{Ha}{Re}$, where Re is the Reynolds number. This has been observed in many experiments, starting from Murgatroyd [12], and in all the experiments with insulating ducts performed in Riga and Purdue in the 1960s. It was observed that the friction law in duct flows with a transverse magnetic field was exactly the same as in a laminar regime, even when turbulence was still present but was quasi-2D (see the review paper by Lielausis [13], as well as Branover [14]). Second, it implies that the Joule dissipation is essentially located within the Hartmann layer, just like the viscous dissipation. According to Eq. (4), which implies an exact balance between viscous friction and electromagnetic force, those two dissipations are equal to each other. This means that the total damping supported by any quasi-2D vortical structure, is located within the Hartmann layers present at its two ends. The relevant form of the kinetic energy theorem applied to a fluid domain of such a structure, at the leading order (the terms disregarded are Ha times smaller), is then

$$\frac{\partial}{\partial t} \int_{-h}^{+h} \frac{\rho u^2}{2} dz \approx - \int_{-h}^{+h} \frac{j^2}{\sigma} dz. \tag{9}$$

Clearly the predominant contribution to the integral on the left-hand side comes from the core, whereas the predominant contribution to the other integral comes from the Hartmann layers. Then, it is straightforward to derive an estimate of these terms, which yields the following expression for the Hartmann damping time τ_H:

$$\frac{\rho U_0^2 h}{\tau_H} \approx \sigma B^2 U_0^2 \delta. \tag{10}$$

---

[2] The authors suggest that these boundary layers present along the walls parallel to the magnetic field, which were named "side layers", be now named "Shercliff layers", since they were first analysed by Shercliff [5].

This demonstrates that the damping of this quasi-2D structure (an eddy, for instance, turbulent or not, or any other flow structure with a non-zero vorticity, like a free-shear layer) is much weaker than the usual Joule damping, whose time scale is

$$\tau_J = \frac{\rho}{\sigma B^2}.$$

It is given by the following expression (Sommeria and Moreau [15]) when the Hartmann walls are insulating:

$$\tau_H = \frac{h}{B}\sqrt{\frac{\rho}{\sigma\nu}} = Ha\tau_J. \tag{11}$$

On the basis of the above ideas, which yield specific properties for the core flow and for the two Hartmann layers, Sommeria and Moreau [15] introduced the z-averaged velocity

$$\mathbf{v}_\perp = \frac{1}{2h}\int_{-h}^{+h}\mathbf{u}_\perp dz, \tag{12}$$

which is very close to the core velocity when Ha is very large, and derived the following equation for the quasi-2D flow:

$$\frac{d\mathbf{v}_\perp}{dt} = -\frac{1}{\rho}\nabla_\perp p_0 + \nu\nabla_\perp^2\mathbf{v}_\perp - \frac{\mathbf{v}_\perp}{\tau_H}, \tag{13}$$

where viscous friction and Lorentz force present in the Hartmann layer are both included in and modelled by the last linear damping term. The predictions derived from this equation have been investigated by Messadek and Moreau [16] in the MATUR experiment. The agreement is fair if the magnetic field is large enough to make inertia negligible within the Hartmann layer. This requires that $N \gg 1$, but in such conditions inertia may remain non-negligible within the core flow.

Therefore, for flows at moderate Hartmann number (say in the range 10–200), where the concept of a Hartmann layer separated from the core flow is well justified, the question of introducing inertial effects in the Hartmann layer theory is quite relevant. This question has recently been addressed by Pothérat et al. [17], who expressed the core velocity $\mathbf{u}_{\perp 0}$ in the plane perpendicular to the magnetic field as a series expansion in terms of the two small parameters $1/N$ and $1/Ha$:

$$\mathbf{u}_{\perp 0} = \mathbf{u}_{\perp 0}^0 + \frac{1}{N}\mathbf{u}_{\perp 0}^{1,0} + \frac{1}{Ha}\mathbf{u}_{\perp 0}^{0,1} + O\left(\frac{1}{NHa}, \frac{1}{N^2}, \frac{1}{Ha^2}\right). \tag{14}$$

At the leading order, the basic averaged equations reduce exactly to the inviscid version of Eq. (13). But at the following order ($1/N$, which is usually much larger than $1/Ha$), the z-averaged velocity obeys a more complex equation involving a new and complex non-linear term:

$$\frac{d\mathbf{v}_\perp}{dt} = -\frac{1}{\rho}\nabla_\perp p_0 - \frac{\mathbf{v}_\perp}{\tau_H} + \frac{\tau_H}{Ha^2}\left(\frac{7}{36}D + \frac{1}{8}\frac{\partial}{\partial t}\right)(\mathbf{v}_\perp \cdot \nabla_\perp)\mathbf{v}_\perp, \quad (15)$$

where the operator D has the following definition:

$$D : \mathbf{F} \rightarrow D\,\mathbf{F} = (\mathbf{v}_\perp \cdot \nabla_\perp)\mathbf{F} + (\mathbf{F}\cdot\nabla_\perp)\mathbf{v}_\perp.$$

Equations (13) and (15) have been solved numerically by Pothérat et al. [18]. At moderate values of Ha, the results obtained with Eq. (15) agree more satisfactorily with the experimental data, namely in the case of isolated vortices [19] and in the case of the MATUR experiment [16], than those obtained with the simple model, Eq. (13). It is clear from the expression of the last term in (15) that the results obtained at large Ha with the two model equations coincide.

From a physical point of view, according to Pothérat et al. [17, 18], the introduction of inertia within the Hartmann layer theory results in a kind of extra diffusion, the vorticity distribution predicted by Eq. (15) being smoother than that predicted by Eq. (13). This is also suggested by the experimental data mentioned above. And, the z-dependence of the actual core velocity $\mathbf{u}_{\perp 0}$ is essentially controlled by a sort of Ekman secondary flow driven by centrifuge effects, which take place within the Hartmann layer. This would be exactly an Ekman flow if the vortical structures would be exactly circular. But their complex shape, due to the shear present at large scale in the quasi-2D flow, is responsible for the complex form of the last term in Eq. (15). The fluid, which enters the structure within the Hartmann layer, is released within the core, almost uniformly. This results in a swelling of the structure in the centre and in the z-distribution of $\mathbf{u}_{\perp 0}$, which is parabolic and gives the eddies a kind of "barrel shape". A similar effect comes from the electric circuit associated with each eddy, due to the current entering the eddy at each end according to Eq. (8), and exiting in the middle part. It also yields a parabolic z-distribution of the velocity $\mathbf{u}_{\perp 0}$. But this second effect scales as Ha^{-1} and is therefore usually much smaller than the inertial effect, which scales as N^{-1}.

The stability of the Hartmann layer at an insulating solid wall has been investigated theoretically by Lock [20], Pavlov and Simkovich [21, 22], Takashima [23], and Lingwood and Alboussière [24] among others, and experimentally in many experiments in duct flows in Riga and Purdue (see reviews by Lielausis [13] and Tsinober [25]) and more recently by Moresco and Alboussière [26].

Lock's analysis predicted that, with infinitely small initial disturbances, the Hartmann layer remains stable until the critical value $R_L = 48,250$ of the only non-dimensional parameter

$$R = \frac{U_0\delta}{\nu} = \frac{Re}{Ha}.$$

Pavlov and Simkovich [21, 22] stressed the importance of finite amplitude perturbations for the stability of the Hartmann layer, and initiated non-linear stability analysis. Lingwood and Alboussière [24] performed a detailed non-linear analysis. According to Lingwood and Alboussière, when the amplitude of the initial disturbance A_0 is finite, the critical value of R decreases when A_0 increases, until an asymptotic limit $R_G = 328$. This global stability limit is fairly well confirmed by the measurements performed by Moresco and Alboussière [26] in the Grenoble High Magnetic Field Facility, using magnetic fields up to 13 T. They obtained a critical value of 380 for R. This has also been confirmed by a recent numerical analysis by Krasnov et al. [27], who find a critical value close to 350.

It is noticeable that these values differ significantly from those previously available (around 250) after the first experiments on duct flows, performed by Murgatroyd [12] and in Riga and Purdue. This may probably be explained by the fact that previous experiments have been performed in ducts, which involved the entrance effect into the magnetic field with other sources of turbulence.

Another experimental result, which is worth to be mentioned here, is the role of the wall roughness on the stability threshold [13,14,26]. In Moresco and Alboussiere's experiment with a rough wall, this threshold slightly decreased from 380 to about 320 when Ha increased, whereas it is Ha-independent when the wall is smooth.

3 The Hartmann layer at a free surface

Let us now turn attention to free-surface flows. The situation here is far more complicated as we will see below. Despite the existence of certain theoretical results (see a review by Molokov and Reed [28]), the work on the active role of the Hartmann layers at the free surface has been initiated only recently.

Concerning experimental work, it has been noticed that in many situations it is very difficult to stabilise free-surface flows with the magnetic field. A good example is the experiment by Bucenieks et al. [29], who investigated a film flow in a trough inclined to gravity in the presence of a magnetic field parallel to backing plate (Fig. 2). This and other similar experiments reveal a very non-uniform film thickness in the transverse cross section of the trough with maximum at the vertical sidewalls. This maximum may exceed the film thickness at the centre of the trough by a factor of 2 or even 5. Bucenieks et al. [29] observed two flow regimes simultaneously: slow one at the sidewalls, and fast one in the centre. The flow pattern was very complicated, exhibiting "dolphins", i.e., time-periodic humps appearing at the sidewalls and disappearing downstream. The origin of these effects is still unclear. However, possible triggers may be inertia, three-dimensional effects, poor wetting of the trough's walls or dimensions of the meniscus (in liquid metals it is quite high, of the order of 0.3–1.3 cm [28]). These effects ultimately result in a highly non-uniform electromagnetic braking in various regions of the flow, in which the Hartmann layers and/or electrically conducting walls play a decisive role.

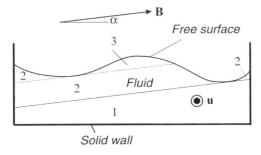

Fig. 2. Schematic diagram of a free-surface flow in a trough, showing regions of different type in the transverse cross section. The magnetic field lines intersect two solid walls in region 1, solid wall and free surface in region 2, and free surface at two positions in region 3

We will be interested here in the Hartmann layers, which are formed not only along all the solid walls, but along the free surface as well (Fig. 2). Outside the Hartmann layers there are regions of inviscid fluid (the cores). There is also a free shear layer originating at the left bottom corner of the trough and extending along the magnetic field, which will not be discussed here.

Let us turn attention to the Hartmann layers at the free surface. These layers may be active or passive. It depends mainly on whether a magnetic field line crosses a free surface at two points (such as in region 3 in Fig. 2), or only at one, crossing a solid wall as well (region 2).

The effect of the Hartmann layer on the core flow is weaker than that exerted by the layer at the solid wall. Indeed, for the existence of the Hartmann layer, traction at the adjacent surface is needed. However, the free surface is traction-free, and the conclusion might be that the primary, active boundary layer, which strongly affects the core flow, cannot exist. In this case, the layer at the free surface would be passive, and thus may be ignored. Then, all the boundary conditions (kinematic, dynamic, and vanishing of the normal component of current) may be applied directly to the core variables. Indeed, this is the case for regions such as 2 in Fig. 2, in which the Hartmann layer at the corresponding solid wall retains its controlling influence on the core. An example of such a flow is discussed in § 3.1.

There is another situation, however, when a magnetic field line crosses the free surface at two points such as in region 3. Then this argument fails, the Hartmann layer at the free surface becomes active, and in § 3.2 we explain the reasons for this.

3.1 Passive free-surface layer

Here we discuss a peculiar example, which demonstrates controlling features of the Hartmann layer at a solid wall in free-surface flows [30]. Consider a horizontal fluid layer on a solid infinite plate in the presence of a strong, vertical

($\alpha = 90°$), uniform magnetic field. The plate is in general electrically conducting, but we again consider an insulating plate here for the sake of simplicity. Now, suppose that an initial, finite-amplitude disturbance $z = h(x, y, t)$ is created, where h is a single-valued function. The disturbance will evolve owing to gravity and surface tension, while the Lorentz force will prevent its spreading. In the inertialess approximation, and for $Ha \gg 1$, the flow splits into the core and the Hartmann layers at the plate and at the free surface. As each magnetic field line crosses both the wall and the free surface, Hartmann layer at the free surface is passive and is not discussed further. The evolution of this disturbance may or may not be guided by the Hartmann layer at the plate as we will see below.

The dimensionless equations for the core are:

$$p_0 = -z + p_w, \qquad \varphi_0 = -zj_{z,0} + \varphi_w, \qquad (16)$$

$$\mathbf{j}_{\perp 0} = \mathbf{e_B} \times \nabla_\perp p_w, \qquad j_{z,0} = Ha^{-1}\nabla_\perp^2 \phi_w, \qquad j_{z,0} = \Upsilon p_w, \qquad (17)$$

$$\mathbf{u}_{\perp 0} = \mathbf{e_B} \times \nabla_\perp \varphi - \nabla_\perp p_w, \qquad v_{z,0} = z\nabla_\perp^2 p_w, \qquad (18)$$

$$h_t = \nabla_\perp \cdot (h\nabla_\perp p_w) - \Upsilon\varphi_w + h\Upsilon^2 p_w. \qquad (19)$$

Here $p_w = h - Bo^{-1}\kappa$ and φ_w are the pressure and the electric potential at the plate, respectively, and

$$\kappa = \nabla_\perp \cdot \frac{\nabla_\perp h}{\sqrt{1 + (\nabla_\perp h)^2}}$$

is the curvature of the free surface. These functions are independent of z.

In Eqs. (16)–(19) the scaling is based on the interaction between gravity and the electromagnetic force. In particular, the fluid velocity and time are scaled with $U_0 = \rho g/\sigma B^2$ and $\tau_g = \sigma B^2 L/\rho g$, respectively, where g is acceleration due to gravity and L is typical lengthscale of the disturbance. We note that the scales of velocity and time are proportional to $1/B^2$ and B^2, respectively, i.e., the magnetic field slows down propagation of disturbances significantly. The Bond number, $Bo = L^2\rho g/\gamma$, represents the ratio of gravity to surface tension. Here γ is the surface tension coefficient.

The operator, acting on a function $s(x, y)$ is the Jacobian Υ defined as follows:

$$\Upsilon s = [s, h] = \partial_x s\partial_y h - \partial_y s\partial_x h. \qquad (20)$$

It should be noted that this differential operator acts along instantaneous isolines of h. Thus, $\Upsilon h \equiv 0$, and $\Upsilon p_w \equiv -Bo^{-1}\Upsilon\kappa$.

Using these results and eliminating vertical component of current between Eqs. (17–2) and (17–3), gives the system of coupled equations for two unknown functions, h and φ_w, as follows:

$$h_t = \nabla_\perp \cdot \left\{ h \nabla_\perp \left(h - Bo^{-1}\kappa \right) \right\} - \Upsilon\varphi_w - Bo^{-1}h\Upsilon^2\kappa, \tag{21}$$

$$\frac{Bo}{Ha} \nabla_\perp^2 \varphi_w = -\Upsilon\kappa. \tag{22}$$

These are the evolution equation for the elevation of the free surface and the Poisson equation for the wall potential, respectively.

At this point several observations can be made.

1. The consequence of Eq. (17−3) is that the vertical current is determined by the variation of curvature along the instantaneous isolines of h. If curvature is a constant along the isolines (e.g., a straight line or a circle), i.e.,

$$\kappa = f(h), \tag{23}$$

where f is an arbitrary function, then $\Upsilon\kappa = 0$, and the vertical current vanishes as a result. If it vanishes within the whole fluid layer, the Hartmann layer is passive, as there is no current entering the layer, and the wall potential φ_w vanishes as well (cf. Eq. (22)). The resulting evolution equation is:

$$h_t = \nabla_\perp \cdot (h\nabla_\perp h) - Bo^{-1}\nabla_\perp \cdot (h\nabla_\perp \kappa). \tag{24}$$

If one considers an initially axisymmetric disturbance, for which curvature of h vanishes identically, the flow will be axisymmetric for all times until the disturbance decays completely. The analysis [31] shows that the axisymmetric perturbations are stable in this sense.

2. Vertical current is created by surface tension only. If one considers large-scale, smooth disturbances, for which surface tension may be neglected, i.e., in the limit $Bo \rightarrow \infty$, there is no vertical component of current, the wall potential vanishes, and the Hartmann layer becomes passive. In this case the evolution Eq. (21) reduces to

$$h_t = \nabla_\perp \cdot (h\nabla_\perp h), \tag{25}$$

which is a particular case of Eq. (24), and which has an analogy in porous media flows.

3. From Eq. (22) follows that if curvature varies along the isolines of h, then the parameter which determines the importance of surface tension is Ha/Bo. As typically $Bo \sim 1$ [28], this means that for non-symmetric disturbances the Hartmann layer is always active, and that the surface tension effects cannot be neglected. This has significant implications for modelling MHD flows with free surface.

More generally, for $Ha \gg 1$, two processes take place within the flow. One is purely diffusive, in which the Hartmann layer is passive. This happens on a slow, "gravity" timescale τ_g. The other one is fast, occurring on a timescale $\tau_\gamma = \tau_g Bo/Ha$, and is related to reshaping of the disturbance. This involves surface tension and active Hartmann layer.

Suppose $\Upsilon\kappa \neq 0$, then from Eq. (22) follows that $\varphi_w = O(Ha/Bo) \gg 1$, and thus the term containing ϕ_w on the right-hand side of Eq. (21) dominates the other terms. Introducing the rescaled potential $\Phi = \varphi_w Bo/Ha$ and time $T = tHa/Bo$, to the leading order this gives:

$$\partial_T h = -\Upsilon\Phi, \tag{26}$$

$$\nabla_\perp^2 \Phi = -\Upsilon\kappa. \tag{27}$$

The system of Eqs. (26) and (27) governs the process of reshaping the free surface, which tends to reduce the variation of κ along the isolines of h. Indeed, an equilibrium for this process occurs when $\partial_T h = 0$, i.e., when $\Upsilon\Phi = 0$.

This fast process itself is quite peculiar. The horizontal components of velocity become very high, $O(Ha/Bo)$, compared to those during the purely diffusive stage being $O(1)$, which yields

$$\mathbf{u}_{\perp 0} = \frac{Ha}{Bo}\mathbf{e_B} \times \nabla_\perp \Phi. \tag{28}$$

As function Φ is 2D, the fast regime is characterised by a quasi-2D flow in the horizontal plane with vorticity in the core being $\omega_{z,0} = -\Upsilon\kappa\frac{Ha}{Bo}$. The fluid flows along the instantaneous isolines of the electric potential. This is the effect of the active Hartmann layer.

However, unless as a result of the restructuring the disturbance becomes isolated and axisymmetric, the process is not complete. Indeed, Eq. (23) may be considered as a planform equation with infinite number of equilibrium states for Eqs. (26) and (27). A good example is the Helmholtz equation

$$\nabla_\perp^2 h = -k^2 h,$$

which is obtained from Eq. (23) by linearising curvature and assuming that function f is linear. It is shown in [31] that the process goes through many stages, in which a slow, diffusive regime is followed by a fast regime until a new equilibrium is reached with higher horizontal lengthscale. Then the process of transition to larger and larger scales via a sequence of slow/fast regimes repeats until the amplitude of the disturbance becomes so small that the non-linear effects vanish.

Of course, when inertia is present, it induces additional vorticity in the core, the associated weak vertical current entering the Hartmann layer, and all the other effects discussed in § 2.

More generally, however, the flow is not quasi-2D, as the electric potential and thus all the three components of the fluid velocity are linear functions along the magnetic field (cf. Eqs. (16)–(18)). This is especially so for the electrically conducting plate [30]. Thus, the vorticity in general retains all the three components.

3.2 Active free-surface layer

When the magnetic field line intersects a free surface at two positions, such as in region 3 in Fig. 2, the corresponding Hartmann layers become active. Consider a flow in a rivulet on a plate inclined to gravity in a transverse magnetic field parallel to the plate [32] (see Fig. 3). The main effect of the magnetic field, as with the duct flows, is to eliminate variation of the velocity along the field lines. However, the usual strong damping characteristic for duct flows is absent as there are no solid walls transverse to the magnetic field. As the core velocity is constant along magnetic field lines, it induces weak stresses at the curved free surface. These stresses are eliminated by the Hartmann layers.

The electric currents induced in the core are very weak, which leads to weak Lorentz force, comparable to the viscous force in magnitude. The core becomes viscous, with viscous effects acting in the plane transverse to the field only. The magnitude of the resulting core velocity is the same as in the hydrodynamic flow in the absence of the magnetic field. The electric currents induced in the core must close in the Hartmann layers at the free surface, which thus control the core flow, and together with viscous forces shape the velocity profile in the plane transverse to the field.

Finally, it is tempting to speculate as to what will happen in the trough as shown in Fig. 2. Any non-uniformity in the free-surface elevation, i.e., the appearance of region such as 3, may result in very high velocities in this region due to low damping, producing further instabilities.

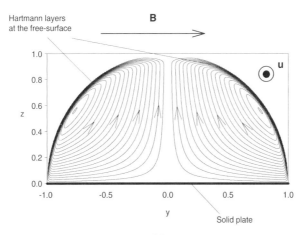

Fig. 3. Active Hartmann layers at the free surface in a rivulet in a magnetic field parallel to the plate (transverse cross section). The magnetic field lines cross free surface only. The arrows show the direction of the electric current, which is induced in the core and passes through the Hartmann layers. (From [32].)

4 Concluding remarks

The main lesson arising from this review is that the Hartmann layer belongs to the class of the *active boundary layers*, like the Ekman layer in rotating fluids. Those active layers differ from the classic Blasius-type layers, which adapt themselves to the properties of the neighbouring core flow in order to satisfy the no-slip condition at the wall, without any reaction on the core flow. In the case of the Ekman layer, the reaction on the core flow comes from the conservation of the flow rate of the secondary centrifuge flow and remains simple. In the case of the Hartmann layer, this character of an active layer is always true along a solid wall and it may be also true along a free surface.

When the Hartmann layer is active, its reaction on the core flow is manifold and quite subtle. First, the conservation of the recirculating current (7) between the core and the layer plays a role similar to the conservation of the recirculating flow rate in the Ekman layer. But, second, it has the other property (8) that any quasi-2D vortical flow is associated with a local electric current, proportional to the local core vorticity, exiting the Hartmann layer and entering into the core, where it forces some open electric circuit. Of course, these two conditions are not independent, as the presence of the j_z component (8) implies some divergence of the current sheet $\mathbf{J}_\perp$ (7). Third, it is worth to underline that most of the drag (including both viscous friction and electromagnetic drag) is located within the Hartmann layer (as well as within the Hartmann wall when it is non-insulating). And, remarkably, when Ha is still larger (say above 200), the theory of this Hartmann damping becomes quite simple, since inertia is negligible within the layer (although it may be significant within the core flow, as in a turbulent regime). Then the Hartmann damping may be modelled in the equation of motion (13) with a linear damping term behind which viscous friction and $\mathbf{j} \times \mathbf{B}$ are hidden.

Among the main consequences of these strong and specific properties of the Hartmann layer, we must mention the difficulty to develop numerical software capable to compute accurately flows at very large Ha numbers. Indeed, there are two options, either to mesh the Hartmann layer, or to model it analytically. As shown by Tagawa et al. [33], the first option requires at least five meshes across the depth of the layer and this may imply enormous computing resources. For instance, to compute flows relevant to the liquid metal blanket of fusion nuclear reactors, where $Ha \approx 10^4$–10^5, this condition makes any attempt quite unrealistic. On the contrary, the second option may save the main part of these resources, but, so far, it has only been checked on simple geometries (plane walls).

We conclude by emphasizing that surface tension effects acting via the active Hartmann layers play a very important role in the dissipation of energy. If they are not taken into account in flow models, the dissipation may be too weak.

Acknowledgements. The authors are indebted to Dr. K.A. Moerch, from the Department of Applied Physics of the Technological University of Denmark, Copenhagen, who provided them with the Hartmann portrait and some reminiscences of the Hartmann achievements [1].

References

1. Moerch KA (2000) Private communication. Physics Department, Technical University of Denmark, DK-2800 Lyngby, Denmark
2. Hartmann J (1937) Hg–dynamics I. Theory of the laminar flow of an electrically conductive liquid in a homogeneous magnetic field. Det Kgl Danske Videnskabernes Selskkab Math–fys Medd 15(6):1–28
3. Hartmann J, Lazarus F (1937) Hg–dynamics II. Experimental investigations on the flow of mercury in a homogeneous magnetic field. Det Kgl Danske Videnskabernes Selskkab Math–fys Medd 15(7):1–45
4. Ludford GSS (1961) The effect of a very strong magnetic cross-field on steady motion through a slightly conducting fluid. J Fluid Mech 10:141–155
5. Shercliff JA (1953) Steady motion of conducting fluids in pipes under transverse magnetic fields. Proc Camb Phil Soc 49:136–144
6. Braginskii SI (1960) Magnetohydrodynamics of weakly conducting fluids. Sov Phys JETP 37(10):1005
7. Hunt JCR, Ludford GSS (1968) Three-dimensional MHD duct flows with strong transverse magnetic fields. 1. Obstacles in a constant area channel. J Fluid Mech 33:693–714
8. Kulikovskii A (1968) Slow steady flows of a conducting fluid at high Hartmann numbers. Izv Akad Nauk SSSR Mekh Zhidk Gaza 3:3–10
9. Hunt JCR, Shercliff JA (1971) Magnetohydrodynamics at high Hartmann number. Ann Rev Fluid Mech 3:37–62
10. Moffatt HK (1964) Electrically driven steady flows in magneto-hydrodynamics. In: Görtler H (ed) Proceedings of the XIth International Congress of Applied Mechanics, Munich, Springer-Verlag, pp:946–953
11. Hunt JCR, Williams WE (1968) Some electrically driven flows in magnetohydrodynamics. 1. Theory. J Fluid Mech 31:705–722
12. Murgatroyd W (1953) Experiments in magnetohydrodynamic channel flow. Phil Mag 44:1348–1354
13. Lielausis O (1975) Liquid metal magnetohydrodynamics. Atomic Energy Rev 13:527–581
14. Branover H (1978) Magnetohydrodynamic Flow in Ducts. Wiley, New York
15. Sommeria J, Moreau R (1982) Why, how and when MHD turbulence becomes two-dimensional. J Fluid Mech 118:507–518
16. Messadek K, Moreau R (2002) An experimental investigation of MHD quasi-2D turbulent shear flows. J Fluid Mech 456:137–159
17. Pothérat A, Sommeria J, Moreau R (2000) An effective two-dimensional model for MHD flows with transverse magnetic field. J Fluid Mech 424:75–100
18. Pothérat A, Sommeria J, Moreau R (2005) Numerical simulations of an effective two-dimensional model for flows with a transverse magnetic field. J Fluid Mech 534:115–143

19. Sommeria J (1988) Electrically driven vortices in a strong magnetic field. J Fluid Mech 189:553–569
20. Lock RC (1955) The stability of the flow of an electrically conducting fluid between parallel planes under a transverse magnetic field. Proc Roy Soc London Ser A 233:105–125
21. Pavlov KB, Simkhovich SL (1972) Stability of Hartmann flow with respect to two-dimensional perturbations of finite amplitude. Magnetohydrodynamics 8(1):58–64
22. Simkhovich SL (1974) Influence of three-dimensional finite perturbations on the stability of Hartmann flow. Magnetohydrodynamics 10(3):17–22
23. Takashima M (1996) The stability of the modified plane Poiseuille flow in the presence of a transverse magnetic field. Fluid Dyn Res 17:293–310
24. Lingwood RJ, Alboussière T (1999) On the stability of the Hartmann layer. Phys Fluids 11:2058–2068
25. Tsinober A (1990) MHD flow drag reduction. In: Bushnell DM, Hefner JN (eds) Viscous Drag Reduction in Boundary layers. AIAA 123:327–349
26. Moresco P, Alboussière T (2004) Experimental study of the instability of the Hartmann layer. J Fluid Mech 504:167–181
27. Krasnov DS, Zienicke E, Zikanov O, Boeck T, Thess A (2004) Numerical study of the instability of the Hartmann layer. J Fluid Mech 504:183–211
28. Molokov S, Reed CB (2000) Review of free-surface MHD experiments and modelling. Technical Report ANL/TD/TM99–08, Argonne National Laboratory
29. Bucenieks I, Lielausis O, Platacis E, Shishko A (1994) Experimantal study of liquid metal film and jet flows in a strong magnetic field. Magnetohydrodynamics 30:219–230
30. Molokov S (2002) Evolution of a free surface in a strong vertical magnetic field. In: Proceedings of the 5th International Conference on Fundamental and Applied MHD (PAMIR), Ramatuelle, France, 16–20 Sept 2002 2:65–70
31. Molokov S (2007) Evolution of finite amplitude disturbances in a fluid layer in a strong vertical magnetic field (in preparation)
32. Molokov S, Reed CB (2000) Fully developed magnetohydrodynamic flow in a rivulet. Technical Report ANL/TD/TM00–12, Argonne National Laboratory Report
33. Tagawa T, Authié G, Moreau R (2002) Buoyant flow in long vertical enclosures in the presence of a strong horizontal magnetic field. Part 1. Fully-established flow. Eur J Mech B Fluids 21(4):383–398

Liquid Metal Magnetohydrodynamics for Fusion Blankets

Leo Bühler

Forschungszentrum Karlsruhe, Postfach 3640, 76021 Karlsruhe, Germany
(leo.buehler@iket.fzk.de)

1 Introduction

The realization of controlled thermonuclear fusion could lead to a significant contribution to future energy demands. The reaction between the fuel components tritium and deuterium requires temperatures above 10^8 K so that any confinement using solid walls is excluded. At these temperatures the fuels are ionized and form an electrically highly conducting plasma that can be confined by strong magnetic fields to a defined volume. During the past decades different concepts of magnetic confinement have been investigated and a number of conceptual designs for commercial or experimental fusion reactors have been studied.

One of those is the linear magnetic confinement in a very long solenoid (up to 130 m [1]). For strong enough magnetic fields, charged plasma particles are spiraling freely along the field lines from which they cannot escape. At both ends of the linear solenoid the plasma is compressed by stronger magnetic fields created by coils of special shape (called the mirror). Particles which move not primarily parallel to the field lines are reflected at the ends, others may leave the confinement at the ends. A major part of the kinetic energy of escaping charged particles can be directly converted into electric energy.

In order to avoid the very complicated mirrors it is a straightforward idea to bend the former geometry and to connect both ends. Magnetic field lines, along which charged particles move, become closed and the confined plasma fills a torus. The primary magnetic field required for plasma confinement has toroidal orientation. The best-established machine with toroidal confinement is the *Tokamak* in which a toroidal electric current is driven around the doughnut-shaped plasma. The toroidal currents induce a secondary, poloidal magnetic field, which in superposition with the primary one generates nested toroidal magnetic surfaces. The field lines follow a helical path on those surfaces as they wind around the torus. Tokamaks have proven their capabilities for fusion confinement in a number of experiments in several countries, from which the most prominent is probably the Joint European Torus experiment

S. Molokov et al. (eds.), Magnetohydrodynamics – Historical Evolution and Trends,
171–194. © 2007 *Springer.*

(JET) in Culham, UK, in which in 1991 the worlds first controlled release of deuterium–tritium fusion power of 1.7 MW was realized. Two years later the US experiment, the Tritium-fueled Fusion Test Reactor (TFTR) at Princeton generated 6 and later 10 MW. In 1997, JET was able to produce more that 16 MW and the upcoming international thermonuclear experimental reactor (ITER) will beat all prior experiments by orders of magnitude in fusion power, confinement time, and capital investment. Primarily designed as a physics experiment, ITER will also provide the opportunity to test a number of engineering components, which are essential elements of future thermonuclear fusion power plants, including liquid metal breeding blankets.

A blanket is the solid structure that is positioned between the plasma and the magnetic coils in order to shield or protect the latter ones from intolerable radiation doses. Moreover, the blanket has two other functions which are the absorption of fast neutrons, conversion of their energy into heat, and breeding of tritium, one of the fuel components. The plasma-facing wall called the *first wall* receives a high heat flux emitted from the fusion plasma. The major heat input to the blanket occurs by volumetric heating due to strong neutron radiation. To ensure safe, reliable operation, all heat released in the blanket has to be removed at such rates that wall temperatures do not exceed critical values.

Various engineering concepts have been discussed in the past with the aim to achieve sufficient cooling by using liquid metals such as lithium or the eutectic lithium–lead alloy as possible coolants. In principle, liquid metals are prime candidates for coolants. They can be operated at high temperature, they have high thermal conductivity, and due to the lithium content the coolant serves simultaneously as a material for the tritium breeding. Blankets which rely exclusively on the heat transfer capabilities of liquid metals are known as *self-cooled liquid metal blankets.* In other concepts the liquid metal serves only as breeder material while the heat is removed by coolants like water or helium gas at high pressure. We will call those concepts *separately-cooled liquid metal blankets.* A combination of both ideas results in the so-called *dual coolant blankets,* where the strong heat flux from the first wall (and from other walls) is removed by helium, while the volumetrically deposited heat is removed by the liquid metal flow. It would be impossible to address here all individual concepts that appeared during the last three decades. In the following we will outline some specific features of different blanket types illustrated by examples.

The magnetohydrodynamic (MHD) issues for applications in fusion have been discussed in the past by numerous authors and one should recall here, for example, the reports by Hunt and Hancox [2], Lielausis [3], Hunt and Holroyd [4]. These reports and references cited there highlight all major aspects concerning liquid-metal MHD in a fusion environment and most results presented there serve up to now as a fundamental basis for the evaluation of new designs.

2 Formulation

The presentation of governing equations follows closely the textbook by Müller and Bühler [5] focusing on fusion-specific aspects. Details of solution procedures may be found in that reference or in original papers cited therein. The discussion concentrates exclusively on MHD-related issues and omits such important topics as heat transfer, tritium breeding, and corrosion.

2.1 Governing Equations

For applications in fusion blankets MHD equations in the *inductionless limit* are preferred. With this assumption the magnetic field is a known quantity that does not depend on the flow. The nondimensional inductionless equations for an incompressible viscous fluid consist of the momentum balance

$$\frac{1}{N}\left[\frac{\partial \mathbf{v}}{\partial t} + (\mathbf{v} \cdot \nabla)\,\mathbf{v}\right] = -\nabla p + \frac{1}{Ha^2}\nabla^2 \mathbf{v} + \mathbf{j} \times \mathbf{B}, \tag{1}$$

and Ohm's law

$$\mathbf{j} = -\nabla\phi + \mathbf{v} \times \mathbf{B}, \tag{2}$$

with conservation of mass and charge

$$\nabla \cdot \mathbf{v} = 0, \quad \nabla \cdot \mathbf{j} = 0. \tag{3}$$

Here, $\mathbf{v}$, $\mathbf{B}$, $\mathbf{j}$, p, and ϕ stand for velocity, applied magnetic field, current density, pressure, and electric potential, scaled by the reference velocity v_0, the magnitude of the applied magnetic field B_0, $j_0 = \sigma v_0 B_0$, $p_0 = L\sigma v_0 B_0^2$, and $\phi_0 = L v_0 B_0$, respectively. The typical geometric dimension of the duct cross section is denoted by L and the fluid properties like density ρ, electric conductivity σ, and kinematic viscosity ν are assumed to be constant. Inductionless models are valid in general for low magnetic Reynolds numbers $R_m = \mu\sigma v_0 L \ll 1$, with magnetic permeability μ, but they can be applied even to cases with higher R_m provided that $R_m\, O\,(\mathbf{j}) \ll 1$, where $O\,(\mathbf{j})$ represents the order of magnitude of nondimensional current density in the fluid (as outlined, e.g., by Walker (1986) [6]).

The two nondimensional groups are the *interaction parameter* and the *Hartmann number*,

$$N = \frac{\sigma L B_0^2}{\rho v_0} \quad \text{and} \quad Ha = L B_0 \sqrt{\frac{\sigma}{\rho\nu}}, \tag{4}$$

which characterize the ratios (electromagnetic/inertia forces) and (electromagnetic/viscous forces)$^{1/2}$, respectively. The hydrodynamic Reynolds number is given in terms of these groups as $Re = Ha^2/N$.

The boundary conditions at the fluid–wall interface Γ are

$$\mathbf{v} = 0 \quad \text{and} \quad \mathbf{j} \cdot \mathbf{n} = \mathbf{j}_w \cdot \mathbf{n} \text{ at } \Gamma, \tag{5}$$

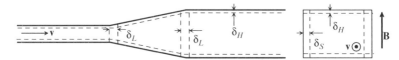

Fig. 1. Sketch of layers for $Ha \gg 1$

where **n** is the inward unit normal to the wall and $\mathbf{j}_w$ the current density in the wall. For smooth walls of conductivity σ_w, whose thickness t is small compared with L, the local current entering the wall is discharged into the thin wall in a quasi two-dimensional (2D) way. To describe this behavior the charge conservation equation is integrated across the wall which leads to the *thin-wall condition*

$$\mathbf{j} \cdot \mathbf{n} = -\frac{\partial \phi}{\partial n} = \nabla \cdot (c \nabla_\tau \phi_w) \quad \text{at } \Gamma, \tag{6}$$

where $c = \frac{\sigma_w t}{\sigma L}$ is the *wall conductance parameter*, ϕ_w stands for the dimensionless wall potential defined at the fluid–wall interface and the subscript τ denotes components tangential to the wall [7]. Currents leaving the fluid enter the wall, turn in the wall into a tangential direction, and create in the wall a distribution of wall potential. For applications in fusion blankets, the metal walls are usually assumed to be perfectly wetted by the fluid with $\phi_w = \phi$. However, in order to minimize pressure drop some design concepts propose the use of thin insulating coatings of thickness δ_i with resistivity ρ_i between the fluid and the wall. As reported by Bühler and Molokov [8], the wall potential is then related to the fluid potential by

$$\mathbf{j} \cdot \mathbf{n} = \frac{1}{\kappa} (\phi_w - \phi) \quad \text{at } \Gamma, \tag{7}$$

with the nondimensional contact resistance $\kappa = \rho_i \delta_i \, \sigma / L$.

2.2 Conditions in fusion blankets

Prime candidates for breeder and/or coolants in fusion blankets are pure lithium and the eutectic lithium lead alloy Pb-17Li. The material properties are given in Table 1 for a temperature of 450°C which is a typical temperature for applications in fusion blankets. For comparison, the values of FLIBE (LiF-34BeF$_2$) at 500°C, which has been considered as another liquid breeder or coolant, has been added to the table. The liquid metals are superior to the latter material in breeding ratio, thermal conductivity and lower viscosity so that FLIBE never was a real alternative. With these properties we may estimate the controlling nondimensional groups for geometries with $L = 0.05$ m, $B = 10$ T, and $v_0 = 0.5$ m/s.

We observe that for the liquid metals used in fusion blankets the Hartmann number is very high, $Ha \gg 1$, i.e., the electromagnetic forces dominate over the

Table 1. Physical properties of liquid breeder materials

	$\rho \left[\text{kg}/\text{m}^3\right]$	$\nu \left[\text{m}^2/\text{s}\right]$	$\sigma \left[1/\Omega\,\text{m}\right]$	Ha	N	Re
Li	4.9×10^2	7.1×10^{-7}	2.9×10^6	4.5×10^4	6.0×10^4	3.2×10^4
Pb-17Li	9.2×10^3	1.4×10^{-7}	7.5×10^5	1.1×10^4	8.2×10^2	1.5×10^5
FLIBE	2×10^3	0.7×10^{-5}	1.5×10^2	50	0.8	3.4×10^3

viscous ones. As a consequence, the fluid moves quasi inviscidly through cores which occupy most of the blanket. The core flow establishes a balance between pressure gradient and Lorentz force. Viscous effects are confined to thin layers, as shown in Fig. 1. The viscous layers at walls to which the magnetic field has a normal component are the *Hartmann layers*. They are very thin and scale as $\delta_H = O\left(Ha^{-1}\right)$. Layers at walls parallel to the field are the *parallel layers* or the *side layers* whose thickness scales as $\delta_s = O\left(Ha^{-1/2}\right)$. One could dedicate such layers to *Arthur Shercliff* who investigated first the flows in parallel layers for insulating rectangular ducts in 1953 [9], but one should also remember in this context the name of *Julian Hunt* who demonstrated in 1965 the possible occurrence of high-velocity jets in such layers, depending on the conductivity of the walls [10].

Layers between different cores are known as *Ludford layers*. They originate from discontinuities of either geometrical or electrical properties at the walls like corners, different conductivities or thicknesses of walls, etc. and spread into the fluid along magnetic field lines. In the viscous-electromagnetic regime these layers scale in thickness like Shercliff's layers, i.e. $\delta_L = O\left(Ha^{-1/2}\right)$, provided that $Ha \gg Re^2$ [11]. The latter condition is hardly met in self-cooled fusion applications.

Inspection of the interaction parameter N in Table 1 tells us that inertia forces are fairly small compared with electromagnetic forces, at least in the cores. However, inertia may play a role for flows of heavy alloys like Pb-17Li or in pure Pb, preferentially in side layers of Hunt's type, where the velocity in the jets may exceed the mean velocity by orders of magnitude. Moreover, inertia will affect essentially the Ludford layers between cores and as a consequence they will change their thickness to $\delta_L = O\left(N^{-1/3}\right)$ in the inertial-electromagnetic regime, if $Re^{1/2} \ll Ha \ll Re^2$ [11].

The values for N and Re given in Table 1 have been evaluated for a velocity that is typical for self-cooled blankets. Velocities in separately cooled blankets can be smaller at least by two orders of magnitude which increases N or decreases Re by more than two orders of magnitude compared to the values shown above. Under such conditions one comes closer to a viscous-electromagnetic balance in Ludford layers of Li blankets but flows of Pb-17Li in such layers will remain inertial.

Finally, it should be mentioned that the main effect of a strong magnetic field is the formation of uniform flow conditions in the cores and that turbulent fluctuations, if they were carried into the field, are damped out very quickly. This leads to laminar MHD flows even far beyond the hydrodynamic thresholds for the onset of turbulent motion. This observation facilitates the analysis on one hand, but reduces the heat transfer capabilities to that of laminar flows.

2.3 Analysis

For numerical analyses of three-dimensional (3D) MHD flows it is often useful to eliminate currents from the momentum equation and Ohm's law which yields

$$\frac{1}{N}\left[\frac{\partial}{\partial t} + (\mathbf{v} \cdot \nabla)\right]\mathbf{v} = -\nabla p + \frac{1}{Ha^2}\nabla^2\mathbf{v} - B^2\mathbf{v}_\perp + \mathbf{B} \times \nabla\phi, \qquad (8)$$

$$\nabla^2\phi = \nabla \cdot (\mathbf{v} \times \mathbf{B}), \qquad (9)$$

where $\mathbf{v}_\perp$ represents the velocity components in the plane perpendicular to the direction of the externally applied magnetic field. The first equation is a standard one present in many commercially available fluid dynamics codes with a source term $\mathbf{B} \times \nabla\phi$ to be modeled by the user. The second equation is a diffusion-type equation for determining the potential ϕ as an additional scalar variable with the source term $\nabla \cdot (\mathbf{v} \times \mathbf{B})$. Examples of the use of commercial software for solving MHD problems for fusion applications are described, e.g., by DiPiazza and Ciofalo (2002) [12] or by Kharicha et al. (2004) [13]. A number of academic codes has been developed, which make use of the MHD equations in the form (8) and (9). Among them are, e.g., Myasnikov and Kalyutik (1997) [14], Sterl (1990) [15], and the list goes back at least to Aitov et al. (1979) [16] and Schumann (1976) [17].

However, until today, neither commercial nor academic codes are able to simulate pressure-driven 3D MHD flows in the relevant parameter range for *Ha* and *N* with sufficient accuracy. A major problem here is that important phenomena happen on different scales but have to be resolved simultaneously. For example, the flow in the core varies on length scales of the order of one, while essential properties of the flow are determined in the very thin Hartmann layers or in the parallel layers which are a bit thicker. The proper resolution of these layers is a key issue for the correct computation of pressure drop, since these layers, together with the adjacent walls, determine the electric resistance for current loops, limit the current density in the core and the pressure drop. Moreover, together with the thin-wall condition (6) for $c \ll 1$, and thus relevant in fusion applications, the discretized numerical problem in terms of linear algebraic equations looses its diagonal dominance and becomes finally ill-conditioned with severe consequences on convergence and numerical stability of the codes.

2.4 Approximations for strong magnetic fields

For high Hartmann numbers, $Ha \gg 1$, one may consider the flow in the cores as being inviscid and exclude viscous Hartmann layers from the numerical simulation of the core flow. The solution in Hartmann layers is well known from asymptotic considerations (see, e.g., Moreau who describes in 1990 the detailed properties of Hartmann layers [18]) so that we can take them into account in an integral manner. The most important aspect of the Hartmann layers is their ability to provide a path where currents may close. This modifies the thin-wall condition (6) to

$$\mathbf{j}_c \cdot \mathbf{n} = -\frac{\partial \phi}{\partial n} = \nabla \cdot [(c + \delta) \nabla_\tau \phi_w] \quad \text{at } \Gamma, \tag{10}$$

where $\delta = (Ha \, |\mathbf{n} \cdot \mathbf{B}|)^{-1}$ stands for the local dimensionless thickness of the Hartmann layer. This formulation has been used by Bühler (1995) [19]. A condition valid for insulating walls has been given by Walker et al. (1971) [20]. The kinematic boundary condition applied to the inviscid core now requires $\mathbf{v}_c \cdot \mathbf{n} = 0$, which is a sufficient condition for inviscid flows. This approximation is valid at walls which are not parallel to the magnetic field. Applications to circular ducts become inaccurate in *Roberts layers*, in the neighborhood of lines where the magnetic field is tangential to the wall. However, since this region is small, the influence on the core velocity, flow rate, or pressure drop is negligible at high Hartmann numbers [21, 22]. Equations (8)–(10) are frequently used either for numerical simulation of channel flow or as the basis of asymptotic analyses. A general modeling of near-wall layers for numerical simulations of MHD flows has been outlined by Widlund (2003) [23].

Inertial forces in the cores of fusion blankets are often negligible compared to Lorentz forces for $N \gg 1$. This leads us to Kulikovskii's *magnetostatic approximation* published in 1968 [24]. This ingenious approach has inspired a number of researchers in the past for evaluation of MHD flows in fusion blankets, where it was sometimes called the *core flow approximation* (see, e.g., [22, 25, 26]). Bühler [19] describes in 1995 the implementation of Kulikovskii's approach into a numerical code using tensor notation with boundary fitted coordinates. The coordinate transformation

$$\mathbf{x} = \bar{\mathbf{x}}\left(u^1, u^2\right) + h\left(u^1, u^2\right) u^3 \hat{\mathbf{z}}, \tag{11}$$

which is illustrated in Fig. 2, maps the computational volume defined by the coordinates u^1, u^2, u^3 onto the physical space in terms of coordinates x, y, z. The momentum equation with mass conservation and Ohm's law with conservation of charge now read as

$$\partial_i p = V b_{ik} j^k, \qquad \partial_k \left(V v^k\right) = 0, \tag{12}$$

$$j_i = -\partial_i \phi + V b_{ik} v^k, \quad \partial_k \left(V j^k\right) = 0, \tag{13}$$

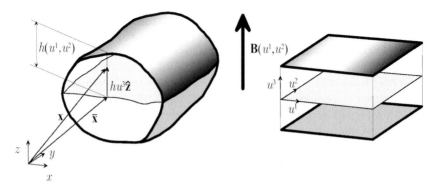

Fig. 2. Sketch of the mapping used to describe boundary- and magnetic field- fitted coordinates

where $Vb_{ik}j^k$ and $Vb_{ik}v^k$ stand for the Lorentz force and induced electric field, and the antisymmetric tensor b_{ik} represents the interaction with the magnetic field. These equations can be integrated analytically along field lines according to Kulikovskii's approach, where the two integration functions, aside from pressure $p\left(u^1, u^2\right)$, are taken as the wall potentials $\phi_\pm\left(u^1, u^2\right)$ at the upper and lower walls at $u^3 = \pm 1$. The wall potentials are determined from the thin-wall condition (10) in the form

$$Vj^3 = \mp\partial_k\left[(c + \delta)\, Ag^{ki}\partial_i\phi_\pm\right]. \tag{14}$$

Once the 2D pressure and wall potentials are obtained it is straightforward to evaluate all the 3D properties of the MHD flow by analytical relations. The numerical code based on Eqs. (12)–(14) can be applied to calculate MHD flows in nearly arbitrary geometries, for spatially varying magnetic fields, conducting and insulating walls, walls with contact resistance, and it performs best for high Hartmann numbers.

3 Self-cooled liquid-metal blankets

In the following, some liquid-metal blankets are outlined that have been promoted during the last decades of fusion research and key issues concerning the liquid-metal flow are briefly discussed. Let us start the presentation with self-cooled blankets. The use of the same fluid as both tritium breeder and coolant greatly simplifies design and materials considerations. There are important constraints related to the use of liquid metals in the blanket of a fusion reactor. For example, compatibility between the coolant and structural material limits the allowable coolant–wall interface temperature and the reactivity of lithium with air and water is an important design hazard [27].

3.1 Poloidal blankets

One of the simpler self-cooled liquid-metal blankets is the blanket designed for the linear confinement in the central part of a tandem mirror reactor (TMR) as outlined in MARS [1]. This blanket, as shown in Fig. 3, consists of two staggered rows of circular pipes which are curved around the plasma. Behind the pipes there are ducts of more or less rectangular cross section in the so-called beam zone. The fluid enters the ducts and pipes through a manifold at the top, flows once through the blanket and leaves it via a collector at the bottom.

For MARS the Hartmann number is $Ha = 10^4$ and the Reynolds number is $Re = 10^5$. Therefore, it is realistic as well as conservative to assume laminar flow through the blanket. The heat transfer is then dominated by conduction. The wall conductance parameter is about $c \approx 10^{-2}$, with the consequence that walls are much better conducting than the Hartmann layers. The velocity profile for the fully developed flow in conducting circular pipes is uniform, $v = \hat{x}$, and the pressure gradient evaluates according to Chang and Lundgren (1961) [28] as

$$\frac{\partial p}{\partial x} = -\frac{c}{1+c}. \tag{15}$$

If we return to dimensional quantities (indicated by *) we find that the dimensional pressure gradient,

$$\frac{\partial p^*}{\partial x^*} = -\sigma_w v_0 B_0^2 \frac{t}{L} \frac{1}{1+c}, \tag{16}$$

becomes independent of the fluid conductivity for $c \ll 1$. Moreover, the pressure drop and therefore the maximum pressure in the blanket depend linearly

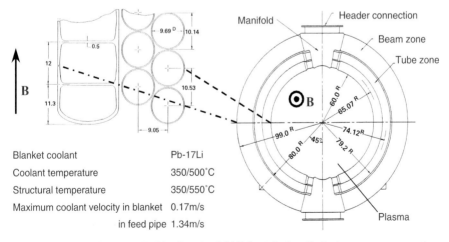

Blanket coolant   Pb-17Li

Coolant temperature   350/500°C

Structural temperature   350/550°C

Maximum coolant velocity in blanket   0.17m/s

in feed pipe   1.34m/s

Fig. 3. Sketch of the TMR blanket for MARS with detailed view on cross sections of rectangular ducts (called beams) and circular pipes facing the plasma [1]

on the thickness t of the wall, which results in wall stresses that are independent of L and t but depend only on the wall conductivity σ_w, velocity v_0, and magnetic field B_0. It is therefore impossible to allow larger blanket pressure by using thicker walls. It is worth to notice here that the MHD pressure drop at magnetic fields of $B_0 = 4.7$ T exceeds the hydrodynamic pressure drop by nearly four orders of magnitude.

In the MARS study also "end-of-loop" effects caused by gradients of $\mathbf{v} \times \mathbf{B}$ in flow direction with resulting extra pressure drop have been taken into account by empirical correlations. The pressure drop in the whole blanket becomes $\Delta p^* = 1.57$ MPa.

If we apply the same ideas to a tokamak reactor we arrive at so-called poloidal blankets. The simplest case is the pure poloidal flow design as shown in Fig. 4b. For a surface heat flux of 0.5 MW / m², a neutron wall loading of 5 MW / m², and magnetic fields up to 7.5 T a pressure drop of 2.6 MPa is required to remove the fusion heat at a mixed-mean temperature rise of 150°C. When the velocities are increased to reduce the first wall temperature to acceptable levels the pressure drop becomes prohibitively high. Abdou et al. [29] showed that the average velocity required to keep the first wall temperature at an acceptable level is too high from either a thermal efficiency point of view (too low average temperature) or from an MHD pressure drop point of view (wall stress and pumping power). Thus, a poloidal flow concept for fusion blankets with conducting walls does not look attractive as a result of the relatively poor heat transfer performance. The use of electrically insulating coatings at the walls of poloidal channels could reduce the pressure drop below 1 MPa as outlined by Kirillov et al. (2000) [30].

A method to reduce pressure drop in poloidal flows, that was preferentially promoted by the Russian fusion team (e.g., [31]), was the use of so-called *slotted channels*, which have a large dimension L along field lines. Inspection of Eq. (16) shows that increasing L reduces pressure drop. On the other hand, if L is increased too much, the solid structure looses its mechanical strength so that designs using anchorlinks become necessary.

The *helical flow concept* shown in Fig. 4a intends to help reduce the interface temperature by mechanical stirring. The stirring is obtained at the expense of higher pressure drop [32] and only a detailed analysis can show if the positive aspects associated with this idea prevail. First estimates show that the pressure drop of 3.5 MPa results in hoop stress equal to the limiting value of the wall material [29].

Another method, called MHD *flow tailoring*, relies on heat transfer improvement by exploiting salient features of MHD flows in strong magnetic fields to create desirable velocity profiles in poloidal ducts. This can reduce blanket complexity and costs as well as enhance thermal hydraulic performance (see, e.g., [33]). A particular form of flow tailoring, involving ducts with alternating expansions and contractions, lends itself to the design of first-wall coolant ducts. As a result of successive contractions and expansions, strong jets are created periodically along the sidewalls, where one of those walls constitutes

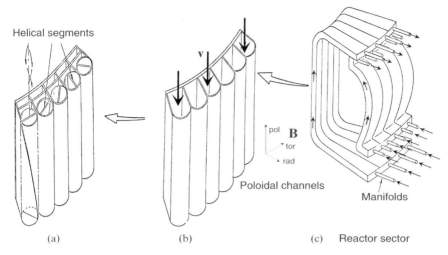

Helical segments

V

pol **B**

tor

rad

Poloidal channels

Manifolds

(a)   (b)   (c)   Reactor sector

Fig. 4. Poloidal blankets for a tokamak reactor; (a) helical vanes in circular pipes improve heat transfer at the first wall, (b) simple poloidal channels, (c) view of reactor sectors [29]

the first wall in the blanket. The jets should take the heat from the first wall and mix it periodically with the colder bulk of the flow. By this method it should be possible to increase the velocity locally at the first wall up to a factor of 3 at the expense of some extra pressure drop.

Heat transfer improvement in rectangular poloidal ducts can be achieved also by unstable or turbulent side layers. It has been shown experimentally by Reed and Picologlou (1989) [34] that the high-speed side layers become unstable at a critical Reynolds number between 2,600 and 5,100. Similar results have been obtained by Burr et al. (2000) [35] who measured additionally the improvement in heat transfer. At strong enough magnetic fields the destabilizing effect of strong shear generated at the sidewalls wins the competition with the damping effect by Joule's dissipation, and turbulent side layers are created. Due to the strong nonisotropic character of the electromagnetic forces, the turbulence structure is characterized by large-scale quasi-2D vortices with their axes aligned in the direction of the magnetic field. A surprising result in the latter reference is the fact that even if the side layers become unstable or turbulent, the pressure drop still obeys the laminar law over the whole parameter range investigated. The discrepancy between experimentally observed critical Reynolds numbers and theoretical predictions by linear stability theory, published by Ting et al. (1991) [36], is still an open question.

The most advanced design of a poloidal blanket is the *self-cooled lead-lithium blanket* (SCLL), investigated in the European Power Plant Conceptual Study [37], derived from an earlier design called TAURO [38]. The SCLL is a blanket of a fusion power reactor based on advanced plasma physics assumptions and large technological extrapolation compared with present-day knowledge. The design is based on the principle of coaxial flow, proposed in the ARIES-AT study [39]. In-vessel components are based on the use of ceramic fiber enforced SiC_f/SiC composite structure and the use of high-temperature Pb-17Li (at temperature above $1000\,^{\circ}C$) both as coolant and breeder. SiC is an excellent candidate for a low-activation wall material. Being nearly an electrical insulator, it yields high electrical resistance, reduces currents, and associated pressure drop. The pressure drop in fully developed poloidal flows in the SCLL blanket seems to be not an issue for the design. 3D effects at the blanket ends or in supply lines, however, give their extra contribution to flow resistance and should be investigated in the future.

Finally it is appropriate to mention here the so-called *dual coolant blanket* presented in 1993 by Malang et al. [40]. Although it is intrinsically not a self-cooled blanket, its similarity with poloidal blankets justifies a brief discussion here. In a dual coolant blanket the high surface heat flux at the first wall is removed by fast helium flow while the volumetrically deposited heat in poloidal breeding channels is removed by the liquid metal flow. The dual coolant blanket avoids MHD problems encountered in self-cooled concepts for cooling of the first wall. Helium cooling of the first wall provides a double containment of the liquid metal and leads therefore to improved safety and reliability. These basic ideas have been further advanced during the European Power Plant Conceptual Study by Norajitra et al. (2002) [41], derived from Sze et al. (2000) [42]. In the latest design all walls of poloidal channels are helium-cooled and the fluid in these channels is insulated from the walls both thermally and electrically by the use of SiC inserts. Electrical insulation yields small pressured drop and thermal insulation allows liquid-metal temperatures above the usual critical interface temperature, since the latter one is kept low by helium cooling of the walls. MHD issues are typically the same as for the ceramic-wall SCLL blanket. However, since the inserts at conducting walls are not perfectly electrically insulating one has to account for leakage currents and their consequences on flow redistribution as already described in 1993 by Bühler and Molokov [43].

3.2 Cellular blankets

Another design concept is the *cellular blanket* considered in the ORNL/ Westinghouse Tokamak Blanket Study, cited, e.g., by Hunt and Holroyd (1977) [4] and Walker and Wells (1979) [44], who performed analyses of the MHD flow in breeder cells under the combined action of the strong stationary toroidal magnetic field and a weaker time-dependent poloidal field. Here, lithium is used as fluid. The key parameters in feeding pipes are $Ha = 2.7 \times 10^4$, $N = 1.4 \times 10^5$, $c = 8.5 \times 10^{-3}$. The main design feature of this blanket, as shown in Fig. 5, is that the fluid flows along the first wall only over a short distance so that the heat can be removed from the first wall with moderate velocity and pressure drop, without exceeding critical temperature margins. The key MHD

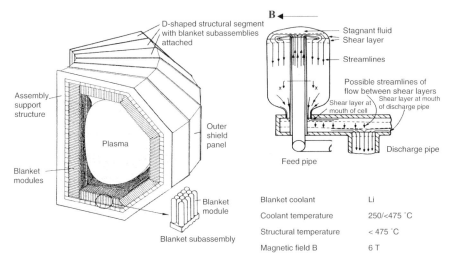

Fig. 5. View on cellular blanket modules [46] and sketch of a single cell [4]

features here are the flow in coaxial circular pipes forming the cells, flows at the plasma-facing hemispheres and the mechanical integrity of the cells caused by Lorentz forces due to the pulsed poloidal magnetic field and by plasma disruptions (accidental rapid breakdown of toroidal plasma current). The formation of shear layers near the first wall will leave a volume of stagnant fluid trapped in the dome of the cell which might deteriorate heat transfer there. An optimization of parameters for best performance of a cellular blanket was given by Holroyd and Mitchell (1984) [45].

3.3 Toroidal blankets

In all studies of self-cooled flow concepts the key issue is the pressure drop of the MHD flow that is preferentially oriented perpendicular to the magnetic field. The idea to use a toroidal flow direction near the first wall appears already in 1971 in a report by Hunt and Hancox [2], in which the toroidal cooling ducts are circular tubes (see Fig. 6a). A toroidal flow minimizes the magnetohydrodynamic interaction. A perfect alignment of the flow with the magnetic field could ideally yield pressure drop as in hydrodynamic pipe flow. The only effect which a magnetic field exerts upon the moving fluid is then the suppression of turbulent fluctuations. This leads to a reduction of heat transfer comparable to that expected for laminar flows, which, however, could be acceptable for the fast flows at the first wall.

A design that is closer to an engineering application has been developed at the Argonne National Laboratory and published, e.g., by Smith (1985) [27]. In this concept (Fig. 6b) the first wall is cooled by a fast toroidal flow of Pb-17Li in narrow channels. The fluid is supplied to the first wall ducts through

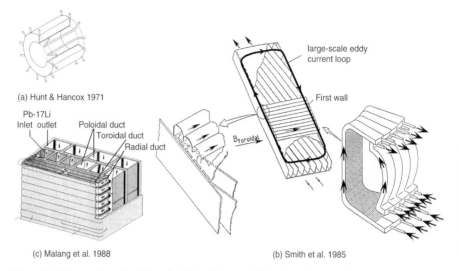

(a) Hunt & Hancox 1971

(c) Malang et al. 1988   (b) Smith et al. 1985

Fig. 6. Schematics of self-cooled liquid metal blankets for a tokamak reactor with toroidal flow near the first wall

larger, slightly slanted poloidal manifolds (perpendicular to the field). The mean velocity in poloidal ducts can be kept at relatively low values which reduces the MHD pressure drop through the manifold. A second advantage of this design is that walls of poloidal channels can take higher stress levels than the first wall since they are not exposed to the surface heat flux and receive less radiation dosage. The flow in toroidal channels is perpendicular only to the poloidal field which is much weaker than the toroidal one. Thus, the velocity in the toroidal channels can be increased considerably over that in poloidal ones without increasing significantly the overall pressure drop. Estimates for pressure drop presented by Abdou et al. (1983) [29] are near 3 MPa for the inboard blanket and near 1.7 MPa for the outboard blanket.

Later it was discovered that there is a potential current loop for large-scale eddy currents across common dividing walls of radial ducts which connect the poloidal manifold and the toroidal channels at the first wall (see Fig. 6b). The additional pressure drop due to this so-called multichannel effect (also known as *Madarame effect* according to its discoverer; he made approximations which allowed solutions for geometries more complex than could be analyzed from first principles [47]) was estimated to be more than 3 MPa, but it has been shown that some kind of electric insulation could reduce this effect considerably. It should also be mentioned that the influence of eddy currents is not uniformly distributed over all coolant ducts, i.e., ducts in the center of the module suffer more from this unfavorable behavior. This leads to nonuniform cooling of the first wall.

The toroidal flow concept was attractive enough that it had been considered as a candidate for Next European Torus (NET, now ITER). A description of the blanket proposed by the Forschungszentrum Karlsruhe was published in 1988 by Malang et al. [48]. In this concept the pressure drop in poloidal and radial channels is further reduced by the use of so-called flow-channel-inserts (FCI), which are loosely fitted into the ducts for electrical insulation and decoupling of neighboring fluid domains. The feasibility of this approach has been demonstrated experimentally by Barleon et al. (1989) [49], who showed a reduction of pressure drop by one order of magnitude.

The bright prospects offered by toroidal concepts and the challenge in predicting associated MHD flows stimulated a number of research activities related to radial-toroidal MHD bend flow like Molokov and Bühler [50] who show that the flow is very sensitive to parameters like the wall conductance and aspect ratio of toroidal cross section. Depending on these parameters the flow exhibits a variety of flow patterns including helical or vortex-type structures. For nonperfect alignment of toroidal ducts with the field, the bends are called forward and backward elbows. Such flows, treated by Walker and coauthors [51, 52], significantly differ from those obtained for perfect alignment. Stieglitz et al. [53] find good agreement between their bend experiments and asymptotic analysis for pressure drop and surface potentials for high Ha and N. The viscous and inertial contributions were detected to scale like $Ha^{-1/2}$ and $N^{-1/3}$, thus indicating that parallel layers in the viscous and inertial regime play a role for fusion relevant parameter values. Multichannel effects and flow coupling have been investigated by Stieglitz and Molokov [54] and Reimann et al. [55], who show a tremendous increase of pressure drop and inertial influence with the increasing number of electrically coupled channels. But they also find that electrical decoupling of radial channels is sufficient to significantly reduce this effect and reestablish uniform pressure drop in neighboring channels.

4 Separately cooled blankets

In separately cooled liquid metal blankets Li or Pb-17Li are used as breeder materials while the heat is removed by helium or water. These electrically nonconducting coolants do not suffer from MHD interactions. For that reason it is possible to circulate them through the magnetic field at high enough flow rates without intolerable MHD pressure drop. Heat transfer in the liquid metal relies on heat conduction, although the liquid breeder is far from being stagnant. It may move freely due to buoyant convection in the blanket, but it is also circulated at small flow rates to external facilities for tritium removal. Due to the small liquid metal velocities the MHD pressure drop in the blanket is small compared to that in self-cooled blankets.

4.1 Water-cooled blankets

The *water-cooled lead-lithium blanket* that was investigated within the European fusion research community is based on a geometry which is closely

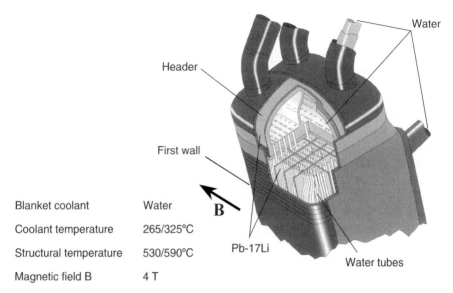

Blanket coolant	Water
Coolant temperature	265/325°C
Structural temperature	530/590°C
Magnetic field B	4 T

Fig. 7. Water-cooled lead–lithium blanket (WCLL) [57]

related to poloidal flow concepts. The liquid metal breeder is filled in rectangular poloidal channels from which the heat is removed by a large number of inserted water tubes. A first design had been shown by Giancarli et al. [56]. The blanket displayed in Fig. 7 was presented in 2000 [57]. Engineering reasons required further modifications, especially concerning the supply of the liquid breeder into the poloidal boxes and connections of the water pipes with the upper plenum. In the latest design (not shown here) the available space for liquid metal feeding pipes was really small so that the liquid metal could be supplied to and removed from the breeder channels only through small circular holes machined in a massive header structure, where the wall was much thicker than the bore diameter. Although the liquid metal velocity in the large breeder channels is as small as 5 mm / s the velocity in the smaller access bores is considerably larger so that the major pressure drop is created in the headers. Important MHD issues in the context of this blanket concept are magnetoconvection in long vertical containers, flows in ducts with internal obstacles, and pressure drop in ducts with very thick walls. The blanket is based on the use of ferromagnetic steel which raises the question on MHD flows in ferromagnetic pipes.

The use of water as coolant requires a high-pressure containment and limits the upper coolant temperature to values below 325°C, although liquid metals offer the potential for operating a reactor at much higher temperatures with better thermal conversion efficiency. Another non-MHD issue of water-cooled blankets is the permeation of tritium from the breeder into the water from where its removal is difficult.

4.2 Helium-cooled blankets

Helium as a coolant in separately cooled blankets resolves some disadvantages compared to water. Helium can be operated at higher temperatures, it is inert and does not react with liquid lithium. For these reasons helium was already considered as a potential coolant in cellular lithium blankets of the type shown in Fig. 5. In a concept addressed briefly in 1985 by Smith et al. [27], lithium was contained in tube bundles cooled by a crossflow of helium.

More recently a *helium-cooled lead-lithium* blanket has been investigated within the European fusion research program. This blanket consists of liquid-breeder-filled rectangular boxes that are arranged around the plasma similar to the cellular blanket (see Fig. 8). All walls including the first wall are cooled by helium, which flows inside the walls in small channels. The boxes, called the breeder units, are fed and drained from the back side through narrow gaps that connect the breeder units with poloidal manifolds. It is necessary to insert in each breeder unit a number of five cooling plates to remove the volumetrically released fusion power and to keep the wall temperature below critical values. As indicated in Fig. 8 the fluid enters one unit coming from the manifold through a small distributing gap, flows radially inwards between the cooling plates towards the first wall, turns at the first wall in poloidal direction, changes into the neighboring unit through a narrow gap, flows back in outward radial direction and leaves the second unit at the back through another narrow gap towards the poloidal manifold. The velocity of the flow in the breeder units required for tritium removal is near 1 mm/s so that MHD pressure drop should not be a serious issue for the flow in the breeder units. The flow through the

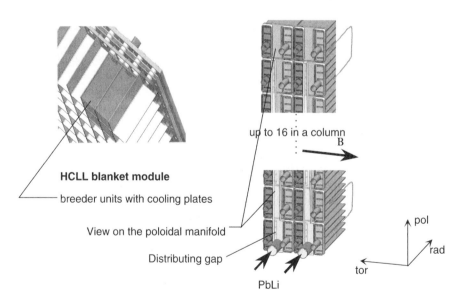

Fig. 8. Helium-cooled lead–lithium blanket (HCLL) [58]

distributing gaps, however, moves at higher velocities. Here, the flow contracts and expands preferentially in the plane of the magnetic field, thus creating the strongest MHD interactions. Moreover, one poloidal manifold feeds up to eight breeder units which leads to the highest velocities and pressure drop in these manifolds.

MHD issues to be studied for this type of blanket are flows in slender ducts formed by the cooling plates, and expansions and contractions at the entrance and exit to the breeder units. The liquid metal is heated volumetrically and cooled at all boundaries. This leads to strong temperature gradients responsible for buoyancy-driven magnetoconvective flows, whose velocity may even exceed that of the imposed forced flow. Depending on the orientation of individual breeder units (near the equatorial plane or near top poloidal positions) Kharicha et al. [13] find either one or two recirculating convection loops between the cooling plates. Another important point is the electrical coupling between neighboring fluid domains at both sides of common cooling plates or at common walls between two breeder units. Those walls are electrically conducting, without surface insulation, and currents may pass through these walls from one fluid region into the other.

5 Exotic blankets

The blanket types shown above (with the exception of the SCLL based on ceramic walls) rely on available materials and known fabrication technologies, so that their realization seems feasible. During the US Advanced Power Extraction (APEX) Study researchers tried to identify and explore novel, possibly revolutionary, concepts for chamber technology. Two ventured concepts investigated during that study are shown in Fig. 9.

The first one is EVaporation Of Lithium and Vapor Extraction (*EVOLVE*), fabricated from high-temperature refractory alloys. It is based on a transpiration-cooled first wall with evaporation of lithium similar to heat pipes. The rear part of the blanket consists of liquid lithium-filled trays from which heat is removed by evaporation of lithium by pool boiling at 1,200–1,400°C. The vapor is removed from the blanket through large pipes. Calculations indicate that an evaporating system with Li can remove a first wall surface heat flux $>2\mathrm{MW/m^2}$ with an accompanying neutron wall load of $>10 \mathrm{\ MW/m^2}$ [59]. Critical MHD issues for this concept are MHD flows through porous structures at the first wall, MHD pool boiling where Lorentz forces inhibit the movement of raising bubbles and thereby deteriorate heat transfer.

The other concept shown in the figure is a blanket with a thick liquid first wall. *Liquid wall concepts* set solid first walls aside and rely instead on heat transfer capabilities of a fast-moving free-surface flow around the plasma that is pressed to the walls by centrifugal forces when it moves down along the curved back wall. Concepts using Lorentz forces to attach the liquid to

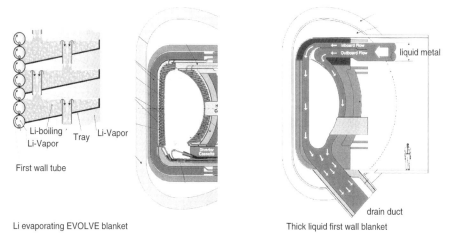

Fig. 9. Conceptual designs for EVOLVE and for a thick liquid first wall blanket [59]

the walls have been also discussed. The design is conceptually simple and provides a renewable first liquid wall that avoids replacing solid structures at regular intervals. For ideal film flows in a closed torus with toroidal magnetic field Hartmann walls perpendicular to the field are absent. Therefore, current closure is excluded and the MHD resistance to such flows should be low. On the other hand, any fusion reactor must have penetrations through the blanket for plasma diagnostics and heating which introduces again Hartmann walls and associated Hartmann braking. Moreover, the flow happens across magnetic fields that vary along the flow direction. This induces axial potential differences that drive axial currents which, depending on their orientation, could generate Lorentz forces that might detach the film from the wall. A complete review on all types of free surface MHD flows and in particular those relevant for fusion applications, until 1999 was given by Molokov and Reed [60] (for a particular application see [61]). Plasma pollution by evaporating Li is an open issue in addition to free-surface stability and interaction with the plasma. Extraction of the liquid metal from the evacuated plasma chamber could be another prohibiting design issue.

6 Conclusions

Liquid metal MHD flows in fusion blankets considered during the past decades cover a number of phenomena like pressure-driven duct flow, flows in expansions and contractions, in bends of different orientation, electrical flow coupling, flows in ducts covered with insulating coatings, buoyant convection, free surface flows, and even evaporation and boiling. So far there is nothing really special in fusion MHD flows, except that the magnetic fields are much higher than in any other known engineering application. All phenomena occurring in 3D flows in single ducts seem well understood and can often be predicted with sufficient accuracy for designing engineering components of fusion reactor blankets. Asymptotic theory is the preferred method for predicting MHD flows for fusion applications but with increasing progress in computational resources complete numerical solutions could reach the desired parameter range in a couple of years. If so, asymptotic solutions will still serve as a tool for code validation and they will continue to provide insight into physical phenomena involved. Academic computational tools will contribute to results at the frontiers of applications which are not yet assessable for commercial fluid dynamics codes. In principle, many commercial codes allow the user to implement such items like Lorentz forces or electric potentials (if not already existing). However, experience shows that the user depends exclusively on the policy of the software companies. Any change in the code's version may require serious updates of self-developed user subroutines and the effort to do that might be as large as to proceed in research. Moreover, the given types of boundary conditions are not always suited for efficient calculations of MHD flows. On the other hand, academic codes are strongly linked to personnel that develops and handles these tools.

For designing fusion blankets with such complexity as outlined above it will not be sufficient to accurately predict MHD flows in single components. Since Madarame we know that global effects, that do not arise until the full system is considered, can deteriorate the overall performance. For that reason engineering approaches are required in addition, which take into account flow coupling in global complex 3D systems, even at the expense of some flow details.

All progress made in fusion-related MHD research was supported by experiments, which confirmed often the validity of theoretical solutions, or which yielded results for cases where theoretical predictions did not yet exist. The enormous costs of operating ITER and testing of liquid metal blankets in a nuclear environment, will make MHD experiments and pretesting of components necessary in order to minimize potential risks during the development. The future MHD research for fusion blankets may have to focus on one particular reference concept concerning modeling and experiments. Nevertheless, a certain part of the research activity should be dedicated to fundamental aspects of MHD flows which are not restricted to a specific design. This could be necessary in order to attract and to keep qualified scientists in this challenging field. Moreover, during the past three decades we have seen a number of different blanket concepts appearing and disappearing on the scene, from which only the most prominent ones were presented in this review. What has remained from all these investigations for future applications are the fundamental research results but not concept specific details.

References

1. Mirror Advanced Reactor Study (MARS) (1984) Technical Report UCRL-53480, Lawrence Livermore National Laboratory, Livermore, CA
2. Hunt JCR, Hancox R (1971) The use of liquid lithium as coolant in a toroidal fusion reactor. Technical Report CLM-R 115, Culham Laboratory, England
3. Lielausis O (1975) Liquid-metal magnetohydrodynamics. Atomic Energy Rev 13(3):527–581
4. Hunt JCR, Holroyd RJ (1977) Applications of laboratory and theoretical MHD duct flow studies in fusion reactor technology. Technical Report CLM-R169, Culham Laboratory, England
5. Müller U, Bühler L (2001) Magnetofluiddynamics in Channels and Containers. Springer, Wien/New York
6. Walker JS (1986) Laminar duct flows in strong magnetic fields. In: Branover H, Lykoudis PS, Mond M (eds) Liquid-metal Flows and Magnetohydrodynamics. American Institute of Aeronautics and Astronautics, Monterey, CA
7. Walker JS (1981) Magnetohydrodynamic flows in rectangular ducts with thin conducting walls. J de Mécanique 20(1):79–112
8. Bühler L, Molokov S (1994) Magnetohydrodynamic flows in ducts with insulating coatings. Magnetohydrodynamics 30(4):439–447
9. Shercliff JA (1953) Steady motion of conducting fluids in pipes under transverse magnetic fields. Proc Camb Phil Soc 49:136–144
10. Hunt JCR (1965) Magnetohydrodynamic flow in rectangular ducts. J Fluid Mech 21:577–590
11. Hunt JCR, Leibovich S (1967) Magnetohydrodynamic flow in channels of variable cross-section with strong transverse magnetic fields. J Fluid Mech 28(2):241–260
12. Di Piazza I, Ciofalo M (2002) MHD free convection in a liquid-metal filled cubic enclosure. I. Differential heating. Int J Heat Mass Transf 45(7):1477–1492
13. Kharicha A, Molokov S, Aleksandrova S, Bühler L (2004) Buoyant convection in the HCLL blanket in a strong uniform magnetic field. Technical Report FZKA 6959, Forschungszentrum Karlsruhe
14. Myasnikov MV, Kalyutik AI (1997) Numerical simulation of incompressible MHD flows in channels with a sudden expansion. Magnetohydrodynamics 33(4): 342–349
15. Sterl A (1990) Numerical simulation of liquid-metal MHD flows in rectangular ducts. J Fluid Mech 216:161–191
16. Aitov TN, Kalyutik AI, Tananaev AV (1979) Numerical investigation of three-dimensional MHD flow in a curved channel of rectangular cross section. Magnetohydrodynamics 15(4):458–462
17. Schumann U (1976) Numerical simulation of the transition from three- to two-dimensional turbulence under a uniform magnetic field. J Fluid Mech 74:31–58
18. Moreau R (1990) Magnetohydrodynamics. Kluwer Academic, Dordrecht
19. Bühler L (1995) Magnetohydrodynamic flows in arbitrary geometries in strong, nonuniform magnetic fields. Fusion Technol 27:3–24
20. Walker JS, Ludford GSS, Hunt JCR (1971) Three-dimensional MHD duct flows with strong transverse magnetic fields. Part 2. Variable-area rectangular ducts with conducting sides. J Fluid Mech 46:657–684
21. Roberts PH (1967) Singularities of Hartmann layers. Proc R Soc Lond 300(A):94–107

22. Hua TQ, Walker JS (1989) MHD flow in insulating circular ducts for fusion blankets. Fusion Technol 15:699–704
23. Widlund O (2003) Wall functions for numerical modeling of laminar MHD flows. Eur J Mech B/Fluids 22(3):221–237
24. Kulikovskii AG (1968) Slow steady flows of a conducting fluid at large Hartmann numbers. Fluid Dynamics 3(1):1–5
25. McCarthy KA, Tillack MS, Abdou MA (1989) Analysis of liquid metal MHD flow using an iterative method to solve the core flow equations. Fusion Eng Des 8:257–264
26. Hua TQ, Walker JS (1991) MHD considerations for poloidal-toroidal coolant ducts of self-cooled blankets. Fusion Technol 19:951–960
27. Smith DL, Baker CC, Sze DK, Morgan GD, Abdou MA, Piet SJ, Schultz SR, Moir RW, Gordon JD (1985) Ovierview of the blanket comparison and selection study. Fusion Technol 8(1):10–113
28. Chang C, Lundgren S (1961) Duct flow in magnetohydrodynamics. Zeitschrift für angewandte Mathematik und Physik XII:100–114
29. Abdou MA et al. (1983) Blanket comparison and selection study-interim report. Technical Report ANL/FPP-83-1 I, Argonne National Laboratory
30. Kirillov IR, RF DEMO team (2000) Lithium cooled blanket of RF DEMO reactor. Fusion Eng Des 49–50:457–465
31. Lavrent'ev IV (1989) MHD-flow at high Rm, N and Ha. In: Lielpetris J, Moreau R (eds) Liquid Metal Magnetohydrodynamics, Kluwer Academic, Dordrecht 21–43
32. Madarame H, Tillack MS (1986) MHD flow in liquid metal blankets with helical vanes. J Fac Eng, Univ Tokyo (B) 38(4):1–18
33. Picologlou BF, Reed CB, Hua TQ, Barleon L, Kreuzinger H, Walker JS (1989) MHD flow tailoring in first wall coolant channels of self-cooled blankets. Fusion Eng Des 8:297–303
34. Reed CB, Picologlou BF (1989) Side wall flow instabilities in liquid metal MHD flow under blanket relevant conditions. Fusion Technol 15:705–715
35. Burr U, Barleon L, Müller U, Tsinober AB (2000) Turbulent transport of momentum and heat in magnetohydrodynamic rectangular duct flow with strong side wall jets. J Fluid Mech 406:247–279
36. Ting AL, Walker JS, Moon TJ, Reed CB, Picologlou BF (1991) Linear stability analysis for high-velocity boundary layers in liquid-metal magnetohydrodynamic flows. Int J Engng Sci, 29(8):939–948
37. Giancarli L, Bühler L, Fischer U, Enderle R, Maisonnier D, Pascal C, Pereslavtsev P, Poitevin Y, Portone A, Sardain P, Szczepanski J, Ward D (2003) In-vessel component designs for a self-cooled lithium-lead fusion reactor. Fusion Eng Des 69(1-4):763–768
38. Pérez AS, Giancarli L, Molon S, Salavy JF (1995) Progress on the design of the TAURO breeding blanket concept. Technical Report DMT 95/575 (SERMA/LCA/1829), CEA
39. Raffray AR, Jones R, Aiello G, Billone M, Giancarli L, Golfier H, Hasegawa A, Katoh Y, Kohyama A, Nishio S, Riccardi B, Tillack MS (2001) Design and material issues for high performance SiCf/SiC-based fusion power cores. Fusion Eng Des 55(1):55–95
40. Malang S, Bojarsky E, Bühler L, Deckers H, Fischer U, Norajitra P, Reiser H (1993) Dual coolant liquid metal breeder blanket. In: Ferro C, Gasparotto M,

Knoepfel H (eds) Fusion Technol 1992, Proceedings of the 17th Symposium on Fusion Technol, Rome, Italy, 14–18 September 1992. Elsevier, Amsterdam 1424–1428

41. Norajitra P, Bühler L, Fischer U, Malang S, Reimann G, Schnauder H (2002) The EU advanced dual coolant blanket concept. Fusion Eng Des 61–62:449–453

42. Sze DK, Tillack M, El-Guebaly L (2000) Blanket system selection for the ARIES-ST. Fusion Eng Des 48(3–4):371–378

43. Bühler L, Molokov S (1993) Magnetohydrodynamic flows in ducts with insulating coatings. Technical Report KfK 5103, Kernforschungszentrum Karlsruhe

44. Walker JS, Wells WM (1979) Stress in liquid lithium modules in a TOKAMAK blanket due to changing poloidal magnetic field. In: Proceedings of the 8th Symposium on Engineering Problems of Fusion Research, San Francisco, California, November 13–16. IEEE Pub No 79CH1441-5 NPS, pp 394–398

45. Holroyd RJ, Mitchell JTD (1984) Liquid lithium as a coolant for TOKAMAK fusion reactors. Nuclear Engng Des/Fusion 1:17–38

46. Walker JS, Wells WM (1979) Forces on liquid lithium modules in a TOKAMAK blanket due to the pulsed poloidal magnetic field. Technical Report ORNL/TM-6907, Oak Ridge National Laboratory

47. Madarame H, Taghavi K, Tillack MS (1985) The influence of leakage currents on the MHD pressure drop. Fusion Technol 8:264–269

48. Malang S, Arheidt K, Barleon L, Borgstedt HU, Casal V, Fischer U, Link W, Reimann J, Rust K, Schmidt G (1988) Self-cooled liquid-metal blanket concept. Fusion Technol 14:1343–1356

49. Barleon L, Lenhart L, Mack HJ, Sterl A, Thomauske K (1989) Investigations on liquid metal MHD in straight ducts at high Hartmann numbers and interaction parameters. In: Müller U, Rehme K, Rust K (eds) Proceedings of the 4th International Topical Meeting on Nuclear Reactor Thermal-Hydraulics, Karlsruhe, October 10–13. G Braun, Karlsruhe, pp 857–862

50. Molokov S, Bühler L (1994) Liquid metal flow in a U-bend in a strong uniform magnetic field. J Fluid Mech 267:325–352

51. Moon TJ, Walker JS (1990) Liquid metal flow through a sharp elbow in the plane of a strong magnetic field. J Fluid Mech 213:397–418

52. Moon TJ, Hua TQ, Walker JS (1991) Liquid-metal flow in a backward elbow in the plane of a strong magnetic field. J Fluid Mech 227:273–292

53. Stieglitz R, Barleon L, Bühler L, Molokov S (1996) Magnetohydrodynamic flow through a right-angle bend in a strong magnetic field. J Fluid Mech 326:91–123

54. Stieglitz R, Molokov S (1997) Experimental study of magnetohydrodynamic flows in electrically coupled bends. J Fluid Mech 343:1–28

55. Reimann J, Barleon L, Bühler L, Lenhart L, Malang S, Molokov S, Platnieks I, Stieglitz R (1995) Magnetohydrodynamic investigations of a self-cooled Pb-17Li blanket with poloidal-radial-toroidal ducts. Fusion Eng Des 27:593–606

56. Giancarli L, Severi Y, Baraer L, Leroy P, Mercier J, Proust E, Quintric-Bossy J (1992) Water-cooled lithium-lead blanket design studies for demo reactor: Definition and recent developments of the box-shaped concept. Fusion Technol 21:2081–2088

57. Giancarli L, Ferrari M, Fütterer MA, Malang S (2000) Candidate blanket concepts for a European fusion power plant study. Fusion Eng Des 49–50:445–456

58. Rampal G, Poitevin Y, Li-Puma A, Rigal E, Szczepanski J, Boudot C (2005) HCLL TBM for ITER – design studies. Fusion Eng Des 75–79:917–922

59. Abdou MA, the APEX TEAM (2001) On the exploration of innovative concepts for fusion chamber technology. Fusion Eng Des 54:181–247
60. Molokov S, Reed CB (2000) Review of free-surface MHD experiments and modelling. Technical Report ANL/TD/TM99-08, Argonne National Laboratory
61. Smolentsev S, Abdou MA (2005) Open-surface MHD flow over a curved wall in the 3-d thin-shear-layer approximation. Appl Math Model 29(3):215–234

Geostrophic Versus MHD Models

Thierry Alboussière

Laboratoire de Géophysique Interne et Tectonophysique, CNRS, Observatoire de Grenoble, Université Joseph Fourier, Maison des Géosciences, BP 53, 38041 Grenoble Cedex 9, France (`thierry.alboussiere@obs.ujf-grenoble.fr`)

1 Introduction

Low magnetic Reynolds number magnetohydrodynamic (MHD) flows and low Rossby number rotating flows share a number of common features. Both are subjected to a strong linear force: Lorentz or Coriolis. From an energetic point of view, they are very different, as Lorentz forces are dissipative in nature (Joule dissipation adding to viscous dissipation) while Coriolis forces are purely conservative. Both forces however tend to favour a two-dimensional (2D) flow, independent of the direction of the applied magnetic field or rotation axis. This can be seen most easily on steady solutions, where three-dimensional (3D) structures are absent in the bulk of the flow. Exactly how these MHD or rotating flows become 2D is very interesting. In MHD flows, due to Joule dissipation, this takes the form of a pseudo-diffusion ([1], or [2]), whereby 3D features are diffusively stretched in the direction of the magnetic field, with a pseudo-diffusion coefficient proportional to the square of their size in the perpendicular direction. In rotating flows the process is dominated by inertial waves. 3D structures give rise to dispersive fast (small Rossby number) inertial waves, whereas nearly 2D structures correspond to inertial waves with a wave-number perpendicular to the rotation axis, hence a vanishing pulsation according to their dispersion relationship (see [3] for a complete exposition). Those 2D structures are thus not rapidly dispersed and remain alone eventually. Historically, the tendency towards 2D flows in rotating systems has been suggested and demonstrated experimentally by Proudman [4] and Taylor [5,6].

Having established the 2D nature of these flows, it was then natural to derive 2D flow equations. For rotating flows, Montgomery (1938) [7] pointed out the role of the so-called geostrophic contours (constant depth in the direction of rotation in the case of uniform density): this follows from the conservation of the background angular momentum. If the flow departs slightly from

S. Molokov et al. (eds.), *Magnetohydrodynamics – Historical Evolution and Trends*, 195–209. © 2007 *Springer*.

those contours, this is a source of vorticity in the rotating frame of reference. The contribution of Ekman layers was recognized following Ekman [8]: for the case of a solid boundary, they are also a source (or rather a sink) of vorticity associated to Ekman pumping. Finally, the core two-dimensional viscosity is also "dissipating" enstrophy. When the material derivative of the vorticity is expressed in terms of these above-mentioned three contributions, the equation of the so-called homogeneous model for rotating flows is obtained. This has been used for modelling oceanic circulation [3]: some typical boundary layers arising from this model have been invoked to represent western streams such as the Gulf stream or the Kuroshio stream. Stewartson [9,10] played an important role in analysing shear layers developing on singular surfaces parallel to the rotation axis.

In parallel, progress has been made in the understanding of the behaviour of electrically conducting fluids in the presence of an imposed magnetic field. Hartmann and Lazarus (1937) [11] discovered Hartmann layers, the boundary layer analogue to the Ekman layer and responsible for an increase in wall friction. Another type of layers, parallel to the direction of the magnetic field and an analogue to Stewartson layers, has been put in evidence by Shercliff (1953) [12] first. Kulikovskii (1958) [13] made a major contribution when he launched the concept of "characteristic surfaces", analogous to the geostrophic contours. Characteristic surfaces are made of magnetic lines with a constant ratio of magnetic field intensity divided by the line length. Magnetic circulation is conserved for flows following these surfaces. Otherwise, in case of crossflow, vorticity is generated by the electrical currents created by the variations in magnetic circulation. Holroyd and Walker (1978) [14] have written inertialess 2D equations, which were only recently put under a form similar to the rotating homogeneous model and analysed in terms of potential 2D structures [15].

The similarity between rotating and MHD flows will be emphasized here as much as it is possible, so as to benefit from all advances in either field. More efforts have gone into rotating flows, and they are perhaps slightly easier to handle as there is only one equation to consider (Navier–Stokes) while Ohm's law has to be combined with Navier–Stokes in MHD studies. There has seemed that rotating flows could show a greater variety of shear layers. Using the analogy between both types of flows, one can either find the corresponding MHD layers or find a good reason why the corresponding layer is not physical.

In § 2, the fundamental lengthscales arising in rotating or MHD flows will be introduced. Section 3 will be devoted to the derivation of the homogeneous model of rotating flows. An MHD 2D model is derived in § 4. The next § 5 will provide an example of a 2D MHD flow calculated from the model just derived. Finally, conclusions and perspectives will be presented in § 6.

2 Fundamental length scales

Starting from the governing equations of rotating and MHD flows, respectively, fundamental scales will be identified and used to justify the subsequent 2D models to be introduced in the second stage.

2.1 Rotating flows

Fluid motion in a reference system rotating with a steady angular velocity $\mathbf{\Omega}$ is subjected to apparent forces, Coriolis and centrifuge. While centrifuge forces can be absorbed in a modified pressure term, Coriolis forces have to be considered specifically as they play a distinct role. Momentum, or Navier–Stokes equation, with Coriolis forces, is the governing equation for the solenoidal velocity field $\mathbf{u}$. This equation takes the following dimensionless form:

$$\frac{\partial \mathbf{u}}{\partial t} + (\mathbf{u} \cdot \nabla)\mathbf{u} = -\nabla p + 2\,E^{-1}\,\mathbf{u} \times \mathbf{e}_z + \nabla^2 \mathbf{u}, \tag{1}$$

where dimensionless vector position $\mathbf{x}$, time t, velocity field $\mathbf{u}$, and pressure field p are derived from their corresponding dimensional quantities using the scales H, H^2/ν, ν/H and $\rho \nu^2/H^2$, respectively. Here H is a length scale of the fluid domain, ρ the density of the fluid and ν its kinematic viscosity. The Ekman number appearing in the equation is defined as $E = \nu/(\Omega H^2)$. Reference axes have been chosen so that the rotation axis lies in the z-direction and $\mathbf{e}_z$ denotes the unit vector along the z-direction.

It is possible to extract fundamental time and length scales from the governing equation. Taking the curl of Eq. (1) eliminates the pressure:

$$\frac{\partial \nabla \times \mathbf{u}}{\partial t} + \nabla \times [(\mathbf{u} \cdot \nabla)\mathbf{u}] = 2\,E^{-1}\,\frac{\partial \mathbf{u}}{\partial z} + \nabla^2\,(\nabla \times \mathbf{u}). \tag{2}$$

In the asymptotic limit of strong rotation ($E \longrightarrow 0$), the first term on the right-hand side of Eq. (2) is dominant and can be balanced by the first term (time derivative) of the left-hand side when a short timescale E is invoked, i.e., the time for the reference system to rotate an angle of one radian. These two terms are responsible for the inertial waves, which satisfy the following dispersion relationship:

$$\omega = \pm 2E^{-1} \cos\theta, \tag{3}$$

where the velocity of inertial modes is defined as $\mathbf{u} = \mathbf{u}_0\,e^{i(\omega t + \mathbf{k} \cdot \mathbf{x})}$ and where θ is the angle between $\mathbf{k}$ and $\mathbf{\Omega}$. After a transient period of inertial wave propagation, the final state of the flow corresponds to a quasi-steady 2D state ($\theta \simeq \pi/2$, hence $\omega \simeq 0$). These inertial waves can be disrupted by non-linear inertial terms (the second term on the left-hand side), provided the Rossby number is not small compared to unity. In our dimensionless formulation, the Rossby number is the product of the dimensionless velocity with the Ekman number, while it is $Ro = U/(\Omega H)$ when U denotes a dimensional velocity scale. Viscous effects (second term on the right-hand side) can also affect inertial waves by viscous damping and are responsible for thin shear layers in the steady or quasi-steady state: $E^{1/2}$ Ekman layers along boundaries not parallel to the rotation direction and $E^{1/3}$ Stewartson layers developing along surfaces containing the rotation direction. Finally, the condition for the existence of a quasi-steady 2D flow is that of a small Rossby number only for regions outside Ekman and $E^{1/3}$ Stewartson layers. These conditions will be used a posteriori to assess the validity of 2D solutions.

2.2 MHD flows

In MHD flows, our attention will be restricted here to the case of low Lundquist and magnetic Reynolds numbers. Momentum equation and Ohm's law are the two governing equations, expressed in a dimensionless form:

$$\frac{\partial \mathbf{u}}{\partial t} + (\mathbf{u} \cdot \nabla)\mathbf{u} = -\nabla p + Ha^2 \mathbf{j} \times \mathbf{B} + \nabla^2 \mathbf{u}, \tag{4}$$

$$\mathbf{j} = -\nabla \phi + \mathbf{u} \times \mathbf{B}. \tag{5}$$

Dimensional scales already defined above for rotating flows are still valid. In addition, the dimensionless electric current density $\mathbf{j}$ and electric potential field ϕ are obtained from their dimensional counterpart using the scales $\nu\sigma B_0/H$ and νB_0, where B_0 is a typical value of the imposed magnetic field.

To obtain an equation for the velocity field $\mathbf{u}$ suitable for analysis, one can take the curl of the momentum equation twice and substitute $\nabla \times \mathbf{j}$ using the curl of Ohm's law:

$$\frac{\partial}{\partial t}\left(-\nabla^2 \mathbf{u}\right) + \nabla \times \nabla \times [(\mathbf{u} \cdot \nabla)\mathbf{u}] = Ha^2 \frac{\partial^2 \mathbf{u}}{\partial z^2} - \left(\nabla^2\right)^2 \mathbf{u}. \tag{6}$$

This equation is obtained using the approximation of a locally uniform magnetic field. This is not exact but this does not affect the following scaling analysis. The asymptotic strong MHD regime is characterized by a large value of the Hartmann number ($Ha \longrightarrow \infty$). The dominant first term on the right-hand side of Eq. (6) can only be balanced by the first term (time derivative) on the left-hand side on a short timescale of order Ha^{-2}, i.e., the so-called Joule time $\rho/(\sigma B^2)$. These two terms define an equation of pseudo-diffusion for the velocity field. Its dispersion relationship takes the following form:

$$\omega = iHa^2 \frac{(\mathbf{k}.\mathbf{B})^2}{k^2}, \tag{7}$$

for elementary solutions defined as $\mathbf{u} = \mathbf{u}_0 \, e^{i(\omega t + \mathbf{k} \cdot \mathbf{x})}$. This is similar to the dispersion relationship of an equation of diffusion, but with a diffusion coefficient D dependent on the length scale l of the velocity disturbance $D \simeq Ha^2 l^2$. The ultimate state of pseudo-diffusion is a quasi-steady 2D flow. The non-linear inertial term (the second one on the right-hand side of Eq. (6)) can disrupt this pseudo-diffusion process provided the interaction parameter N is small compared to unity. In our dimensionless formulation, the interaction parameter is defined as the square of the ratio of the Hartmann number by the dimensionless velocity, or $N = \sigma B_0^2 H/(\rho U)$ using the dimensional scale U for the velocity. Viscous terms (the last term in Eq. (6)) provide additional diffusion to the MHD pseudo-diffusion and are responsible for the existence of thin shear layers in the steady state: Ha^{-1}-thick Hartmann layers on walls non-parallel to the magnetic field and $Ha^{-1/2}$-thick "parallel" layers developing along surfaces made of magnetic lines. 2D models will be valid for a large value of the interaction parameter and will apply outside Hartmann or parallel layers.

3 The "homogeneous model" of rotating flows

This model has been developed initially to model oceanic circulation for which the thin aspect ratio (depth over horizontal scales) gives another reason, in addition to rotation effects, to focus on 2D flows. This aspect ratio condition is not necessary and the homogeneous model has also been applied to other geometries, e.g., to the thick atmosphere of Jupiter or to the earth's liquid inner core.

Let us assume for simplicity that the fluid domain is symmetrical with respect to a plane perpendicular to the axis of rotation. A 2D model arises naturally for the long time evolution of flows in rotating systems if one considers Eq. (2) and the fact that the fluid domain is bounded in the direction of the rotation axis. From this last condition, it is concluded that the strongest flow components will be perpendicular to the rotation axis. This follows from continuity as Ekman layers cannot accept a jump in the normal component but only a jump in the tangential components of velocity. Hence, the 2D flow is a flow with 2D flow components in the direction perpendicular to the axis of rotation. From Eq. (2), the flow component parallel to the rotation axis is odd with respect to the plane of symmetry and small compared to perpendicular components. The main component of vorticity is parallel to the axis of rotation and is 2D.

Traditionally, the homogeneous model is derived from the equation of vorticity, projected on the direction of the rotation axis. Sources of vortex stretching are due to two causes: geometrical effects and Ekman pumping. If the depth of the fluid domain changes in the direction of the 2D flow, axial stretching or compression must occur so that the core flow remains tangent to the upper and lower boundaries. Regarding Ekman layers developing at a solid boundary, there is a fundamental linear relationship between the normal flow velocity entering the layer and the vorticity of the 2D flow.

We are going to present a slightly different derivation based on global conservation, Ekman layer properties and analysis of the 2D core flow. The geometry of the cavity consists of the space situated between two symmetrical surfaces, defined in the orthonormal coordinate system (x, y, z) by the two functions, $z^u(x, y)$ and $-z^u(x, y)$, respectively, where z is still referring to the direction of the axis of rotation. The "depth" of the cavity is defined as the distance between both surfaces as a function of x and y, $d(x, y) = 2z_u(x, y)$. Assuming the existence of a core flow were x and y components of the flow are independent of z, a purely 2D flow $\mathbf{u}_0$ with no z component is introduced:

$$\mathbf{u}_0 = \begin{bmatrix} u_{0x}(x, y) \\ u_{0y}(x, y) \end{bmatrix}. \tag{8}$$

One must be careful as this is not the real flow and, for instance, this flow need not be divergence-free. However, in the absence of flow injection through the upper and lower boundaries, global volume conservation implies that the 2D mass flow rate, denoted $\mathbf{Q}(x, y)$, is a divergence-free 2D vector field. It is attributed to a streamfunction, ψ:

$$\mathbf{Q} = \nabla \psi \times \mathbf{e}_z. \tag{9}$$

One has to link $\mathbf{Q}$ to $\mathbf{u}_0$, taking into account the effect of the upper and lower Ekman layers. These layers contribute a small flow deficit in the direction of the core flow, but more importantly, they generate a component of crossflow due to Ekman pumping. This crossflow is small but has important consequences in terms of mass conservation: a vortex core flow creates a purely divergent crossflow in Ekman layers. In our dimensionless formulation, the crossflow is equal to $E^{1/2}\mathbf{e}_z \times \mathbf{u}_0$, so that the global 2D flow can be written:

$$\mathbf{Q} = d\,\mathbf{u}_0 + E^{1/2}\mathbf{e}_z \times \mathbf{u}_0. \tag{10}$$

Finally, one must write the restriction of the momentum Eq. (1) to the x and y components in the core of the flow, in terms of the 2D vector field $\mathbf{u}_0$ and the pressure field $p = p_0(x, y)$ at $z = 0$:

$$\frac{\partial \mathbf{u}_0}{\partial t} + (\mathbf{u}_0 \cdot \nabla)\mathbf{u}_0 = -\nabla p_0 + 2\,E^{-1}\mathbf{u}_0 \times \mathbf{e}_z + \nabla^2 \mathbf{u}_0. \tag{11}$$

Let us denote the single component (in the direction of the rotation axis) of the curl of $\mathbf{u}_0$ by ω_0. We shall take the curl of the restricted momentum Eq. (11), bearing in mind that all variables are 2D:

$$\frac{\partial \omega_0}{\partial t} + \mathbf{u}_0 \cdot \nabla \omega_0 + (\nabla \cdot \mathbf{u}_0)\omega_0 = -2\,E^{-1}(\nabla \cdot \mathbf{u}_0) + \nabla^2 \omega_0. \tag{12}$$

The divergence of $\mathbf{u}_0$ can be expressed from the 2D divergence of Eq. (10), where $\mathbf{Q}$ is solenoidal:

$$\nabla \cdot \mathbf{Q} = d\nabla \cdot \mathbf{u}_0 + \mathbf{u}_0 \cdot \nabla d - E^{1/2}\omega_0 = 0, \tag{13}$$

hence

$$\nabla \cdot \mathbf{u}_0 = -\frac{1}{d}\mathbf{u}_0 \cdot \nabla d + E^{1/2}\frac{\omega_0}{d}. \tag{14}$$

In the first term of the right-hand side, one can express $\mathbf{u}_0$ with $\mathbf{Q}$ at the leading order: $\mathbf{u}_0 = \mathbf{Q}/d$. Eq. (14) becomes:

$$\nabla \cdot \mathbf{u}_0 = \mathbf{Q} \cdot \nabla \left(\frac{1}{d}\right) + E^{1/2}\frac{\omega_0}{d}, \tag{15}$$

where the small $E^{1/2}\omega_0/d$ term is necessarily retained as $\nabla \cdot \mathbf{u}_0$ will be multiplied by the large E^{-1} factor in Eq. (12). Next, the non-linear term $\mathbf{u}_0 \cdot \nabla \omega_0$ will be approximated using $\mathbf{u}_0 \simeq \mathbf{Q}/d$ so as to retain only consistent orders of magnitude in a $E^{1/2}$ Taylor expansion. Eq. (12) can finally be written as follows:

$$\frac{\partial \omega_0}{\partial t} + \nabla \left(\frac{\omega_0}{d}\right) \times \nabla \psi = -2\,E^{-1}\nabla \left(\frac{1}{d}\right) \times \nabla \psi - 2E^{-1/2}\frac{\omega_0}{d} + \nabla^2 \omega_0. \tag{16}$$

Note that all terms are scalars in the 2D equation above: cross products are identified with their single non-zero z-component. Note also that vorticity transport and geometrical vortex stretching have been combined into a single term: $\nabla \omega_0 \times (\nabla \psi/d) + \omega_0 \nabla (1/d) \times \nabla \psi = \nabla (\omega_0/d) \times \nabla \psi$. These are local vorticity terms, not planetary or background vorticity. Finally, as we have carefully retained dominant terms in the vorticity equation above, we can again use $\mathbf{u}_0 \simeq \mathbf{Q}/d$ to express vorticity in terms of the streamfunction ψ:

$$\omega_0 = -\nabla \cdot \left(\frac{\nabla \psi}{d}\right). \tag{17}$$

The 2D Eqs. (16) and (17) constitute the so-called homogeneous model of quasi-geostrophic flows.

The homogeneous model can be entered into a numerical formulation, and quasi-geostrophic flows can be computed. Here, we shall only have a look at this model to identify some expected features. In a steady or quasi-steady low-Rossby number regime, there can be a competition between the last two terms (dissipation by Ekman layer friction or bulk 2D viscous dissipation), which results in the development of $E^{1/4}$ thick Stewartson layers. This feature is legitimate in the quasi-geostrophic model, as $E^{1/4}$ is thicker than $E^{1/3}$: as it is recalled in § 2.1, the length-scales perpendicular to the rotation axis of the order $E^{1/3}$ or smaller are not associated with two-dimensional structures, but to 3D ones. Apart from Stewartson layers, there can be Munk layers [16] when topography effects exist: the first and last terms at the right-hand side of Eq. (16) compete on a length scale $E^{1/3}\beta^{-1/3}$, where β is the magnitude of $\nabla(1/d)$. When β is very small compared to unity, this length scale is large compared to $E^{1/3}$ and corresponds to a valid quasi-geostrophic flow structure. When unsteady solutions are considered, the easiest case is that of a balance between the time-dependent term and the dominant term in Ekman number expansion, that of planetary vortex stretching. Considering local coordinates (x,y) such that the depth d is constant along the x direction, this balance in Eq. (16) can be written as follows:

$$\frac{\partial \omega_0}{\partial t} = 2\,E^{-1}\beta \frac{\partial \psi}{\partial x}. \tag{18}$$

This equation is generating the so-called Rossby waves. When β is small compared to unity, homogeneous elementary solutions $e^{i\omega t + i\mathbf{k}\cdot\mathbf{x}}$ must satisfy:

$$\omega = 2\,E^{-1}\beta \frac{k_x}{k^2}. \tag{19}$$

These waves live in two dimensions and are slower than inertial waves. Rossby waves are indeed the trace of inertial waves in the 2D plane perpendicular to the axis of rotation.

4 A two-dimensional MHD model

The derivation of a 2D MHD model presented here follows very closely the presentation of the model of homogeneous rotating flows above. Global 2D conservation laws (for mass and electric charge) are combined to the evolution equation for vorticity and electric current in the bulk of the fluid to provide two coupled governing equations.

The magnetic field is supposed to lie in the z-direction, but its intensity can now be a function of x and y, $B_z = B_z(x, y)$: this is known as the "straight magnetic lines approximation". As a first step, we consider that the "horizontal" components (perpendicular to $\mathbf{B}$) of the velocity and also of the electric current density are independent of z in the bulk of the fluid:

$$\mathbf{u}_0 = \begin{bmatrix} u_{0x}(x,y) \\ u_{0y}(x,y) \end{bmatrix}, \qquad \mathbf{j}_0 = \begin{bmatrix} j_{0x}(x,y) \\ j_{0y}(x,y) \end{bmatrix}. \tag{20}$$

The 2D integrated volume flow rate is still denoted $\mathbf{Q}(x, y)$ and we introduce also the 2D integrated electric current density $\mathbf{I}(x, y)$. We shall then consider the 2D mass and electric charge conservation laws after integration along z over the total depth $d(x, y)$ of the cavity (from the lower wall $z = -z_u(x, y)$ to the upper wall $z = z_u(x, y)$). This implies the existence of two streamfunctions:

$$\mathbf{Q} = \nabla \psi \times \mathbf{e}_z, \qquad \mathbf{I} = \nabla h \times \mathbf{e}_z, \tag{21}$$

where the variable name $h(x, y)$ for the electric current streamfunction recalls us that this is indeed the induced magnetic field in the z direction. The next step is to express the 2D, integrated fluxes in terms of a bulk flow contribution (20) and a contribution from Hartmann layers. In MHD, the flow deficit in

Hartmann layers is of little importance. The electric current developing in them is the important feature [12]:

$$\mathbf{Q} = d\,\mathbf{u}_0, \qquad\qquad \mathbf{I} = d\,\mathbf{j}_0 + 2\,Ha^{-1}B_z\mathbf{e}_z \times \mathbf{u}_0, \qquad (22)$$

written here for the case of a slowly varying depth $\partial d/\partial x << 1, \partial d/\partial y << 1$. We then consider the restriction of the momentum Eq. (4) and of Ohm's law (5) to the x and y components in the core of the flow, in terms of the 2D vector fields $\mathbf{u}_0$, $\mathbf{j}_0$ and pressure and electric potential fields at $z = 0$, $p = p_0(x, y)$, $\phi = \phi_0(x, y)$:

$$\frac{\partial \mathbf{u}_0}{\partial t} + (\mathbf{u}_0 \cdot \nabla)\mathbf{u}_0 = -\nabla p_0 + Ha^2 B_z\mathbf{j}_0 \times \mathbf{e}_z + \nabla^2\mathbf{u}_0, \qquad (23)$$

$$\mathbf{j}_0 = -\nabla\phi + B_z\mathbf{u}_0 \times \mathbf{e}_z. \qquad (24)$$

Taking the curl of these 2D equations leads to:

$$\frac{\partial \omega_0}{\partial t} + \mathbf{u}_0 \cdot \nabla\omega_0 + (\nabla \cdot \mathbf{u}_0)\omega_0 = -Ha^2\nabla \cdot (B_z\mathbf{j}_0) + \nabla^2\omega_0, \qquad (25)$$

$$\nabla \times \mathbf{j}_0 = -\nabla \cdot (B_z\mathbf{u}_0). \qquad (26)$$

From Eq. (22), divergence terms in the equations above can be written:

$$\nabla \cdot (B_z\mathbf{j}_0) = (\mathbf{I} \cdot \nabla)\left(\frac{B_z}{d}\right) + 2Ha^{-1}\mathbf{e}_z \cdot \nabla \times \left(\frac{B_z}{d^2}\mathbf{Q}\right), \qquad (27)$$

$$\nabla \cdot (B_z\mathbf{u}_0) = (\mathbf{Q} \cdot \nabla)\left(\frac{B_z}{d}\right), \qquad (28)$$

which can be substituted into the 2D governing Eqs. (23) and (24). Using the streamfunctions ψ and h defined by Eq. (21), the governing equations are finally written:

$$\frac{\partial \omega_0}{\partial t} + \nabla\left[\frac{\omega_0}{d}\right] \times \nabla\psi = Ha^2\nabla\left[\frac{B_z}{d}\right] \times \nabla h + 2Ha\nabla \cdot \left[\frac{B_z}{d^2}\nabla\psi\right] + \nabla^2\omega_0, \qquad (29)$$

$$-\nabla \cdot \left[\frac{1}{d}\nabla h\right] = \nabla\left[\frac{B_z}{d}\right] \times \nabla\psi, \qquad (30)$$

where Eq. (17) relating ω_0 and ψ is still valid here, in the MHD context.

These governing Eqs. (17), (29), and (30) will be discretized and solved numerically in § 5 for a particular configuration. Let us here first analyse these equations from a general point of view in the same way as for the homogeneous model of rotating flows. One can first look at the structures arising from the balance of bulk viscosity and Hartmann layer friction, i.e., between the last two terms in Eq. (29). By scaling analysis, this corresponds to shear layers of thickness $Ha^{-1/2}$. This corresponds to the typical thickness of parallel layers (see § 2). This does not mean however that we are representing parallel layers well with the 2D model. On the contrary, we know that structures developing on a length scale of unity in the direction of the magnetic field and on a length-scale $Ha^{-1/2}$ or less in a direction perpendicular to the magnetic field are essentially 3D. We may obtain a solution looking like a parallel layer, but this is just an approximation. And this approximation does not become any more accurate as the Hartmann number is increased. As a corollary, we may just remove the last term (bulk viscosity) from Eq. (29) as it will always be negligible compared to the term before (Hartmann layer friction), as long as legitimate 2D structures are modelled. The situation has to be contrasted with that of rotating flows for which a similar balance had provided a legitimate $E^{1/4}$ length scale (2D Stewartson layers), larger than 3D $E^{1/3}$ Stewartson layers[1].

For steady, high interaction parameter flows, one may examine the balance between Hartmann layer friction and the first term on the right-hand side of Eq. (29), representing a source of vorticity as the electric current passes across characteristic surfaces (B_z/d constant). To do so, we must use the electric Eq. (30), expressing the fact that curl is generated for the 2D current when the 2D flow passes across characteristic surfaces. In order to combine more easily Eqs. (29) and (30), it will be useful to write them in local coordinates (x, y), such that the characteristic function B_z/d is constant along the x axis. In the steady, inertialess regime, denoting the magnitude of $\nabla(B_z/d)$ by G and neglecting other variations of B_z and d for simplicity, the governing 2D equations take the following form:

$$0 = -Ha^2 G \frac{\partial h}{\partial x} + 2Ha \frac{B_z}{d^2} \nabla^2 \psi, \qquad (31)$$

$$-\frac{1}{d} \nabla^2 h = -G \frac{\partial \psi}{\partial x}. \qquad (32)$$

Taking the Laplacian of Eq. (31) and substituting h using Eq. (32), one gets a local equation for ψ, bearing in mind that all variations of B_z and d have been discarded, except obviously when building G:

$$-HaG^2 d \frac{\partial^2 \psi}{\partial x^2} + 2\frac{B_z}{d^2} \nabla^4 \psi = 0. \qquad (33)$$

This balance shows that shear layers of thickness $Ha^{-1/4}G^{-1/2}$ and length unity can develop along characteristic surfaces, as illustrated in [15]. This equation can also be used to derive the development length $Ha^{1/2}G$ observed for instance in pressure-driven flows in cylinders with a step transverse magnetic field. Another outcome of Eq. (33) takes the form of boundary layers along walls cutting characteristic surfaces. They are equivalent to Munk layers in rotating fluids, or rather Stommel layers [17], as the balance in Eq. (33) involves topographic effects and Hartmann layer friction. In this case, scaling analysis of Eq. (33) provides a typical thickness $Ha^{-1/2}G^{-1}$. When $G << 1$,

---

[1] The fact that both structures have been discovered by Stewartson and named after him does not help to distinguish them.

this scaling is legitimate with respect to the 2D nature of the solution. In reality, there is probably a double layer structure in most practical cases, with a genuine $Ha^{-1/2}$ (3D) parallel sublayer within this thicker $Ha^{-1/2}G^{-1}$ 2D layer. The steady features above have been described in more details in [15,18].

Finally, one can take a look at the unsteady 2D regime in the simplest regime of negligible non-linear inertial terms. The time-dependent term in Eq. (29) is balanced by the dominant term in terms of Hartmann number expansion, $Ha^2\nabla[B_z/d]\times\nabla h$. Using again the local coordinates defined above, and neglecting variations of B_z and d, the following equation can be derived:

$$\frac{1}{d}\frac{\partial}{\partial t}\left(\nabla^2\right)^2\psi = Ha^2G^2d\frac{\partial^2\psi}{\partial x^2}.$$

Solutions to this equation are the counterpart of Rossby waves in rotating fluids. Starting from elementary $e^{i\omega t + i\mathbf{k}\cdot\mathbf{x}}$ the dispersion equation is:

$$\omega = iHa^2G^2d^2\frac{k_x^2}{k^4},\tag{34}$$

typical of a pseudo-diffusion equation. This is obviously reminiscent of the 3D pseudo-diffusion in the direction of the magnetic field. For 2D flows, this has become pseudo-diffusion in the direction of the characteristic surfaces, with a pseudo-diffusion coefficient, $Ha^2G^2d^2l^4$, where l is the typical size of the structure considered. When G is very small compared to unity, this pseudo-diffusion coefficient is small compared to that of 3D MHD pseudo-diffusion.

5 A flow in a spatially varying magnetic field

This section will be devoted to an example of an MHD 2D flow, calculated numerically from Eqs. (17), (29), and (30). As can be seen in Fig. 1, the configuration is that of a duct of a rectangular cross section with a (smooth) step transverse magnetic field. Two apparently slightly different configurations will be considered: one in which the shape of the duct is strictly rectangular and another one in which, downstream the magnetic field change, the top and bottom duct walls are slightly curved. The difference lies in the fact that characteristic surfaces are degenerate in the first case, while characteristic surfaces are well defined and aligned with the mean flow direction in the second case, downstream the magnetic field step.

To be more specific, the width of the duct is equal to 2 in dimensionless terms while its length is 24, from $x = -6$ to 18, the change in magnetic field intensity taking place between $x = -2$ and $+2$:

$$\begin{aligned} B_z &= 0.2, & for \quad & x \geq 2, \\ B_z &= 1 - 0.6x + 0.05x^3, & for \quad & -2 \leq x \leq 2, \\ B_z &= 1.8, & for \quad & x \leq -2, \end{aligned}\tag{35}$$

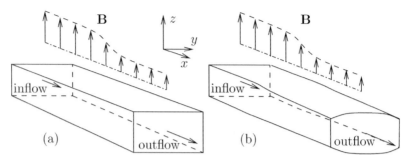

Fig. 1. Rectangular ducts with transverse step magnetic field: (a) flat case, (b) curved case

ensuring a smooth magnetic field and first derivative for all values of x. The depth of the duct in the direction of the magnetic field is uniform and equal to 1 for the first case (a), and varies in the following way for case (b):

$$
\begin{aligned}
d &= 1, & for \quad & x \le -2, \\
d &= 1 - (0.25 + 0.1875x - 0.015625x^3)y^2, & for \quad & -2 \le x \le 2, \\
d &= 1 - 0.5y^2, & for \quad & x \ge 2.
\end{aligned}
\tag{36}
$$

Figure 2 shows lines of constant value for the characteristic function B_z/d for both configurations (a) and (b), in the (x, y) plane. Numerical simulations have been run for a single value of the Hartmann number, $Ha = 5 \times 10^3$, and increasing values of the Reynolds number (see Fig. 1). In these simulation, the small bulk viscous term has been retained but boundary conditions at $y = \pm 1$ are free-slip conditions for the velocity field. A no-slip condition would be numerically more demanding as the parallel layers would cause considerable mesh refinement. Moreover, we are not particularly interested in $Ha^{-1/2}$ parallel layers in this study. The smallest value of Reynolds number corresponds to an inertialess solution. In the region of varying magnetic field, the flow goes through boundary layers, as expected [15], scaling like $Ha^{-1/2}G^{-1}$ as discussed in the previous section. For larger values of the Reynolds number, a clear asymmetry appears between upstream and downstream regions on each side of the magnetic field change. Upstream the flow is quite similar to the inertialess case, while downstream the wall jets, generated in the non-uniform region of magnetic field, remain and undergo instability. These instabilities lead to vortices, eventually damped by Hartmann layer friction.

The difference between cases (a) and (b) is visible in the behaviour of the wall jets. In Fig. 2, snapshots of the vorticity isovalues are shown in the final statistically steady regime. For the case (a) of a strictly rectangular duct, the jets undergo shear instability and mixing is rather efficient. For the case (b) of non-uniform duct depth, the characteristic surfaces channel the jets and delay their instability, so that these jets survive further downstream. As discussed in the previous section, vortices are stretched along the x-direction by pseudo-

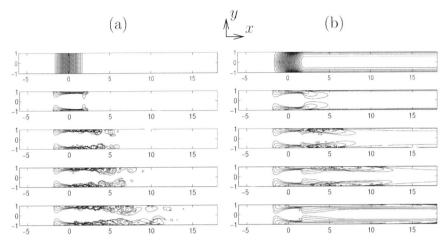

Fig. 2. Isovalues of the characteristic function B_z/d (*top*), vorticity isolines for increasing Reynolds, $Re = 5$, 10^3, 2×10^3 and 4×10^3: (a) flat case, (b) curved case

diffusion, which is probably at the origin of the observed stabilization of the jets. Their flow direction coincides with the orientation of the surface layers which helps them to propagate further downstream.

6 Conclusions and perspectives

Rotating and MHD flows have been compared for a long time. The comparison is extended here to show that the 2D models can be derived within the same framework in both cases. The 2D equations are different indeed but they both indicate that the evolution of (local) vorticity is subjected to four effects: (1) non-linear transport and stretching of local vorticity; (2) "topographic" constraint, related to the conservation of background circulation or magnetic circulation; (3) Ekman or Hartmann layer friction; and (4) 2D bulk viscosity. These terms involve different derivation orders and their magnitude have different scalings with respect to the Ekman or Hartmann number. Hence, various shear layers can arise with a range of different thicknesses. Attention must be paid to the lower acceptable size for these 2D structures, $Ha^{-1/2}$ for MHD flows, and $E^{1/3}$ for rotating flows, under which structures are necessarily 3D.

We have also discussed the unsteady behaviour of 2D flows. Rossby waves dominate 2D rotating flows unless there is no topographic (or beta-plane) effect. Correspondingly, we have shown here that 2D unsteady MHD flows are subjected to pseudo-diffusion in the direction of characteristic surfaces, unless there is no topography and the magnetic field is uniform[2]. These unsteady 2D

---

[2] By the way, this shows that the case of a uniform transverse magnetic field and parallel walls constitutes a very singular case.

features can be linked to the basic unsteady features of the 3D flows. Inertial waves responsible for the 2D nature of rotating flows reappear as Rossby waves in the (imperfectly) 2D beta-plane configuration. Similarly, pseudo-diffusion responsible for the 2D nature of MHD flows has a pseudo-diffusion reminiscence in the 2D MHD equations.

The analysis of the MHD equations has been restricted here to the case of electrically insulating boundaries. It would be interesting to extend this work to electrically conducting boundaries for practical applications, as done by Bühler and Molokov [19]. Another interesting extension for practical applications (e.g., cooling by liquid metal films of fusion reactors) would be to study the effect of a free-surface flow, as initiated by Molokov [20]. Combining MHD 2D modelling and free-surface analysis should result in a "shallow water" model for MHD flows.

Let us mention other effects not included in the present work which can affect MHD flows. Inertial effects can become significant in the Hartmann layers themselves at sufficiently high Reynolds number. This causes the appearance of Ekman pumping within the Hartmann layers. In consequence, the 2D equations contain an extra term. This effect has been studied by Pothérat et al. [21] and by Dellar [22] in the absence of topography. One can even go a step further and consider the effects of instability [23] and transition to turbulence [24] in Hartmann layers. One can envisage the possibility of a 2D flow with turbulent Hartmann layers. This is analogous to the case of atmospheric or oceanic studies where a turbulent Ekman layer is adjacent to a 2D large-scale flow.

From a theoretical point of view, an important question is related to the nature of the MHD 2D turbulence under the effect of the pseudo-diffusive effects described above. As we know, beta-plane turbulence, studied initially by Rhines [25] departs very significantly from pure 2D turbulence. This turbulence is characterized by a mixture of non-linear inertial terms and linear Rossby waves. Not much has been done so far to characterize 2D MHD turbulence, in the presence of topography or non-uniform magnetic fields.

References

1. Moffatt HK (1967) On the suppression of turbulence by a uniform magnetic field. J Fluid Mech 28(3):571–592
2. Sommeria J, Moreau R (1982) Why, how, and when, MHD turbulence becomes two-dimensional. J Fluid Mech 118:507–518
3. Greenspan HP (1968) The Theory of Rotating Fluids. Cambridge University Press, Cambridge
4. Proudman J (1916) On the motion of solids in liquids possessing vorticity. Proc Roy Soc A 92:408–424
5. Taylor GI (1921) Experiments with rotating fluids. Proc Roy Soc A 100:114–121
6. Taylor GI (1923) Experiments on the motion of solid bodies in rotating fluids. Proc Roy Soc A 104:213–218

7. Montgomery RB (1937) A suggested method for representing gradient flow in isentropic surfaces. Bull Am Meteorol Soc 18:210–212

8. Ekman VW (1905) On the influence of the Earth's rotation on ocean-currents. Arkiv för Metamatik Astronomi och Fysik 2(11):1–52

9. Stewartson K (1957) On almost rigid rotations. J Fluid Mech 3:17–26

10. Stewartson K (1966) On almost rigid rotations. Part 2. J Fluid Mech 26:131–144

11. Hartmann J, Lazarus F (1937) Experimental investigations on the flow of mercury in a homogeneous magnetic field. K Dan Vidensk Selsk Mat-Fys Medd 15(7):1–45

12. Shercliff JA (1953) Steady motion of conducting fluids under transverse magnetic fields. Proc Camb Phil Soc 49:136–144

13. Kulikovskii AG (1968) Slow steady flows of a conducting fluid at high Hartmann numbers. Izv Akad Nauk SSSR Mekh Zhidk i Gaza 3:3–10

14. Holroyd RJ, Walker JS (1978) A theoretical study of the effect of wall conductivity, non-uniform magnetic fields and variable-area ducts on liquid-metal flows at high Hartmann numbers. J Fluid Mech 84(3):471–495

15. Alboussière T (2004) A geostrophic-like model for large-Hartmann-number flows. J Fluid Mech 521:125–154

16. Munk WH (1950) On the wind-driven ocean circulation. J Meteor 7(2):79–93

17. Stommel H (1948) The westward intensification of wind-driven ocean currents. Trans Am Geophys Union 29:202–206

18. Walker JS, Ludford GSS (1972) Three-dimensional MHD duct flows with strong transverse magnetic fields. Part 4. Fully insulated, variable-area rectangular ducts with small divergences. J Fluid Mech 56(3):481–496

19. Bühler L, Molokov S (1994) Magnetohydrodynamic flows in ducts with insulating coatings. Magnetohydrodynamics 30:439–447

20. Molokov S (2002) Evolution of a free surface in a strong vertical magnetic field. In: Proceedings of the 5th International Conference on Fundamental and Applied MHD (PAMIR), Ramatuelle, France, 16–20 Sept 2002 2:65–70

21. Pothérat A, Somméria J, Moreau R (2005) Numerical simulations of an effective two-dimensional model for flows with a transverse magnetic field. J Fluid Mech 534:115–143

22. Dellar PJ (2004) Quasi-two-dimensional liquid-metal magnetohydrodynamics and the anticipated vorticity method. J Fluid Mech 515:197–232

23. Lingwood RJ, Alboussière T (1999) On the stability of the Hartmann layer. Phys Fluids 11(8):2058–2068

24. Alboussière T, Lingwood RJ (2000) A model for the turbulent Hartmann layer. Phys Fluids 12(6):1535–1543

25. Rhines PB (1975) Waves and turbulence on a beta plane. J Fluid Mech 69:417–443

Part III

Turbulence

The Birth and Adolescence of MHD Turbulence

Keith Moffatt

Trinity College, Cambridge, CB2 1TQ, United Kingdom
(H.K.Moffatt@damtp.cam.ac.uk)

Summary. This essay provides a personal account of the development of the subject of magnetohydrodynamic (MHD) turbulence from its birth in 1950 to its "coming-of-age" in 1971, following the development of mean-field electrodynamics, a major breakthrough of the 1960s. The discussion covers the early ideas based on the analogy with vorticity, the passive vector problem, the suppression of turbulence by an applied magnetic field, and aspects of the turbulent dynamo problem.

1 Birth pangs of the 1950s

The conception of the state of magnetohydrodynamic (MHD) turbulence dates from G.K. Batchelor's seminal paper "On the spontaneous magnetic field in a conducting liquid in turbulent motion", published in the Proceedings of the Royal Society in 1950 [1]. At that time, it was already recognised that, just as vortex lines are, under ideal circumstances, frozen in the fluid (i.e., transported with conservation of flux), so the magnetic lines of force are similarly *frozen* in a conducting fluid (again with conservation of flux) in the ideal perfect-conductivity limit. The evolution equations for magnetic field $\mathbf{B}$ in a perfectly conducting fluid, and for $\boldsymbol{\omega}$ in an ideal fluid (with no magnetic effects), are then superficially identical. The word *superficially* is here deliberate, the analogy between $\mathbf{B}$ and $\boldsymbol{\omega}$ being imperfect, in that $\boldsymbol{\omega}$ is constrained, by its very definition, to be equal to the curl of the velocity field $\mathbf{u}$ that transports it, whereas $\mathbf{B}$ of course suffers no such constraint. There is thus far greater freedom in the choice of initial conditions for the pair of fields $(\mathbf{u}, \mathbf{B})$ than for the pair $(\mathbf{u}, \boldsymbol{\omega})$. This imperfection was, I believe, recognised by Batchelor, but dismissed as irrelevant; what was important for him was the physical fact that the velocity $\mathbf{u}$ transports both $\boldsymbol{\omega}$ and $\mathbf{B}$ in a similar way, and that the statistics of these fields may therefore be expected to evolve in a correspondingly similar way, perhaps after the decay of transients associated with initial conditions. In this regard, as would emerge much later, Batchelor was at best only partially correct.

S. Molokov et al. (eds.), Magnetohydrodynamics – Historical Evolution and Trends, 213–222. © 2007 *Springer.*

Fig. 1. Portrait of G.K. Batchelor FRS by artist Rupert Shepherd (1984). The portrait hangs in DAMTP, Cambridge

Of course both ω and $\mathbf{B}$ are subject to diffusive effects, viscous in the case of ω, with diffusivity the kinematic viscosity ν, and resistive in the case of $\mathbf{B}$, with diffusivity the magnetic resisitivity η. However, if $\eta = \nu$, Batchelor's analogy (albeit imperfect for the above reason) persists, and it was this analogy that Batchelor sought to exploit in the context of turbulence. Here, the recognised and very contemporary scenario was that in homogeneous turbulence in an incompressible fluid, the rate of viscous dissipation of mean square vorticity (or *enstrophy* as it is now called) always adjusts itself to be in approximate equilibrium with its rate of production by the all-pervasive process of stretching of vortex lines. If ν were to suddenly decrease (through the agency of some Maxwell demon), then the enstrophy would *increase*, but at the same time, the characteristic length scale of the vorticity field would *decrease* till a new equilibrium at a higher enstrophy level is established. Batchelor argued that, if $\eta < \nu$, the magnetic energy (the analogue of enstrophy) will similarly *increase* through stretching of magnetic lines of force, and he inferred an exponential increase of this energy on the Kolmogorov timescale $(\nu/\epsilon)^{1/2}$ characteristic of the small scales of the turbulence where the enstrophy spectrum is maximal. Batchelor's condition $\eta < \nu$ is satisfied in the interstellar medium where the density is low and the kinematic viscosity ν is correspondingly large.

Batchelor further argued that, when this condition is satisfied and dynamo action occurs, then, as a result of the back-reaction on the turbulence of the growing Lorentz force distribution, the mean magnetic energy density will saturate at a level of order $(\nu\epsilon)^{1/2}$, this being the energy density characteristic of the smallest scales of motion. It was arguable however that, even if saturation is quickly achieved on the Kolmogorov scale, magnetic modes could continue to grow through the familiar stretching mechanism on larger scales l (where the characteristic velocity is $u \sim (\epsilon l)^{1/3}$) provided simply that the local magnetic Reynolds number $R_m = ul/\nu$ is larger than some critical value of order unity. Saturation would then be progressively established at all scales satisfying this criterion, at a level of magnetic energy (*equipartition*) equal to the local (in scale) kinetic energy of the turbulence. This was the alternative turbulent dynamo scenario proposed also in 1950 by Schlüter and Biermann [2], a scenario that was more readily accepted by the astrophysical community. It was a view further developed in the review article of Syrovatsky 1957 [3], which revealed for the first time the high current level of interest and activity in MHD in the former Soviet Union. The two standpoints were considered in Cowling's influential monograph *Magnetohydrodynamics* [4], published in 1957, who however concluded his penetrating discussion with the statement *"These remarks serve to illustrate how unsatisfactory is the present state of the theory of magnetohydrodynamic turbulence. ... Work decisively distinguishing between these standpoints is still to be awaited"*. Within the fluid mechanics community, Batchelor's theory, based on the above *analogy with vorticity*, undoubtedly retained its appeal, but there seemed little prospect of providing convincing proof of its validity by theoretical argument. Nor of course was there at that time any prospect of either numerical simulation or laboratory experiment that could even remotely approach the range of parameters where (on either theory) turbulent dynamo action might be anticipated. Only quite recently (see, e.g., [5]) are numerical simulations at sufficiently high R_m becoming possible.

I was fortunate to start my own research in this field, under Batchelor's supervision and guidance, in 1958. Batchelor had just completed his study of the *passive scalar problem*, and he gave me an advance copy of two famous papers on this topic [6, 7] published in JFM one year later. I had been much influenced by Cowling's monograph, and also by lectures on *Cosmical Electrodynamics* given that year by Leon Mestel in Part III of the Cambridge Mathematical Tripos. It seemed to me that the techniques that Batchelor had used for the passive scalar problem might be adapted to the *passive vector problem*, which is of course just the *kinematic dynamo problem* as we now understand it. Batchelor had originally suggested that I work on the problem of the effect of turbulence on the rate of evaporation of droplets in a turbulent airflow; but he readily agreed to this change of focus. Thus, it was that in 1959 I started to think about the detailed nature of the back-reaction of Lorentz forces under the condition $\eta \ll \nu$, when Batchelor's criterion for dynamo growth of magnetic energy is well satisfied. This was to provide the core of my Ph.D. thesis *Magnetohydrodynamic Turbulence* (1962); my paper on this topic appeared 1 year later [8].

2 Marseille 1961: a definitive moment

An important Colloquium *Mécanique de la Turbulence*, sponsored by CNRS, was held in August 1961 on the occasion of the inauguration of the Institut de Mécanique Statistique de la Turbulence in Marseille. The meeting was distinguished by the presence of the great pioneers of the subject, G.I. Taylor, Th. Von Karman, and A.N. Kolmogorov himself. It was the occasion when the first reliable observational evidence in support of the Kolmogorov $k^{-5/3}$ spectrum was first presented by R.W. Stewart (later published by Grant et al. (1962) [9]), only to be followed by Kolmogorov's remarkable contribution *Précisions sur la structure locale de la turbulence dans un fluide visqueux aux nombres de Reynolds élevés* [10] in which he addressed the problem of the intermittency of the local rate of dissipation $\epsilon(\mathbf{x}, t)$, and showed how this could be expected to modify the (-5/3) exponent of the energy spectrum; thus did Kolmogorov undermine the very foundations of the study of turbulence that he had himself laid 20 years previously; it was indeed a revolutionary moment for the subject!

The Colloquium included a section chaired by L.S.G. Kovasznay on *Turbulence in Compressible and Electrically Conductive Media*, to which I was privileged to contribute. It is perhaps an indication of the primitive state of the subject that, apart from myself, only Kovasznay spoke on the subject of turbulence in conducting fluids. He drew attention to the evidence for the presence of turbulence in plasma experiments, and of the need to take account of terms analogous to Reynolds stress in the time-averaged equations of MHD, namely $\langle \mathbf{u} \times \mathbf{B} \rangle$ in the mean induction equation, and $\langle \mathbf{j} \times \mathbf{B} \rangle$ in the mean momentum equation; this seems so absolutely natural now that it is difficult to appreciate how novel, and indeed daring, such a suggestion still appeared at that time. Kovasznay [11] had been primarily concerned with situations typical of plasma experiments in which the source of energy is electromagnetic, and energy flows via MHD instabilities of various kinds to the turbulence with resulting enhancement of the rate of Joule dissipation of energy. In this respect, his approach was complementary to that of the dynamo theoreticians, who were concerned with circumstances when the source of energy was purely dynamic, and the flow of energy was from the resulting turbulence to the magnetic field. I did my best in my contribution [12] to distinguish the main features of these contrasting situations.

3 The low-R_{m} situation

It was recognised by Golitsyn (1960) [13] that, when R_{m} is small, an applied magnetic field is weakly perturbed by turbulence, and the induction equation may therefore be linearised in order to obtain the fluctuating field **b** in terms of the velocity field **u**. The spectrum $\Gamma(k)$ of **b** may then be obtained in terms of the spectrum of **u**, with the result that $\Gamma(k) \sim k^{-11/3}$ (a result that may be compared with the corresponding $k^{-17/3}$ law obtained by Batchelor et al. [7] for the passive scalar case).

On the assumption that Batchelor's criterion $\eta < \nu$ for dynamo action was correct, I considered at about the same time [14] the situation of moderate conductivity, when η is large compared with ν but η still small enough that $R_{\mathrm{m}} \gg 1$; i.e., when the Reynolds number Re of the turbulence satisfies $Re \gg R_{\mathrm{m}} \gg 1$. I argued the case for a $k^{1/3}$ spectrum (like that of vorticity) up to a *conduction cut-off*, $k_c = (\epsilon/\eta^3)^{1/4}$, and a $k^{-11/3}$ spectrum, like Golitsyn's result (and for similar reasons) above k_c. The $k^{-11/3}$ result has been found by Odier et al. (1998) [15] in experiments on turbulence in liquid gallium, a welcome and long-awaited validation of ideas that were both rudimentary and tentative in those early days of the subject.

4 The high-R_{m} situation

The situation when $\eta \ll \nu$ is very different. Here, magnetic fluctuations persist on sub-Kolmogorov scales where the velocity gradient may reasonably be assumed to be approximately uniform. Batchelor [6] had exploited this idea to determine a k^{-1} law for the spectrum of a passive scalar. The same arguments applied to magnetic field (treated as a passive vector) led to an unacceptably divergent k^{+1} spectrum, possibly a symptom of dynamo instability. In fact, it had been shown at about the same time by Pearson (1959) [16] that if a weak random vorticity field is subjected to uniform irrotational strain, then the associated enstrophy in general increases exponentially, and this in spite of the effect of viscosity. This result carried over by analogy to the effect of a similar uniform straining motion on a random magnetic field: the mean magnetic energy increases exponentially, despite the effect of Joule dissipation. This surprising result is perhaps attributable to the unphysical assumption of a strain field that is uniform to infinity, but nevertheless it suggested that stretching of field lines could persist until the growing Lorentz force reacted back upon this strain field, in a way that might lead to structures in which the straining process was exactly compensated by this back-reaction. I did indeed find such structures [8], although it seemed inevitable that finite diffusivity would lead to some leakage of magnetic flux, causing slow decay. It was argued at about the same time by Saffman (1963) [17] that the decrease of scale associated with the stretching process during the kinematic phase would ultimately lead to decay of magnetic energy; this prediction has not been substantiated by later developments.

There has been a recent renewal of interest in such "small-scale dynamo action" stimulated by the work of Kulsrud (1999) [18], with reference to processes in the interstellar medium (see also Schekochihin et al. (2004) [19] and the extensive bibliography therein). The availability of high-speed computer power opens up new possibilities for the investigation of this regime. Together with Y. Hattori, I have recently returned to the study of isolated "magnetic eddies" in the perfect conductivity limit [20]. Even without imposed strain, the behaviour of such eddies under the action of the Lorentz force distribution is of interest! It turns out that, in the simplest case, an axisymmetric magnetic eddy can contract towards the axis of symmetry and split into two nearly spherical eddies which propagate away from each other along the axis of symmetry. These are candidate "coherent structures" of MHD turbulence in the high conductivity limit.

5 Suppression of turbulence by a strong applied field

My first research student at Cambridge in the 1960s was Jacques Nihoul from Liège. One of the problems that he worked on was the effect of a suddenly applied magnetic field on a field of homogeneous turbulence at low R_m. The fact that a magnetic field could suppress turbulence had been demonstrated experimentally by Murgatroyd (1953) [21]; and it was already recognised from the work of Lehnert (1955) [22] and Shercliff (1965) [23] that vorticity components perpendicular to a uniform applied field tend to be preferentially suppressed; this process is effectively linear, and Nihoul (1965) [24] found that the turbulent energy decays as t^{-3} during this suppression phase. I carried this work somewhat further [25] and showed the manner in which anisotropy develops from an initially isotropic state: in fact, the anisotropy ratio $\langle u^2+v^2\rangle/\langle w^2\rangle$ decreases from the isotropic value of 2 asymptotically to 1 during this phase, where u, v are the velocity components perpendicular to the field, and w is the component parallel to the field. This result holds only insofar as the fluid can be regarded as unbounded; much work has since been done on the non-linear effects which resist the anisotropisation process (Sommeria and Moreau (1982) [26]) and on the effect of fluid boundaries perpendicular to the applied field (Pothérat et al. (2000) [27]); but the fact that a strong field induces a state of 'nearly two-dimensional' turbulence having very weak variation parallel to the field seems to be reasonably well established.

6 Helicity and the α-effect

The great breakthrough in dynamo theory came in the mid-1960s, with the work of Steenbeck et al. (1966) [28] and their subsequent series of papers, work that only became widely known some years later when distributed in English translation by Roberts and Stix (1972) [29], and again much later with the publication of the book of Krause and Rädler (1980) [30]. The key idea of the theory lay in recognition of the fact that, within the framework of the kinematic dynamo problem for which the statistics of the velocity field are regarded as "given", the mean electromotive force (or *emf*) $\langle \mathbf{E} \rangle = \langle \mathbf{u} \times \mathbf{b} \rangle$ is linearly related to the mean magnetic field $\langle \mathbf{B} \rangle$, assumed to vary on a length-scale large compared with that of the turbulence. At leading order in the ratio of these scales, and assuming isotropic turbulence, this gives the famous relationship $\langle \mathbf{E} \rangle = \alpha \langle \mathbf{B} \rangle$. This astonishing result, the appearance of a mean emf parallel, rather than perpendicular, to the mean field, was somewhat arbitrarily described by Steenbeck et al. [28] as the α-*effect*, a description that is now firmly established. It is an effect that appears only when the turbulence *lacks reflexional symmetry* (such turbulence may be described as *chiral*), and, as shown by Steenbeck et al., it is responsible for the exponential growth of the mean field in a variety of planetary and stellar circumstances.

This discovery completely superseded, and rendered almost irrelevant, the previous divergence between the points of view that had been advocated by Batchelor [1] and Biermann and Schlüter [2], as described above. The focus from this point on was to be on length-scales large (rather than small) compared with the scale of the energy-containing eddies of the turbulence; this was entirely appropriate as far as the problem of explaining the observed existence of stellar and planetary fields was concerned.

One of my research students in the late 1960s was Glyn Roberts, who worked on dynamo action associated with space-periodic velocity fields; this work, contained in his 1969 Ph.D. thesis, was published 1 year later [31]. (Roberts's subsequent paper [32] developed this theme further, and provided the basis for the Karlsruhe experiment of Müller et al. [33] which, some 30 years later, was to be one of the first experiments successfully demonstrating dynamo action in a fluid.) I had difficulty initially in understanding Glyn's arguments, and I finally succeeded only through carrying out a parallel treatment for homogeneous turbulence [34]; at this stage, it became clear that turbulent dynamo action could occur even when the magnetic Reynolds number $R_m(l)$ based on the scale l of the turbulence is *small* compared with unity; the sole requirement was indeed that the turbulence should lack reflexional symmetry; a magnetic field similarly lacking reflexional symmetry would then grow on scales L large enough for the associated magnetic Reynolds number $R_m(L)$ to be of order unity or greater. Batchelor's theory [1] therefore turned out to be wrong for the reason that it failed to take account of large-scale modes that are available to the magnetic field but not to the vorticity field. Its relevance to reflexionally symmetric turbulence is still however a matter for debate.

By sheer coincidence, I had just 1 year earlier in 1969 [35] published a paper concerning *the degree of knottedness of tangled vortex lines* and relating this to a new invariant, which I named the *helicity*, of Euler flows. (This was the analogue of Woltjer's (1958) [36] invariant for ideal MHD, which henceforth became known as *magnetic helicity*.) A physical interpretation of the *cross-helicity* $\langle \mathbf{u} \cdot \mathbf{b} \rangle$ as the *degree of mutual linkage* of vorticity and magnetic fields emerged at the same time. I learnt some years later that the helicity invariant had been previously discovered by J.-J. Moreau (1961) [37]; but I think it is fair to claim that it was my 1969 paper that firmly established the bridge between topology and the dynamics of ideal fluids which has since proved so fruitful. In any event, non-zero helicity provided the simplest symptom and measure of the required lack of reflexional symmetry, and has played a central role in the understanding of the dynamo process ever since.

The mean-field electrodynamics of Steenbeck et al. had a precursor in the work of Parker (1955) [38], who had considered the effect of what he described as *random cyclonic events* (i.e., helical upwellings) acting on a locally uniform magnetic field $\mathbf{B}$. The aggregate effect of these upwellings was to provide a mean current parallel to $\mathbf{B}$. The role of magnetic diffusivity in this process remained obscure however, and the theory, as presented by Parker, though physically appealing, lacked the mathematical foundation that followed only 10+ years later. A second precursor lay in the work of Braginskii (1964) [39], who had developed a theory of *nearly axisymmetric dynamo action*; here again, an α-effect, whose origin was to be greatly clarified later by Soward (1972) [40], was the main outcome of the theory; but the sheer complexity of Braginskii's treatment meant that for many years it had less impact on the MHD community than it undoubtedly deserved.

With mean-field electrodynamics, turbulent dynamo theory had come of age. The two-scale technique opened the way to dynamic, as opposed to purely kinematic, models of dynamo action, particularly in circumstances where Coriolis, as well as Lorentz, forces played a dominant role [41]. The next decade was to be a period of consolidation, and of reaping the fruits of the great advances of the 1960s, the 'teenage years' of the subject. This in turn would lead into the modern era when high-powered numerical simulation would play a role of ever-increasing importance. The situation as it appeared to me in 1978 may be found in my research monograph *Magnetic Field Generation in Electrically Conducting Fluids* [42], which treated dynamo theory in both its kinematic and dynamic aspects, as then understood.

References

1. Batchelor GK (1950) On the spontaneous magnetic field in a conducting liquid in turbulent motion. Proc Roy Soc A 201:405–416
2. Schlüter A, Biermann L (1950) Interstellare Magnetfelder. Z Naturforsch 5a:237–251
3. Syrovatsky CJ (1957) Magnetohydrodynamics. Uspekhi Fiz Nauk 62:247
4. Cowling TG (1957) Magnetohydrodynamics. Interscience, New York [2nd edn. 1975, Adam Hilger, Bristol]

5. Boldyrev S, Cattaneo F (2004) Magnetic field generation in Kolmogorov turbulence. Phys Rev Lett 92:144501–1445014
6. Batchelor GK (1959) Small-scale variation of convected quantities like temperature in turbulent fluid. Part I. General discussion and the case of small conductivity. J Fluid Mech 5:113–133
7. Batchelor GK, Howells ID, Townsend AA (1959) Small-scale variation of convected quantities like temperature in turbulent fluid. Part 2. The case of large conductivity. J Fluid Mech 5:134–139
8. Moffatt HK (1963) Magnetic eddies in an incompressible viscous fluid of high electrical conductivity. J Fluid Mech 17:225–239
9. Grant HL, Stewart RW, Molliet A (1962) Turbulent spectra from a tidal channel. J Fluid Mech 12:241–268
10. Kolmogorov AN (1962) Précisions sur la structure locale de la turbulence dans un fluide visqueux aux nombres de Reynolds élevés. In: Favre A et al. (eds) La Mécanique de la Turbulence. CNRS, Paris 108:447–451
11. Kovasznay LSG (1960) Plasma turbulence. Rev Mod Phys 32:815–822
12. Moffatt HK (1962) Turbulence in conducting fluids. In: Favre A et al. (eds) La Mécanique de la Turbulence. CNRS, Paris 108:395–404
13. Golitsyn G (1960) Fluctuations of magnetic field and current density in turbulent flows of weakly conducting fluids. Sov Phys Dokl 132:315–318
14. Moffatt HK (1961) The amplification of a weak applied magnetic field in fluids of moderate conductivity. J Fluid Mech 11:625–635
15. Odier P, Pinton J-F, Fauve S (1998) Advection of a magnetic field by a turbulent swirling flow. Phys Rev E 58:7397–7401
16. Pearson JRA (1959) The effect of uniform distortion on weak homogeneous turbulence. J Fluid Mech 5:274–288
17. Saffman PG (1963) On the fine-scale structure of vector fields convected by a turbulent fluid. J Fluid Mech 16:545–572
18. Kulsrud RM (1999) A critical review of galactic dynamos. Annn Rev Astron Astrophys 37:37–64
19. Schekochihin AA, Cowley SC, Taylor SF, Maron JL, McWilliams JC (2004) Simulations of the small-scale turbulent dynamo. Astrophys J 612:276–307
20. Hattori Y, Moffatt HK (2006) The magnetohydrodynamic evolution of toroidal magnetic eddies. J Fluid Mech 558:253–279
21. Murgatroyd W (1953) Experiments on magnetohydrodynamic channel flow. Phil Mag 44:1348–1354
22. Lehnert B (1954) Magnetohydrodynamic waves under the action of the Coriolis force. Astrophys J 119:647–654
23. Shercliff JA (1965) A textbook of magnetohydrodynamics. Pergamon Press, Oxford
24. Nihoul JCJ (1965) The stochastic transform and the study of homogeneous turbulence. Physica 31:141–152
25. Moffatt HK (1967) On the suppression of turbulence by a uniform magnetic field. J Fluid Mech 28:571–592
26. Sommeria J, Moreau R (1982) Why, how, and when, MHD turbulence becomes two-dimensional. J Fluid Mech 118:507–518
27. Pothérat A, Sommeria J, Moreau R (2000) An effective two-dimensional model for MHD flows with transverse magnetic field. J Fluid Mech 424:75–100

28. Steenbeck M, Krause F, Rädler K-H (1966) Berechnung der mittleren Lorentz-Feldstärke für ein elektrisch leitendes Medium in turbulenter, durch Coriolis-Kräfte beinfluster Bewegung. Z Naturforsch 21a:369–376

29. Roberts PH, Stix M (1971) The turbulent dynamo: a translation of a series of papers by Krause F., Radler K-H., and Steenbeck M. Tech. Note 60, NCAR, Boulder, Colorado

30. Krause F, Rädler K-H (1980) Mean-field magnetohydrodynamics and dynamo theory. Pergamon Press, Oxford/New York/Toronto/Sydney/Paris/Frankfurt

31. Roberts GO (1970) Spatially periodic dynamos. Phil Trans Roy Soc A 266: 535–558

32. Roberts GO (1972) Dynamo action of fluid motions with two-dimensional periodicity. Phil Trans Roy Soc A 271:411–454

33. Müller U, Stieglitz R, Horanyi S (2004) A two-scale hydromagnetic dynamo experiment. J Fluid Mech 498:31–71

34. Moffatt HK (1970) Turbulent dynamo action at low magnetic Reynolds number. J Fluid Mech 41:435–452

35. Moffatt HK (1969) The degree of knottedness of tangled vortex lines. J Fluid Mech 35:117–129

36. Woltjer L (1958) A theorem on force-free magnetic fields. Proc Natl Acad Sci USA 44:489–491

37. Moreau J-J (1961) Constantes d'un îlot tourbillonnaire en fluide parfait barotrope. C R Acad Sci Paris, 252:2810–2813

38. Parker EN (1955) Hydromagnetic dynamo models. Astrophys J 122:293–314

39. Braginskii SI (1964) Theory of the hydromagnetic dynamo. Sov Phys JETP 20:1462–1471

40. Soward AM (1972) A kinematic theory of large magnetic Reynolds number dynamos. Phil Trans Roy Soc A 272:431–462

41. Moffatt HK (1972) An approach to a dynamic theory of dynamo action in a rotating conducting fluid. J Fluid Mech 53:385–399

42. Moffatt HK (1978) Magnetic field generation in electrically conducting fluids. Cambridge University Press, Cambridge

How Analogous is Generation of Vorticity and Passive Vectors (Magnetic Fields)?

Arkady Tsinober

Department of Fluid Mechanics, Tel Aviv University, Tel Aviv 69778, Israel
(tsinober@eng.tau.ac.il)

Summary. A brief account is presented on analogies between the processes of evolution of vorticity and magnetic field and related problems starting from the very beginning and including the most recent results. The emphasis is made on essential differences as contrasted to similarities. This is seen already on a purely kinematic level which is the main theme of this communication.

1 Introduction and brief history

> *What is hardest to accept in Batchelor's discussion is the assumed simlarity between* **B** *and* $\boldsymbol{\omega}$. Lundquist 1952

The criticism of Lundquist (1952) [1] refers to the paper by Batchelor (1950) [2] which followed his presentation in Paris in 1949 at a Symposium on problems of motion of gaseous masses of cosmical dimensions [3]. He proposed an analogy between magnetic field and vorticity based on the observation that the equations for vorticity $\boldsymbol{\omega}$ and magnetic field **B** *are identical in form*

$$\frac{\partial \boldsymbol{\omega}}{\partial t} - \boldsymbol{\nabla} \times (\mathbf{u} \times \boldsymbol{\omega}) = \nu \nabla^2 \boldsymbol{\omega}, \tag{1a}$$

$$\frac{\partial \mathbf{B}}{\partial t} - \boldsymbol{\nabla} \times (\mathbf{u} \times \mathbf{B}) = \eta \nabla^2 \mathbf{B}, \tag{1b}$$

and that *there is thus a formal analogy between the two solenoidal vectors* $\boldsymbol{\omega}$ *and* **B**, *provided* $\boldsymbol{\omega}$ *refers to the motion of non-conducting fluid and* **B** *to the motion of conducting fluid. Many of the results concerning vorticity in classical hydrodynamics can now be interpreted in terms of magnetic field in the electromagnetic hydrodynamic problem* (Batchelor [2], p. 409). On the dynamical level Batchelor put forward a hypothesis as follows: *The ultimate balance between magnetic and hydrodynamic systems is such that the large wave-number components contain comparable amount of kinetic and magnetic*

S. Molokov et al. (eds.), *Magnetohydrodynamics – Historical Evolution and Trends*, 223–230. © 2007 *Springer*.

energy (Batchelor [2], p. 414), i.e., magnetic energy saturation at the viscous-eddy energy and scale[1]. Another hypothesis claimed eventual equipartition between the magnetic and kinetic energy at all scales assuming that interactions only between velocities and magnetic fields at the same scale are important, i.e., locality (Schlüeter and Biermann [6], Biermann and Schlüeter [7]).

The analogy proposed by Batchelor is, in fact, an extension of the popular analogy between vorticity $\boldsymbol{\omega}$ and material line elements $\mathbf{l}$ (proposed by Taylor 1938 [8], and which goes back to Helmholz 1858 [9] and Kelvin 1880 [10]), equations for which in the absence of viscosity *are identical in form* as well:

$$\frac{\partial \boldsymbol{\omega}}{\partial t} + (\mathbf{u} \cdot \boldsymbol{\nabla})\boldsymbol{\omega} - (\boldsymbol{\omega} \cdot \boldsymbol{\nabla})\mathbf{u} = 0, \tag{2a}$$

$$\frac{\partial \mathbf{l}}{\partial t} + (\mathbf{u} \cdot \boldsymbol{\nabla})\mathbf{l} - (\mathbf{l} \cdot \boldsymbol{\nabla})\mathbf{u} = 0. \tag{2b}$$

These equations are identical in form *provided* that the field $\mathbf{l}$ is solenoidal, which is not necessarily the case. It follows from the equation for $\mathbf{l}$ that $D(div\mathbf{l})/Dt = 0$, i.e., $div\mathbf{l}$ is a pointwise Lagrangian invariant and is conserved along fluid particle trajectories in inviscid flow. If initially $div\mathbf{l} = 0$, it will remain such all the time as in the case when initially $\mathbf{l}_{t=0} = \boldsymbol{\omega}_{t=0}$, and subsequently $\mathbf{l}(t) = \boldsymbol{\omega}(t)$. This latter relation is usually used in claims about the analogy between vorticity $\boldsymbol{\omega}$ and material line elements $\mathbf{l}$. However, there are two main problems with this analogy. The first is that, generally, material line elements $\mathbf{l}$ are dynamically passive, whereas vorticity is not a dynamically passive quantity. Second, the material line elements, which initially and thereby consequently coincide with vorticity, are special in the sense that they are not dynamically passive quantities anymore and react back on the flow precisely as does vorticity. In other words, the fact that vorticity is frozen in the inviscid flow field does not mean that vorticity behaves in the same way as material lines, but the other way around: those material lines which coincide with vorticity behave like vorticity, because they are not passive anymore as are all the other material lines.

While the above analogies have since been realized to be flawed,[2] only recently more attention was given to the differences rather than similarities (Kraichnan and Kimura [13], Lüethi et al [14], Ohkitani [15], Tsinober [16], [17], Tsinober and Galanti [18]). It is the main purpose of this communication to address the main of these differences as contrasted to similarities mostly on kinematic level. For a review of dynamical aspects and problems and a variety of related issues see Schekochihin et al. [19].

---

[1] A similar analogy was promoted by Chandrasekhar (1951) [4,5].

[2] In the first edition of *Electrodynamics of Continuous Media*, Landau and Lifshitz 1957 [11] devoted the whole § 55, entitled *Spontaneous magnetic field in turbulent flow of conducting fluid*. It was replaced by § 74 with a different title *Turbulent dynamo*, in which the material of the paper by Batchelor [2] was replaced by later developments mainly due to Zel'dovich and his group [12].

2 Magnetic field versus vorticity

The usual comparison is based on looking at the Eqs. (1a, b) for vorticity $\boldsymbol{\omega}$ and the (solenoidal) passive vector $\mathbf{B}$ as proposed by Batchelor [2]. Though a number of differences are known, these differences are hidden when one looks at the equations for $\boldsymbol{\omega}$ and $\mathbf{B}$, which – as mentioned, *are identical in form*. However, a more "fair" comparison should be made between the velocity field, $\mathbf{u}$, and the vector potential, $\mathbf{A}$, with $\mathbf{B} = \boldsymbol{\nabla} \times \mathbf{A}$ (Tsinober and Galanti [18]). Such a comparison allows to see immediately one of the basic differences between the fields $\mathbf{u}$ and $\mathbf{A}$ (apart of the first being non-linear and the second linear), which is not seen from the Eqs. (1). Namely, the Euler equations conserve energy, since the scalar product of $\mathbf{u} \cdot (\boldsymbol{\omega} \times \mathbf{u}) \equiv 0$ is identically vanishing. In contrast – unless initially and thereby subsequently $\mathbf{u} \equiv \mathbf{A}$ – the scalar product of $\mathbf{A} \cdot (\mathbf{u} \times \mathbf{B}) \not\equiv \mathbf{0}$.[3] It is this term, $\mathbf{A}\cdot(\mathbf{u} \times \mathbf{B}) \equiv -A_i A_k s_{ik} + \partial/\partial x_k \{A_k A_l u_l - \frac{1}{2}u_k A^2\}$, which acts as a production term in the energy equation for $\mathbf{A}$. In other words, when the initial conditions for $\mathbf{u}$ and $\mathbf{A}$ are not identical, the field $\mathbf{A}$ (and $\mathbf{B}$), is sustained by the strain, s_{ik}, of the velocity field – in contrast to the field $\mathbf{u}$, which requires external forcing. The production term $-A_i A_k s_{ik}$ is positively skewed and $\langle -A_i A_k s_{ik} \rangle > 0$. A noteworthy feature is that an analogue of Kolmogorov's 4/5 law[4] is valid for the vector $\mathbf{A}$ (see, e.g., Gomez et al. [20] and references therein)

$$\left\langle \Delta u_{\parallel}(\boldsymbol{\Delta}\mathbf{A})^2 \right\rangle = -4/3 r \epsilon_A, \tag{3}$$

where $\Delta u_{\parallel} \equiv \Delta\mathbf{u} \cdot \mathbf{r}/r \equiv [\mathbf{u}(\mathbf{x} + \mathbf{r}) - \mathbf{u}(\mathbf{x})] \cdot \mathbf{r}/r$, $\boldsymbol{\Delta}\mathbf{A} = \mathbf{A}(\mathbf{x} + \mathbf{r}) - \mathbf{A}(\mathbf{x})$, and ϵ_A is the mean dissipation rate of the energy of $\mathbf{A}$. An important point is that the relation (3) holds for *any* random isotropic velocity field including the Gaussian one, which is not the case for the velocity field itself, since $\left\langle \Delta u_{\parallel}(\boldsymbol{\Delta}\mathbf{u})^2 \right\rangle \equiv 0$ for a Gaussian velocity field[5]. Similarly, there are essential differences in the behaviour of vorticity, $\boldsymbol{\omega}$ and $\mathbf{B}$. First, in statistically stationary velocity field (NSE but not Gaussian) the enstrophy ω^2 saturates to some constant value, since vorticity is not a dynamically passive quantity. In contrast, the energy of magnetic field B^2 grows exponentially without limit: in the kinematic regime magnetic field is a passive vector and the fluid flow does not know anything about its presence. Second, growth of magnetic field is insensitive to the particulars of the random flow, e.g., the velocity field can be artificial such as Gaussian. In such a velocity field the production term

---

[3] The corresponding equation for the vector potential $\mathbf{A}$ has the form

$$\frac{\partial \mathbf{A}}{\partial t} + \mathbf{B} \times \mathbf{u} = -\boldsymbol{\nabla} p_A + \eta \nabla^2 \mathbf{A}$$

[4] It is more convenient to use the 4/3 law for the velocity field in the form $\left\langle \Delta u_{\parallel}(\Delta\mathbf{u})^2 \right\rangle = -\frac{4}{3}\langle \epsilon \rangle r$, which turns into the 4/5 law by isotropy.

[5] Yaglom (1949) [21] has shown that the 4/3 law in the form (3) holds for a passive scalar. Here, again it is valid for any random velocity field including Gaussian one.

$B_i B_k s_{ik}$ is also positively skewed and $\langle B_i B_k s_{ik} \rangle > 0$. This is not the case with vorticity: there is no amplification of vorticity in a Gaussian velocity field, the PDF of $\omega_i \omega_k s_{ik}$ is precisely symmetric and consequently $\langle \omega_i \omega_k s_{ik} \rangle \equiv 0$: to be amplified vorticity needs for this 'its own' genuine turbulent velocity field. For other results concerning differences between $\boldsymbol{\omega}$ and $\mathbf{B}$ see Tsinober [16], Tsinober and Galanti [18], and references therein.

3 Evolution of disturbances

Important aspects of the essential difference between the evolution of fields $\boldsymbol{\omega}$ and $\mathbf{B}$ arising from the non-linearity of the equation of $\boldsymbol{\omega}$ and linearity of the equation for $\mathbf{B}$ are revealed when one looks at how these fields amplify disturbances. In other words, $\mathbf{B}$ and $\boldsymbol{\omega}$ possess essentially different stability properties. The reason is that the equation for the disturbance of vorticity differ strongly from that for vorticity itself due to the non-linearity of the equation for the undisturbed vorticity $\boldsymbol{\omega}$, whereas the equation for the evolution of disturbance of the field $\mathbf{B}$ is the same as that for $\mathbf{B}$ itself due to the linearity of the equation for $\mathbf{B}$. Consequently, the evolution of disturbances of the fields $\boldsymbol{\omega}$ and $\mathbf{B}$ is drastically different. For example, in a statistically stationary velocity field the energy of the disturbance of $\mathbf{B}$ grows exponentially without limit (just like the energy of $\mathbf{B}$ itself), whereas the energy of vorticity disturbance grows much faster than that of $\mathbf{B}$ for some initial period until it saturates at a value which is of the order of the enstrophy of the undisturbed flow. It is noteworthy that much faster growth of the energy of disturbances of vorticity during the very initial (linear in the disturbance) regime is due to additional terms in the equation for the disturbance of vorticity, which have no counterpart in the case of passive vector $\mathbf{B}$. Indeed, the equation for the disturbance of vorticity Δ_i^ω,

$$\frac{D\Delta_i^\omega}{Dt} = \Delta_j^\omega s_{ij} + \omega_j \Delta_{ij}^s - \Delta_j^u \frac{\partial \omega_i}{\partial x_j} + \Delta_j^\omega \Delta_{ij}^s - \Delta_j^u \frac{\partial \Delta_i^\omega}{\partial x_j} + \nu \nabla^2 \Delta_i^\omega, \quad (4)$$

contains three terms $\Delta_j^\omega s_{ij}, \omega_j \Delta_{ij}^s, -\Delta_j^u \frac{\partial \omega_i}{\partial x_j}$ being all *linear* in disturbance, whereas the equation for the disturbance of the magnetic field is just the same as that for the magnetic field itself. It is important to stress that the additional 'linear' terms $\left(\omega_j \Delta_{ij}^s \text{ and } - \Delta_j^u \frac{\partial \omega_i}{\partial x_j} \right)$ in Eq. (4) arise due to the non-linearity of the equations for the undisturbed vorticity. In this sense the essential differences between the evolution of the disturbances of vorticity and the evolution of the disturbance of passive vector $\mathbf{B}$ with the same diffusivity can be seen as originating due to the non-linear effects in genuine NSE turbulence even during the linear regime. For more details and for other results concerning differences between the evolution of disturbances of $\boldsymbol{\omega}$ and $\mathbf{B}$ see Tsinober and Galanti [18].

4 Two dimensional flows

In two dimensions (x, y) the differences between ω and $\mathbf{B}$ are even more drastic. First, vorticity vector in this case has only a z-component and there is no stretching/amplification of vorticity as in three dimensions. Magnetic field can possess all three components and there is a process of stretching of magnetic field in the plane (x, y). This process can lead to substantial transient growth of magnetic field, which at later times is always overcome by diffusion and consequent eventual decay of the magnetic field. However, this transient regime can be very long (Kinney et al. [22], Dar et al. [23]). The importance of the two-dimensional (2D) configuration is that it is a state to which tends any magnetohydrodynamic (MHD) flow in some sense (locally and/or globally) due to the development of anisotropy in such flows (Maron and Goldreich [24], Mueller et al. [25], Tsinober [26], Zikanov and Thess [27], and references therein). This means that in the dynamical case the difference between the behaviour of ω and $\mathbf{B}$ becomes even larger than that at the kinematic level.

5 Concluding remarks

Until recently the emphasis was made on analogies between genuine and "passive" turbulence. Most probably it started with the well known Reynolds analogy on transport of momentum and heat (Reynolds 1874 [28]) and study of fluid motion by means of 'colour bands' (Reynolds 1894 [29]). Since then such analogies were promoted in a number of papers (see references in Antonov et al. [30], Tsinober [16]).

The essential differences in the behaviour of passive and active fields, including those described above, point to serious limitations on analogies between the passive and active fields and show that caution is necessary in promoting such analogies. They also serve as a warning that flow visualizations used for studying the structure of dynamical fields (velocity, vorticity, etc.) of turbulent flows may be quite misleading, making the question "what do we see?" extremely nontrivial. The general reason is that the passive objects may not 'want' to follow the dynamical fields (velocity, vorticity, etc.) due to the intricacy of the relation between passive and active fields and Lagrangian chaos, just like there is no one-to-one relation between the Lagrangian and Eulerian statistical properties in turbulent flows (Tsinober [16]). This does not mean that qualitative and even quantitative study of fluid motion by means of 'colour bands' (Reynolds [29]) is always impossible or necessarily erroneous. However, watching the dynamics of material 'colour bands' in a flow may not reveal the nature of the underlying motion, and even in the case of right qualitative observations the right result may come not necessarily for the right reasons. The famous verse by Richardson belongs to this kind of observation.

It is the right place to remind the outstanding and specific property of genuine turbulence – *self*-amplification of the field of *strain*. This is underlying some of (but not all) main differences between genuine and passive turbulence since there is no counterpart to this process in the behaviour of passive objects. It is a reflection of a more general property of genuine turbulence possessing an intrinsic *dynamical* mechanism generating randomness (intrinsic stochasticity), whereas in case of passive objects randomness is imposed by the velocity field.

There are properties of passive objects which do depend on the details of the velocity field (Tsinober [16], [17], Tsinober and Galanti [18]). For example, though growth of magnetic field is insensitive to the particulars of the random flow, its alignment with the eigenframe of the rate of strain tensor is sensitive to the details of the velocity field. Namely, in a genuine turbulent field (NSE) the magnetic field is primarily aligned with the eigenvector corresponding to the intermediate eigenvalue of the rate of strain tensor, whereas in a Gaussian velocity field the magnetic field is aligned with the eigenvector corresponding to largest (i.e., positive and purely stretching) eigenvalue of the rate of strain tensor. Just these very properties can be effectively used to study the differences between the real turbulent flows and the artificial random fields. More precisely, the essential differences in the behaviour of passive objects in a real and synthetic turbulence may be exploited in order to gain more insight into the dynamics of real turbulence. At present, however, the knowledge necessary for such a use is very far from being sufficient. With few exceptions it is even not clear what can be learnt about the dynamics of turbulence from studies of passive objects (scalars and vectors) in real and 'synthetic' turbulence. This requires systematic comparative studies of both. An attempt at such a comparative study was made by Tsinober and Galanti [18]. This is a relatively small part of a much broader field of comparative study of 'passive' turbulence reflecting the kinematical aspects and genuine turbulence representing also the dynamical processes. It seems that this branch of turbulence research is quite promising.

References

1. Lundquist S (1952) Studies in magneto-hydrodynamics. Arkiv för Fysik 5(15):297–347
2. Batchelor GK (1950) On the spontaneous magnetic field in a conducting liquid in turbulent motion. Proc Roy Soc London A201:405–416
3. Batchelor GK (1951) Magnetic fields and turbulence in a fluid of high conductivity, in Problems of Cosmical Aerodynamics. In: Proceeding of a Symposium on "Problems of motion of gaseous masses of cosmical dimensions", Paris, 16-19 August 1949, Central Air Documents Office, pp 149–155
   It is noteworthy that among the participants of this Meeting were such prominent people as Alfven H., Burgers JM, Hoyle F, Kampé de Fériet J, von Karman Th, Liepmann H, von Neumann J, and Spitzer L
4. Chandrasekhar S (1951) The invariant theory of isotropic turbulence in magnetohydrodynamics. Proc Roy Soc 204A:435–449

5. Chandrasekhar S (1951) The invariant theory of isotropic turbulence in magnetohydrodynamics. Proc Roy Soc 207A:301–306
6. Schlüter A, Biermann L (1950) Interstellare Magnetfelder. Z Naturforsch 5a:237–251
7. Biermann L, Schlüter A (1951) Cosmic radiation and cosmic magnetic fields II. Origin of cosmic magnetic fields. Phys Rev 82:863–868
8. Taylor GI (1938) Production and dissipation of vorticity in a turbulent fluid. Proc Roy Soc London A164:15–23
9. Helmholz H (1858) On integrals of the hydrodynamical equations which express vortex motion. Translated from Germain by Tait PG, 1867 with a letter by Lord Kelvin (Thomson W) in London Edinburgh Dublin Phil Mag J Sci, Fourth series 33:485–512
10. Kelvin Lord (Thomson W) (1880) Vibration of columnar vortex. London Edinburgh Dublin Phil Mag J Sci, Fifth series 33:485–512; also (1910) in Mathematical and physical papers, vol 4, Cambridge University Press
11. Landau LD, Lifshits EM (1957) Electrodynamics of continuous media, 1st edn. Nauka, Moscow. English thanslation by Pergamon in 1960
12. Landau LD, Lifshits EM (1981) Electrodynamics of continuous media, 2nd edn. Nauka, English thanslation by Pergamon in 1984
13. Kraichnan RH, Kimura Y (1994) Probability distributions in hydrodynamic turbulence. Progr Astron Aeronaut 162:19–27
14. Lüthi B, Tsinober A, Kinzelbach W (2005) Lagrangian measurement of vorticity dynamics in turbulent flow. J Fluid Mech 528:87–118
15. Ohkitani K (2002) Numerical study of comparison of vorticity and passive vectors in turbulence and inviscid flows. Phys Rev E65:046304, 1–12
16. Tsinober A (2001) An informal introduction to turbulence. Kluwer Academic, Dordrecht
17. Tsinober A (2005) On how different are genuine and "passive" turbulence. In: Peinke J et al. (eds). Progress in Turbulence, Springer, pp 31–36
18. Tsinober A, Galanti B (2003) Exploratory numerical experiments on the differences between genuine and "passive" turbulence. Phys Fluids 15:3514–3531
19. Schekochihin AA, Cowley SC, Taylor SF, Maron JL, McWilliams JC (2004) Simulations of the small-scale turbulent dynamo. Astrophys J 612(1):276–307
20. Gomez T, Politano H, Pouquet A (1999) On the validity of a nonlocal approach for MHD turbulence. Phys Fluids 11:2298–2306
21. Yaglom AM (1949) Local structure of the temperature field in a turbulent flow. Dokl Akad Nauk SSSR 69(6):743–746
22. Kinney RM, Chandran B, Cowley S, McWilliams JC (2000) Magnetic field growth and saturation in plasmas with large magnetic Prandtl number. I. The two-dimensional case. Astrophys J 545:907–921
23. Dar G, Verma MK, Eswaran V (2001) Energy transfer in two-dimensional magnetohydrodynamic turbulence: formalism and numerical results. Physica D157:207–225
24. Maron J, Goldreich P (2001) Similuations of incompresible MHD turbulence. Astrophys J 554:1175–1196
25. Müller W-C, Biskamp D, Grappin R (2003) Statistical anisotropy of magnetohydrodynamic turbulence. Phys Rev E67:066302/1–4
26. Tsinober A (1990) MHD flow drag reduction. In: Bushnell DM, Hefner JN (eds) Viscous Drag Reduction in Boundary layers. AIAA 123:327–349

27. Zikanov O, Thess A (1998) Direct numerical simulation of forced MHD turbulence at low magnetic Reynolds number. J Fluid Mech 358:299–333
28. Reynolds O (1874) On the extent and action of the heating surface of steam boilers. Proc Lit Phil Soc Manchester 14:7–12
29. Reynolds O (1894) Study of fluid motion by means of coloured bands. Nature 50:161–164
30. Antonov NV, Hnatich M, Honkonen J, Jurchishin M (2003) Turbulence with pressure: Anomalous scaling of a passive vector field. Phys Rev E 68(4):046306–1–25

MHD Turbulence at Low Magnetic Reynolds Number: Present Understanding and Future Needs

René Moreau[1], Andre Thess[2], and Arkady Tsinober[3]

[1] Laboratoire EPM, ENSHMG, BP 95, 38402 Saint-Martin d'Hères, France
   (`r.j.moreau@wanadoo.fr`)
[2] Department of Mechanical Engineering, Ilmenau University of Technology, P.O.
   Box 100565, 98684 Ilmenau, Germany (`thess@tu-ilmenau.de`)
[3] Department of Fluid Mechanics and Heat Transfer, Tel Aviv University, Israel
   (`tsinober@eng.tau.ac.il`)

Summary. This paper is an attempt to summarize the most important results and established ideas on magnetohydrodynamic (MHD) turbulence in flows of liquid metals when the magnetic Reynolds number is significantly smaller than unity. It is written on the basis of the round-table discussion organised during the Coventry meeting, with additions introduced by the authors, coming from their own vision of the subject, or raised during their exchanges with other specialists. It covers the turbulent regimes observable in rather well controlled laboratory experiments as well as in metal processes where electromagnetic devices are used for different purposes (stirring, pumping, refining, etc). A number of still not-understood points are mentioned and some needs of new efforts are underlined.

1 Introduction

This paper is not a pure synthesis of comments and remarks raised during the **round-table discussion on turbulence** organized during the Coventry meeting on the history of magnetohydrodynamics (MHD). It is rather a review paper written on the basis of these comments and remarks, complemented with a few additions, which were raised during the subsequent exchanges between the authors and with significant contributions from other colleagues.

The main conclusion arising from all those experimental investigations is the fact that, in flows bounded by electrically insulating walls, the turbulence is not damped, even when the magnetic field is very high. It may on the contrary be more intense than in a similar flow without any magnetic field. Indeed, as discussed later, it becomes quasi-two-dimensional, or "spiral" in Branover's terminology [74].

Turbulence is a common feature of flows at high Reynolds number ($Re = UL/\nu$). In the case of electrically conducting fluids and in the presence of a

S. Molokov et al. (eds.), Magnetohydrodynamics – Historical Evolution and Trends,
231–246. © 2007 *Springer.*

magnetic field, the magnetic Reynolds number ($R_m = \mu \sigma U L$) becomes also significant. But most of the papers presented at this meeting put the emphasis on liquid metals at scales which are usual in the laboratory or in the industry, where R_m is significantly smaller than unity, so that the disturbance of the magnetic field, although often measurable, remains negligible in comparison with the applied magnetic field. Then the typical conditions, which were considered during the round-table discussion and which still prevail in this paper, are focusing on $Re \gg 1$ and $R_m \ll 1$. Very often, the speakers at the Coventry meeting were referring to the experiments previously performed first in Riga and in Purdue (see the review papers by Lielausis [1] and Tsinober [2, 3]), and more recently in Beer-Sheva, Grenoble, and Dresden. But the emphasis was essentially on the guiding ideas and on the leading mechanisms, which allow to understand and to predict the observations.

In the following, §2 is devoted to the simplest condition to analyse the influence of a magnetic field, when this field is uniform and when the turbulence is homogeneous. Here all the average quantities, including mean velocity, double and higher-order velocity correlations are spatially uniform. Homogeneous turbulence has the advantage of being amenable to analytical description and numerical simulation based on Fourier transforms and pseudo-spectral methods, respectively. Moreover, it may be actually observed in laboratory experiments, in MHD as well as in ordinary hydrodynamics.

In §3, we focus on the influence of the Hartmann walls. As it is well known, in the presence of a uniform magnetic field, two characteristics of these walls are of primary importance: their electrical conductivity and their orientation. Concerning their conductivity, it is generally supposed that the walls are either insulating or thin, so that the conductance ratio ($C = \sigma_w e / \sigma L$) is quite small. As a consequence, the main part of the electric current cannot be short-circuited by the walls and the net effect is a moderate damping (proportional to B^{-1}, whereas it varies as B^{-2} when the walls are good conductors). And concerning their orientation, it is well known that, except when they are parallel to the magnetic field, their vicinity is occupied by a Hartmann layer, which has the property to react on the neighbouring core flow. The walls parallel to the magnetic field cannot have such a strong influence on the core flow. And §4 briefly focuses on the cases where the fluid domain is so wide in the direction of the magnetic field that the fluid flow may be considered without any influence of the Hartmann layers.

Section 5 summarizes the discussion on much more complex problems, since it is related to shear flows, where the turbulence cannot be homogeneous. This class includes the usual duct flows, which have been the subject of many experiments in the past, as well as flows around bodies and shear flows generated in the vicinity of an electrical discontinuity in the Hartmann wall.

Section 6 is devoted to the remarks and comments, which were made on the still more complex cases where the applied magnetic field is either AC or non-uniform. Of course, with an AC field, a necessary non-uniformity is due to the skin effect, but other causes, like the finite length of the inductor, may also generate non-uniformities.

In §7 we have listed a number of other conditions (free surface phenomena, Alfvén waves, transport of scalar quantities, non-zero magnetic Reynolds number, natural convection, etc) where turbulence is also a central feature, but still less understood than in the cases discussed in § 1–6.

Throughout this paper, a number of questions of significant importance are raised and underlined, which are still poorly understood and should therefore deserve new efforts. Finally, this paper ends up with a few more remarks and wishes shared by the authors, even if they were not explicitly mentioned during the round-table discussion.

2 Homogeneous turbulence

Homogeneous turbulence refers to a state whose statistical properties are independent of position (Pope [4]). In practice, homogeneous turbulence is approximately realized far away from walls provided no other source of shear does exist. In ordinary hydrodynamics, homogeneous turbulence has become a paradigm and has been intensively studied both experimentally and theoretically. In MHD at low R_{m} the challenge is to predict the statistical properties of turbulence in the presence of a homogeneous magnetic field. The first theoretical predictions of the decay of velocity components parallel and perpendicular to the magnetic field and of their ratio ($\langle u_{\parallel}^2 \rangle = 2 \langle u_{\perp}^2 \rangle$) are due to Moffatt [5]. They were followed by experimental investigations by Alemany et al. [6] and Caperan and Alemany [7], using grid turbulence in a cylinder of mercury located in a long vertical coil and by direct numerical simulations [8–10], which confirmed the existence of angular transfers of energy in the wave-number space.

Many other experiments, performed in channel flows, also yield important information on the turbulent properties (see the review papers by Lielausis [1] and Tsinober [2, 3], as well as Sukoriansky et al. [75], and Branover et al. [74]). However, those results often remain more qualitative than quantitative, because homogeneity is not well guaranteed, and because other effects, which will be discussed later on, also influence the turbulence (e.g. entry into the fringing magnetic field, presence of walls and boundary layers, etc).

Among the main results, which received significant support, let us mention the k^{-3} spectral law for the kinetic energy, although most of the measurements concern only the velocity component parallel to the magnetic field. The simplest heuristic way to interpret this law is based on the assumption that some quasi-steady equilibrium exists for all k, between the local energy transfer, whose timescale must be of the order of $1/\sqrt{k^3 E(k)}$, and the Joule damping time, which is itself k-independent. This requires that $E \approx k^{-3}$. There is an obvious analogy between these results and the equivalent results in rotating or stratified fluids.

Now, if the Joule damping time is independent of the wave number, it does depend on the orientation of the wave vector $\mathbf{k}$. Let us denote θ the angle between $\mathbf{k}$ and the magnetic field. The local and instantaneous Joule timescale within the Fourier space is $\tau_J / \cos^2 \theta$, where $\tau_J = \rho/\sigma B^2$ is the Joule timescale in an isotropic turbulence. If, as shown in all experiments, the time decay law is of the form t^{-n}, this implies that $\cos^2 \theta \approx \tau_J/\tau$. Then, after a time of the order of a few τ_J, the only $\mathbf{k}$ vectors carrying a significant energy are almost perpendicular to the magnetic field. And it is justified to admit that $\cos^2 \theta \approx \left(l_\perp/l_\parallel\right)^2$, where $l_\parallel$ and $l_\perp$ are the typical length scales parallel and perpendicular to the magnetic field. This shows that, after some initial period of decay, there exists an asymptotic morphological anisotropy, such that $l_\parallel/l_\perp \approx (t/\tau_J)^{1/2}$. This basic law, which expresses how the turbulent eddies elongate as time progresses, is often quoted by the expression *tendency to two-dimensionality*. Remarkably, for the linear regime we can understand and predict both the kinematic anisotropy $\left(u_\parallel/u_\perp\right)$ and the morphological anisotropy $(l_\parallel/l_\perp)$, whereas for the non-linear regime, if the above prediction for $l_\parallel/l_\perp$ may still be accepted, no predictions or measurements are available to be substituted to the Moffatt's prediction.

The above interpretation of the morphological anisotropy, due to Alemany et al. [6], which is valid in homogeneous turbulence, is much more general. It can also be established in terms of the reminiscence of the Alfvén waves at small R_m. As shown by Sommeria and Moreau [11], those waves degenerate into a diffusion along the magnetic field lines, which provides the $(t/\tau_J)^{1/2}$ law. And, more recently, the invariance of the parallel component of the angular momentum, discovered by Davidson [12, 13] also yields an interesting way to understand this prediction and to accept it even in complex shear flows.

Clearly, our picture of homogeneous MHD turbulence is still far from complete. In particular, there is a conspicuous lack of reliable experimental studies providing velocity data with high spatial and temporal resolution. In order to obtain such data it is either necessary to develop new measurement techniques for liquid metals like mercury and gallium or to perform MHD experiments in transparent fluids using strong magnetic fields of the order of 10 T. Both directions offer attractive opportunities for probing MHD turbulence down to the Kolmogorov scale and thereby laying the foundations for the development of turbulence models that could be used in engineering applications.

3 Influence of the Hartmann walls

These walls have at least two major effects. The first one develops as soon as the two-dimensionality of the turbulence is fairly well satisfied or, in other words, as soon as the length of the energy-containing eddies in the magnetic field direction becomes of the same order as the gap between the Hartmann walls. Then these walls suppress the velocity component parallel to the magnetic field $\mathbf{B}$. More precisely, as shown by Pothérat et al. [14], the only mechanism, which still drives some velocity fluctuations parallel to $\mathbf{B}$, is the Ekman pumping within the Hartmann layer. A straightforward order of magnitude analysis shows that such a velocity scales as Re/Ha^3, which is indeed very small in most relevant experiments.

The second important effect due to the presence of the Hartmann walls and boundary layers is the damping, which remains present and important in this region where the two-dimensionality cannot exist. Sommeria and Moreau [11] have shown that, when the Hartmann number ($Ha = Bh\sqrt{\sigma/\rho\nu}$, where h stands for the distance between the walls perpendicular to $\mathbf{B}$) is large enough to justify the classical inertialess theory of the Hartmann layer, the relevant damping timescale is not anymore the usual Joule timescale $\tau_J = \rho/\sigma B^2$. It becomes Ha times larger, since it is now

$$\tau_H = \frac{h}{B}\sqrt{\frac{\rho}{\sigma\nu}} = Ha\frac{\rho}{\sigma B^2}. \tag{1}$$

This expression, valid when the Hartmann wall is insulating, can easily be modified to take into account the electrical conductance of the wall. It is noticeable that this Hartmann damping time includes both the viscous and the ohmic dissipation, together present within the Hartmann layers, and capable to brake the whole two-dimensional (2D) eddies. According to Sommeria and Moreau [11], within the core flow all planes perpendicular to the magnetic field are identical and the actual turbulent core flow in such a plane may be modelled with the simple equation

$$\frac{d\mathbf{u}_\perp}{dt} = -\frac{1}{\rho}\nabla_\perp p + \nu\nabla^2\mathbf{u}_\perp - \frac{\mathbf{u}_\perp}{\tau_H}. \tag{2}$$

In this model Eq. (2) the Lorentz force is hidden behind the last, linear term. The validity of these ideas has been well confirmed in the MATUR experiment (Messadek and Moreau [15]) performed in the presence of a high magnetic field (up to 6 T, which represents the Hartmann number of the order of 1,800). This experiment has also shown that below some value of Ha (close to 300), inertia present in the turbulent core starts to modify the Hartmann layer properties, even when those layers are not yet turbulent.

Quite recently another experiment [16] has demonstrated that the Hartmann layers become unstable for $R = Re/Ha \approx 380$, a value significantly larger than that previously admitted ($R \approx 250$, according to Lielausis [1]). And it is also noticeable that a direct numerical simulation performed at the same time by Krasnov et al. [17] agrees fairly well with this new experimental threshold, although it is a bit smaller ($R \approx 350$ instead of 380).

In Pothérat et al. [14], an amendment to the Sommeria and Moreau [11] model Eq. (2) is introduced, under the form of a rather complex non-linear term, which takes into account the Ekman pumping at the scale of each eddy and implies some departure from the 2D core flow. Since it results in enlarging the quasi-2D structures in their middle part, these authors call this three-dimensional (3D) disturbance the "barrel effect". A similar behaviour has been observed in the direct numerical simulation performed by Zikanov and Thess [9] and by Mück et al. [18].

4 Flows in the presence of azimuthal magnetic field

Flows in such configurations (i.e., *without* Hartmann walls) are of particular interest due to the well-known fact that a pure 2D flow does not interact with a homogenous magnetic field orthogonal to the plane of the flow. In other words, the 2D flow in such a configuration is a solution of the ordinary Navier–Stokes equations at arbitrary R_m. To quote Landau and Lifshitz [19]: *We may say that the two-dimensional flow "does not see" a uniform field. In a strong external field, the turbulence degenerates just into this two-dimensional form.* Hence it has been expected that, in such a configuration, one can realize a "pure" 2D turbulent flow (Kit and Tsinober [20]) in a sufficiently strong magnetic field. There is a number of qualitative experimental observations indicating that this is the case, starting with famous experiments by Hartmann and Lazarus in 1937 in MHD-channel flows with short Hartmann walls and the stability study of a channel flow by Wooler [21].

Experiments on flows in the wake past cylinders were designed to study the difference between the flow past cylinders either aligned with the magnetic field or perpendicular to it (Kit et al. [22]). It was obtained, in the case of a cylinder perpendicular to **B**, that the turbulent fluctuations in the wake become even smaller than the noise, whereas in the case of a parallel cylinder turbulent fluctuations grow and when the magnetic field is large enough they become much larger than in the absence of magnetic field. A number of other related results are reviewed by Tsinober [3, 23, 24].

However, all the above examples are not clean in the sense that Hartmann walls are always present in the experiments at whatever small aspect ratio. Therefore, these experiments must be conceived in such a way that the Hartmann damping time (1) is much larger than the other relevant timescales, like the eddy turnover time. It is noteworthy that the cleanest way to observe the process of two-dimensionalization can be achieved in the total absence of the Hartmann walls. Experimentally, this can be done in an axisymmetric configuration with an azimuthal magnetic field in the form $B \sim r$. Such a not curl-free field can be realized by applying an uniform electric current by means of electrodes located upstream or downstream of the working section with the expectation that their influence on the flow in the working section would be negligible [3]. Numerically, this can be done using periodic boundary conditions in the azimuthal direction [25]. The bottom line is that all the observed flows are quasi-2D only and it is still not clear whether a clean purely 2D turbulent flow is achievable in such a configuration, and it seems worthy to try to get a more definite answer to this question of basic nature.

5 Flows with free shear layers

All wall-bounded flows exhibit some shear, at least in the boundary layers, which always plays a significant role since the mean velocity gradient controls the turbulence production. To be more specific, in ordinary turbulence, one may consider that the main feature of any near-wall turbulence is the global equilibrium between its production by the shear and its diffusion into the neighbouring core flow. In MHD the novelty is twofold: increasing the shear the magnetic field also increases the turbulence production, and besides an increased damping mechanism, concentrated within the Hartmann layers, becomes also relevant. This may result in the destabilization of the Hartmann layer just mentioned above.

However, in the Coventry round-table, the shear, which was at the centre of the discussion, is the one which is present in free shear layers between core flows having different mean velocities, or in high velocity jets often present along the walls parallel to the magnetic field. Such free shear layers, which are also quasi-2D, like the turbulent eddies, are submitted to the classical Kelvin–Helmholtz instability, which rapidly feeds a sort of quasi-2D turbulence, almost insensitive to the Joule damping. Indeed, this damping is essentially the one discussed above in § 3, since these quasi-2D eddies have their extremities imbedded within the Hartmann layers. The recent MATUR experiment [15] demonstrated that most properties of such a flow could be understood in terms of the equilibrium between the eddy turnover time $\tau_{tu} \approx l/u_\perp$ and the Hartmann damping time τ_H (1). This namely explains that the most relevant non-dimensional parameter is the ratio

$$\frac{Ha}{Re} = \frac{\tau_H}{\tau_{tu}} = \frac{hu_\perp}{Bl_\perp} \sqrt{\frac{\rho}{\sigma\nu}}, \tag{3}$$

instead of the interaction parameter $N = Ha^2/Re$. This relevance of Ha/Re in most shear flows under a high magnetic field was known as an experimental result for quite a long time [1, 2], but its explanation related to the 2D organization of the core flow and to the Hartmann damping is quite recent.

Of course, the case of rectangular ducts is the most complex. First, in the fully established regime it involves Hartmann layers as well as shear side layers along the walls parallel to the magnetic field. The authors of this paper think that they should be named "Shercliff layers"[1]. Although this was not mentioned during the round table, it would be worth now to re-examine the available data on turbulent duct flows in terms of the above ideas. But the most striking phenomenon now seems to be the entry (or exit) effect, which was studied first by Shercliff [26] and discussed in § 6.

---

[1] This idea to name the side layers « Shercliff layers » has been discussed and checked with Martin Cowley and Julian Hunt, who were Shercliff's Ph.D. students, during the colloquium « Turbulence, Twist and Treacle », organized in Cambridge (21–22 April 2005) on the occasion of the 70th anniversary of Keith Moffatt

A question of significant importance, already evoked at the end of § 2, which has strong implications for the numerical modelling of any quasi-2D turbulent flows, is whether the small scales have the same anisotropy as the large ones. Since they cannot get 2D during their turnover time, they must suffer a strong Joule damping. Then, are they nevertheless relevant and do they contribute to the eddy viscosity (RANS) or to the sub-grid scale (LES) modelling? A related question, which has just started to receive some attention (Knaepen et al. [27]), is whether the parallel velocity component has the same anisotropy as the perpendicular components. It seems, from this first attempt, that during the transition from an initially 3D state to the quasi-2D regime between the Hartmann walls, where the parallel component is much more rapidly damped out, and its anisotropy remains smaller than that of the perpendicular components.

6 AC and spatially non-uniform fields

Turbulent flows driven by time-dependent (AC) magnetic fields play an important role in industrial applications like electromagnetic stirring in metals processes, electromagnetic flow control during the growth of semiconductor crystals (Davidson [28]), as well as electromagnetic pumping. Two particular problems are of particular interest, namely the flow in a channel driven by a travelling magnetic field (which is relevant to electromagnetic pumps) and the flow in a cylindrical cavity with finite length under the influence of a rotating magnetic field. Up to now, most investigators have explicitly or implicitly assumed that the flow is driven by the mean (time-averaged) component of the electromagnetic force and have neglected the time-dependent part of the Lorentz force. This assumption becomes particularly questionable when applied to situations involving electromagnetic flow control by means of magnetic fields with very low frequency (of the order of 2 Hz). Such low frequencies are currently applied in some commercial flow control systems for continuous casting of steel. Even for higher frequencies, there is no reason to believe that the flow under a real time-dependent Lorentz force should always be close to the flow computed with a steady Lorentz force. This is due to the fact that turbulence is known to be sensitive to small details of initial and boundary conditions. It is a challenge both for experimental and theoretical studies to develop a better understanding of turbulence under time-dependent magnetic fields. This will require local turbulence measurements with high temporal resolution and extensive direct numerical simulations.

Closely related to AC problems are flows under the influence of a spatially non-uniform magnetic field. It should be emphasized that the first theoretical treatment of the entry/exit effects (which later was called "M-shaped velocity profiles") was made by Shercliff [29]. A comprehensive understanding has been developed for laminar flows, based on the existence of characteristic surfaces [30, 31], which has been reviewed by Hunt and Shercliff [32].

Then, Holroyd and Walker [33] and Hua and Walker [34] have extended this analysis, introducing two new assumptions, the "straight magnetic field approximation" and the "core flow approximation". But the comparison between their numerical results and their own measurements exhibit a clear discrepancy. Quite recently, this type of approach has been revisited by Alboussière [35] and extended to channels of varying cross section and arbitrary magnetic fields. The most prominent effect of a non-uniform field is to create the so-called M-shaped velocity profile in the entrance region. However, for the turbulent case our understanding is much less advanced. Although there have been a large number of experimental studies for channel flows in a variety of magnetic fields (see, e.g., [36]), the investigations have almost exclusively dealt with the case of large magnetic interaction parameter.

The recent and still ongoing experiment performed by Andreev et al. [37] exhibits a previously unexpected feature: the upstream turbulence seems to be completely damped out in the fringing region, where the M-shaped velocity distribution forms. And the turbulence may then restart at some distance downstream, fed by the Kelvin–Helmholtz instability of the jets, as discussed above. This damping is almost paradoxical, because, following the Shercliff idea, clearly evoked at the Coventry meeting during the presentation of his movies on MHD (see MD Cowley paper, this volume), one might essentially expect a generation of vortices and a localized source of turbulence. However, the situation is not quite the same as in the Shercliff movies, where the magnet is translated upstream, whereas the walls and the free-surface liquid are initially at rest. Then, in the movies there is a non-zero relative velocity of the wall with respect to the magnetic field, whereas in the usual channel flow this relative velocity is zero. This behaviour seems, nevertheless, related to the key properties of such a flow in a strong non-uniform magnetic field, which are related to the characteristic surfaces. The streamlines, as well as the electric current lines must be lying within those surfaces. In other words, any turbulent motion, namely the quasi-2D turbulent motion not significantly braked by the Hartmann effect, cannot exist, since it would imply that the streamlines escape from the characteristic surfaces.

By contrast, industrial applications such as electromagnetic brakes in metallurgy (see, e.g., [38,39]) are characterized by moderate magnetic interaction parameters. Here, the turbulence stays 3D in contrast to the 2D turbulence at high interaction parameters. Future investigations, both experimental and theoretical will be necessary to shed new light on the structure of the microturbulence in this type of flows. For instance, it would be interesting to better understand the dual role of the magnetic field. On the one hand, the magnetic field dissipates energy and leads to a damping of turbulence. On the other hand the magnetic field creates M-shaped shear layers, which are prone to instabilities and transition to turbulence. This antagonism is not well understood and will require a lot of additional work.

7 Other effects

7.1 Free surfaces

The interaction of MHD turbulence with free surfaces is important in a variety of industrial applications. For instance, in the application of magnetic brakes during continuous casting of steel one would like to know how the effect of the magnetic brake influences the amplitude and frequency of oscillations of the free surface of the steel. Another example is the application of low-frequency AC fields to free-surface flows, which creates strong surface oscillations and internal turbulence. Recently, laboratory experiments were performed, in which liquid metal droplets were exposed to low frequency (Fautrelle and Sneyd [40]) and high frequency (Kocourek et al. [41]) magnetic fields. These experiments revealed a startling variety of instabilities and non-linear phenomena, which (even without internal turbulence) are poorly understood.

7.2 Finite Lundquist numbers

Along with possible engineering applications of flows (Shercliff [42]; Tsinober [43], [44]), turbulence in such a situation is expected to exhibit a significantly different behaviour, since Alfvén waves are likely to be present as soon as the Lundquist number is larger than unity (Roberts [45]; Iwai et al. [46]). In such conditions, these waves propagate many times over the characteristic length of the system before being damped, and feed the system with additional oscillatory degrees of freedom. Consequently, resonant phenomena involving external excitation (e.g., sound, vibrations), non-linear Alfvén waves and turbulence production are likely to be present. Among other things, turbulence is expected to be characterized by the exchange of energy between the magnetic field and the fluid flow. It is noteworthy that at $Lu \gg 1$ transient regimes (e.g., starting the facility) become oscillatory and large flow-rate oscillations and even reversals and resonances are possible (Vatazhin et al. [47]; Antimirov and Tabachnik [48]).

7.3 Transport of passive scalars

It is quite astonishing that the transport of passive scalars in turbulent MHD flow has received little attention in the past. This is particularly surprising in view of the fact that the first unambiguous experimental demonstration of the strong anisotropy of MHD turbulence was given by Kolesnikov and Tsinober [49] measuring the concentration of In injected in a flow of mercury. A good understanding of the propagation of passive scalars in MHD turbulence is not only an interesting fluid dynamical problem in its own right, but it is also important for applications in metallurgy and semiconductors crystal growth. In order to improve our understanding of passive scalar transport, it would be desirable to develop experimental methods for local concentration measurements with high spatial and temporal resolution. Alternatively, it could be useful to use temperature as a passive scalar and carry over the experimental techniques known from ordinary hydrodynamics (see, e.g., Warhaft [50]).

7.4 Natural convection

Turbulent convection in the absence of a magnetic field has been intensively studied in the past. Understanding is particularly advanced for the Rayleigh-Bénard problem, which is the flow in a fluid layer heated from below and cooled from above (Siggia [51], Grossmann and Lohse [52]). For fluids as different as water, air, helium, mercury, and sodium there is consensus that the turbulent convective flow is characterized by a well-mixed interior and by thin thermal and velocity boundary layers in the immediate vicinity of the heating and cooling plates. Moreover, a large-scale circulation (sometimes called "wind") has been consistently observed in virtually all experiments conducted for aspect ratios of order unity. The presence of a vertical magnetic field affects the flow in two ways. On the one hand, Hartmann boundary layers at the walls perpendicular to the magnetic field modify the heat transport and velocity distributions near the heating and cooling plates. On the other hand, the structure of the large-scale turbulent convection in the core is also modified on account of the induced Lorentz forces. Experimental results with magnetic field are scarce (Aurnou and Olsen [53], Burr and Müller [54], Burr et al. [55]) in contrast to numerical simulations (e.g., Hanjalic and Kenjeres [56]). The work of Cioni et al. [57] is virtually the only experimental study extending sufficiently far into the non-linear regime to be able to probe turbulent convection. Significant new experimental and numerical work is necessary in this field. As already mentioned in § 2, an attractive alternative to mercury might be the use of transparent fluids in high magnetic fields of the order of 10 T. Studying convection in such systems would permit to extract high-resolution velocity data using optical flow measurement techniques like laser Doppler anemometry.

But there are other convective situations, such as the case of long horizontal cylinders heated at one end and cooled at the other end, and submitted to an externally applied magnetic field (Alboussière et al. [58]; Davoust et al. [59]). In this case, a flow is always present and submitted to the stabilising influence of the magnetic field. Similarly, in long vertical enclosures submitted to a horizontal magnetic field, the usual 3D flow is stabilized and a laminar flow organization, essentially quasi-2D, develops (Tagawa et al. [60] Authié et al. [61]). Remarkably, in both cases, the net heat flux is first increased when the magnetic field is applied, and then it decreases when the fluid velocity is significantly damped. Those simple flows are not yet completely understood, namely in their specific scenarios towards unsteady and turbulent regimes. And there are many others, still more complex, like those relevant to crystal growth and solidification of liquid metals, which require deeper investigations.

7.5 Finite magnetic Reynolds numbers

The most spectacular phenomenon, which takes place when R_m is sufficiently large is the excitation of a magnetic field by dynamo effect. And the most common belief is that in order to maintain MHD dynamo the turbulent flow should lack reflectional symmetry (e.g., Alpha effect). However, it was recognized quite for a while that dynamo is possible in flows without such an asymmetry as well (e.g., Kazantsev [62]; Gailitis [63]; Novikov et al. [64]; Haugen et al. [65]). Subsequent developments showed that generally the presence of helicity does not matter at least on the qualitative level. It is already known from astrophysical context with finite magnetic Prandtl numbers that magnetic energy grows regardless and independently of whether the turbulent velocity field has a helical component (Schekochihin et al. [66,67] and references therein). In fact it is known from the spectral theory of the kinematic dynamo driven by a random velocity field and also from numerical simulations (Tsinober and Galanti [68]) that the magnetic energy grows exponentially independently of whether the velocity field is helical or not and even for random Gaussian velocity field. At the dynamical level (i.e., taking into account the back reaction of the magnetic field) it is natural to expect the saturation of the dynamo as observed also in liquid sodium experiments (Gailitis et al. [69] and references therein). Among the natural candidates as the reason for such a saturation is the local anisotropization of turbulent flow in case of not small R_m. Indications of such an anisotropization are known from direct numerical simulations (Schekochihin et al. [66,67]; Maron and Goldreich [70]; Müller et al. [71]).

Beside the regimes where R_m is larger than the critical value for a dynamo effect, the domain where R_m is of order unity is also quite challenging. It is characterized by the fact that the applied magnetic field is significantly transported by the fluid flow, like scalar quantities such as temperature. But because the magnetic field is a solenoidal vector quantity, the analogy remains quite limited and specific properties should be observed. This challenge has justified an important experiment, known as the for von Karman sodium (VKS) experiment, using a liquid sodium facility, in which the fluid flow is driven by two co-rotating or contra-rotating disks. Important results have been obtained, first at low R_m, such as the $k^{-11/3}$ spectral law for the turbulent magnetic energy, which is the magnetic signature of the classical Kolmoroff law for the kinetic energy. But, as soon as R_m is not very small, quite specific results appear, such as a k^{-1} law in the domain of small wave numbers (Odier et al. [72]; Bourgoin et al. [73]).

8 Concluding remarks

The main conclusion of this paper is that the understanding of MHD turbulence – though improving – is quite partial, even in simple configurations. However, as soon as some non-uniformity in the applied magnetic field is present, or the magnetic field is unsteady, or the magnetic Reynolds is not small, our knowledge remains extremely modest. One of the consequences is that there exists, at the moment, no available numerical model capable to compute a complete MHD flow. It seems that the main reason for this is the lack of detailed experimental data and that a joint international research program should be encouraged to provide such data.

The ability to measure the velocity field in turbulent MHD flows with high spatial and temporal accuracy is an important prerequisite for the possibility to test theoretical predictions in experiments. In MHD such flow measurements are particularly important because liquid metals are opaque and local measurements are often the only means to obtain reliable information. These problems are discussed in detail by Eckert, Cramer, and Gerbeth in this volume.

References

1. Lielausis O (1975) Liquid metal magnetohydrodynamics. Atomic Energy Rev 13:527–581
2. Tsinober A (1975) Magnetohydrodynamic turbulence. Magnetohydrodynamics 11:5–17
3. Tsinober A (1990) MHD flow drag reduction. In: Bushnell DM, Hefner JN (eds) Viscous Drag Reduction in Boundary layers. AIAA 123:327–349
4. Pope S (2000) Turbulent Flows. Cambridge University Press, Cambridge
5. Moffatt HK (1967) On the suppression of turbulence by a uniform magnetic field. J Fluid Mech 28:571–592
6. Alemany A, Moreau R, Sulem PL, Frisch U (1979) Influence of external magnetic field on homogeneous MHD turbulence. J de Mécanique 18:280–313
7. Caperan P, Alemany A (1985) Homogeneous low-magnetic-Reynolds-number MHD turbulence - Study of the transition to the quasi-two-dimensional phase and characterization of its anisotropy. J de Mécanique Th et Appli 4:175–200
8. Schumann U (1976) Numerical simulation of the transition from three- to two-dimensional turbulence under a uniform magnetic field. J Fluid Mech 74:31–58
9. Zikanov O, Thess A (1998) Direct numerical simulation of forced MHD turbulence at low magnetic Reynolds number. J Fluid Mech 358:299–333
10. Knaepen B, Moin P (2004) Large-eddy simulation of conductive flows at low magnetic Reynolds number. Phys Fluids 16:1255–1261
11. Sommeria J, Moreau R (1982) Why, how and when MHD turbulence becomes two-dimensional. J Fluid Mech 118:507–518
12. Davidson PA (1995) Magnetic damping of jets and vortices. J Fluid Mech 299:153–186
13. Davidson PA (1997) The role of angular momentum in the magnetic damping of turbulence. J Fluid Mech 336:123–150

14. Pothérat A, Sommeria J, Moreau R (2000) An effective two-dimensional model for MHD flows with transverse magnetic field. J Fluid Mech 424:75–100
15. Messadek K, Moreau R (2002) An experimental investigation of MHD quasi-2D turbulent shear flows. J Fluid Mech 456:137–159
16. Moresco P, Alboussière T (2004) Experimental study of the instability of the Hartmann layer. J Fluid Mech 504:167–181
17. Krasnov DS, Zienicke E, Zikanov O, Boeck T, Thess A (2004) Numerical study of the instability of the Hartmann layer. J. Fluid Mech 504:183–211
18. Mück B, Günther C, Müller U, Bühler L (2000) Three-dimensional MHD flows in rectangular ducts with internal obstacles. J Fluid Mech 418:265–295
19. Landau LD, Lifshits EM (1981) Electrodynamics of Continuous Media. 2nd edn. Nauka, Moscow, English translation by Pergamon in 1984
20. Kit LG, Tsinober A (1971) Possibility of creating and investigating two-dimensional turbulence in a strong magnetic field. Magnetohydrodynamics 7:312–318
21. Wooler PT (1961) Instability of flow between parallel planes with coplanar magnetic field. Phys Fluids 4:24–27
22. Kit LG, Turuntaev SV, Tsinober A (1970) Investigation with a conduction anemometer of the effect of a magnetic field on disturbances in the wake of a cylinder. Magnetohydrodynamics 5:331–335
23. Tsinober A (1996) Transition to quasi-two-dimensional turbulence and its transport properties in liquid metal turbulent MHD flows. Workshop on MHD-Turbulence: Experiments and Theory, IATF, FZK, Karlsruhe, 9–10, October 1996
24. Tsinober A (2001) An informal introduction to turbulence. Kluwer Academic, Dordrecht
25. Lee D, Choi H (2001) Magnetohydrodynamic turbulent flow in a channel at low magnetic Reynolds number. J Fluid Mech 439:367–394
26. Shercliff JA (1953) Steady motion of conducting fluids in pipes under transverse magnetic fields. Proc Camb Phil Soc 49:136–144
27. Knaepen B, Dubief Y, Moreau R (2004) Hartmann effect on MHD turbulence in the limit Rm≪1. In: Proceedings of the Summer Program. Stanford Center for Turbulence Research, pp 99–107
28. Davidson P (1999) Magnetohydrodynamics in Materials Processing. Annn Rev Fluid Mech 31:273–300
29. Shercliff JA (1962) The theory of electromagnetic flow-measurement. Cambridge University Press, Cambridge
30. Kulikovski AG (1968) Slow steady flows of a conducting fluid at high Hartmann numbers. Izv Akad Nauk SSSR Mekh Zhidk i Gaza 3:3–10
31. Kulikovski AG (1973) Flows of a conducting incompressible liquid in an arbitrary region with a strong magnetic field. Izv Akad Nauk SSSR Mekh Zhidk i Gaza 8:144–150
32. Hunt JCR, Shercliff JA (1971) Magnetohydrodynamics at high Hartmann numbers. Annn Rev Fluid Mech 3:37–62
33. Holroyd RJ, Walker JS (1978) A theoretical study of the effect of wall conductivity, non-uniform magnetic fields and variable-area ducts on liquid-metal flows at high Hartmann numbers. J Fluid Mech 84:471–495
34. Hua TQ, Walker JS (1989) Three-dimensional MHD flow in a channel with inhomogeneous electrical conductivity. Int J Engng Sci 27:1079–1091

35. Alboussière T (2004) A geostrophic-like model for high Hartmann number flows. J Fluid Mech 521:125–154

36. Gelfgat YM, Lielausis OA, Scherbinin EV (1976) Liquid metals under the influence of electromagnetic fields. Zinatne, Riga (in Russian)

37. Andreev O, Kolesnikov Y, Thess A (2007) Experimental study of liquid metal channel flow under the influence of a nonuniform magnetic field. Phys Fluids 19:039902

38. Garnier M (ed) (1997) Proceedings of the Second International Congress on Electromagnetic Processing of Materials, Paris, 27–29 May 1997

39. Asai S (ed) (2000) Proceedings of the Third International Congress on Electromagnetic Processing of Materials, Nagoya, 3–6 April 2000

40. Fautrelle Y, Sneyd AD (2005) Surface waves created by low frequency magnetic fields. Eur J Mech B/Fluids 24:91–112

41. Kocourek V, Karcher C, Conrath M, Schulze D (2006) Stability of liquid metal drops affected by a high frequency magnetic field. Phys Rev E 74:026303

42. Shercliff JA (1976) Technological Alfven waves. Proc IEEE 123:1035–1042

43. Tsinober A (1974) Liquid metal flow in a strong magnetic field. In: All-Union Conference on Engineering problems of Controlled Thermonuclear Fusion, Leningrad, pp 247–249

44. Tsinober A (1987) An outline of some basic problems related to the MEKKA Programme in Karlsruhe. Internal Report 8

45. Roberts PH (1967) An introduction to Magnetohydrodynamics. Longmans, London

46. Iwai K, Kameyama T, Moreau R (2003) Alfven waves excited in a liquid metal. In: Proceedings of the 4th International Conference on Electromagnetic Processing of Materials. Lyon, 14–17 October 2003: 75–86

47. Vatazhin A, Lyubimov G, Regirer S (1970) MHD-flows in channels. Nauka, Moscow, 672 pp, see 209–211 (in Russian)

48. Antimirov M, Tabachnik E (1976) MHD convection in a vertical channel. Magnetohydrodynamics 12:147–154

49. Kolesnikov YB, Tsinober A (1974) An experimental study of two-dimensional turbulence behind a grid. Fluid Dynamics 9:621–624

50. Warhaft Z (2000) Passive scalars in turbulent flows. Annn Rev Fluid Mech 32:203–240

51. Siggia ED (1994) High Rayleigh number convection. Annn Rev Fluid Mech 26:137–168

52. Grossmann S, Lohse D (2000) Scaling in thermal convection: a unifying theory. J Fluid Mech 407:27–56

53. Aurnou JM, Olsen PL (2001) Experiments on Rayleigh-Bénard convection, magnetoconvection and rotating magnetoconvection in liquid Gallium. J Fluid Mech 430:283–307

54. Burr U, Müller U (2002) Rayleigh-Bénard convection in liquid metal layers under the influence of a horizontal magnetic field. J Fluid Mech 453:345–369

55. Burr U, Barleon L, Jochmann P, Tsinober A (2003) Magnetohydrodynamic convection in a vertical slot with horizontal magnetic field. J Fluid Mech 475:21–40

56. Hanjalic K, Kenjeres S (2000) Reorganisation of turbulence structure in magnetic Rayleigh-Bénard convection: a T-RANS study. J Turbulence 1:008

57. Cioni S, Chaumat S, Sommeria J (2000) Effect of a vertical magnetic field on turbulent Rayleigh-Bénard convection. Phys Rev E 62:R4520–4523

58. Alboussière T, Garandet JP, Moreau R (1993) Buoyancy driven convection with a uniform magnetic field. Part 1. Asymptotic analysis. J Fluid Mech 253:545–563
59. Davoust L, Cowley MD, Moreau R, Bolcato R (1999) Bouyancy driven convection with a uniform magnetic field. Part 2. Experimental investigation. J Fluid Mech 400:59–90
60. Tagawa T, Authié G, Moreau R (2002) Buoyant flow in long vertical enclosures in the presence of a magnetic field. Part 1. Fully-established flow. Eur J Mech B/Fluids 21:383–398
61. Authié G, Tagawa T, Moreau R (2003) Buoyant flow in long vertical enclosures in the presence of a strong horizontal magnetic field. Part 2. Finite enclosures. Eur J Mech B/Fluids 22:203–220
62. Kazantsev AP (1967) On magnetic field amplification in a conducting fluid. Soviet JETP 53:1807–1813
63. Gailitis A (1974) Generation of magnetic field by mirror-symmetric turbulence. Magnetohydrodynamics 10:131–134
64. Novikov VG, Ruzmaikin AA, Sokolov DD (1983) Fast dynamo in reflexionally invariant random velocity field. Soviet JETP 85:909–918
65. Haugen NE, Brandenburg A, Dobler W (2003) Is nonhelical hydromagnetic turbulence peaked at small scales. Astrophys J 597:L141–L144
66. Schekochihin AA, Boldyrev SA, Kulsrud RM (2002) Spectra and growth rates of fluctuating magnetic fields in the kinematic dynamo theory with large magnetic Prandtl numbers. Astrophys J 567:828–852
67. Schekochihin AA, Maron JL, Cowley SC (2002) The small scale-structure of MHD turbulence with large magnetic Reynolds numbers. Astrophys J 576:806–813
68. Tsinober A, Galanti B (2003) Exploratory numerical experiments on the differences between genuine and "passive" turbulence. Phys Fluids 15:3514–3531
69. Gailitis A, Lielausis O, Platacis E, Gerbeth G, Stefani F (2002) Laboratory experiments on hydromagnetic dynamos. Rev Mod Phys 74:973–990
70. Maron J, Goldreich P (2001) Simulations of incompresible MHD turbulence. Astrophys J 554:1175–1196
71. Müller W-C, Biskamp D, Grappin R (2003) Statistical anisotropy of magnetohydrodynamic turbulence. Phys Rev E 67:066302
72. Odier P, Pinton JF, Fauve S (1998) Advection of a magnetic field by a turbulent swirling flow. Phys Rev E 58:7397–7401
73. Bourgoin M, Marié L, Pétrélis F, Gasquet C, Guigon A, Lecciane JB, Moulin M, Burguete J, Chiffaudel A, Daviaud F, Fauve S, Odier P, Pinton JF (2002) MHD measurements in von Karman sodium experiment. Phys Fluids 14:3046–3058
74. Branover H, Eidelman A, Golbraikh E, Moiseev S (1999) Turbulence and structures: chaos, fluctuations and helical self-organization in nature and the laboratory. Academic Press, San Diego
75. Sukoriansky S, Zilberman I, Branover H (1986) Experimental studies of turbulence in mercury flows with transverse magnetic fields. Exp Fluids 4:11–16

Modelling of MHD Turbulence

Bernard Knaepen, Olivier Debliquy, and Daniele Carati

Université Libre de Bruxelles, Boulevard du Triomphe, Campus Plaine CP231, 1050 Brussels, Belgium (bknaepen@ulb.ac.be)

1 Introduction

Numerical simulations of turbulent phenomena in fluids have made considerable progress with the emergence of large parallel computers. For simple geometries, very efficient numerical methods have been developed to provide accurate numerical solutions to the equations of fluid dynamics. These approaches are referred to as direct numerical simulation (DNS) and their predictions are often regarded as almost as reliable as the experimental data.

If the propagation of sound waves can be neglected and if thermal effects are not considered, turbulence in non-conducting fluids is described by the incompressible Navier–Stokes equations. For a given geometry, the turbulence properties are then fully determined by a single dimensionless parameter: the Reynolds number Re. There is however a strong limitation to the use of DNS. Indeed, the number of degrees of freedom required to characterize a velocity field u_i that corresponds to a turbulent flow is known to increase as $Re^{9/4}$ in three-dimensional (3D) turbulent systems [1]. DNS of the Navier–Stokes equations are thus limited to moderately small Reynolds number flows. For electrically conducting fluids, the situation is similar, though even more complex. The flow may be strongly affected by the coupling between the velocity and magnetic fields and is described by the magnetohydrodynamic (MHD) equations. The number of degrees of freedom is expected to be even higher in MHD and the limitations to the use of DNS techniques are at least as severe for MHD as they are for non-conducting fluids.

There is thus an interest in developing modelling techniques in which only a fraction of the total number of degrees of freedom is actually simulated. Among these techniques, large eddy simulation (LES) has attracted much interest in the past few decades [2]. LES can be defined as a computer experiment in which the large scales are simulated directly while the small scales are modelled. This technique has been first developed for simulating Navier–Stokes problems at high Reynolds numbers and has been recently adapted to MHD flow simulations [3–7]. It will be presented in § 2 and tested in § 3.1 using

S. Molokov et al. (eds.), Magnetohydrodynamics – Historical Evolution and Trends,
247–262. © 2007 *Springer.*

the incompressible MHD equations as well as the quasi-static (QS) approximation for low magnetic Reynolds number.

2 Large eddy simulation and MHD

The LES technique has been used to investigate numerically different flavours of the MHD equations. In this presentation, only incompressible and isothermal problems are considered, so that the only relevant equations are for the velocity and magnetic fields. However, even for these specific problems, the Lorentz force in the Navier–Stokes equations may be given by different expressions depending on the range of parameters that characterize the flow. We will thus briefly reproduce the relevant MHD equations before introducing the LES approach in the following subsections.

2.1 Non-linear MHD and quasi-static approximation

For a conducting fluid the Navier–Stokes equations have to include the Lorentz force per unit of mass $\mathbf{f}^L$:

$$\partial_t \mathbf{u} = -\nabla \frac{P}{\rho} - (\mathbf{u} \cdot \nabla)\mathbf{u} + \mathbf{f}^L + \nu \nabla^2 \mathbf{u}, \tag{1}$$

where $\mathbf{u}$ is the velocity field, P is the pressure, ρ is the fluid density, and ν is the kinematic viscosity. Assuming that no separation of charge is observed within the fluid, the Lorentz force reduces to the magnetic part. For a very wide range of problems, assuming that relativistic effects are negligible, the **non-linear MHD equations** are directly derived from Maxwell's equations and read:

$$\mathbf{f}^L = \frac{1}{\rho} \mathbf{j} \times \mathbf{B}, \tag{2}$$

$$\mu \mathbf{j} = \nabla \times \mathbf{B}, \tag{3}$$

$$\partial_t \mathbf{B} = -\mathbf{u} \cdot \nabla \mathbf{B} + \mathbf{B} \cdot \nabla \mathbf{u} + \eta \nabla^2 \mathbf{B}, \tag{4}$$

where $\mathbf{B}$ is the magnetic field and $\mathbf{j}$ is the current density. The magnetic properties are characterized by the magnetic permeability μ, the magnetic diffusivity $\eta = 1/(\sigma\mu)$, and the electric conductivity σ of the fluid. These equations have to be supplemented by the incompressibility condition $\nabla \cdot \mathbf{u} = 0$ and by $\nabla \cdot \mathbf{B} = 0$.

Several dimensionless parameters are usually defined to characterize the regimes of turbulence in a conducting fluid. The kinematic Reynolds number is defined by the ratio:

$$Re = \frac{UL}{\nu}, \tag{5}$$

where U and L are respectively a characteristic velocity and a characteristic length scale of the flow. The kinematic Reynolds number is directly proportional to the ratio between the non-linear convective term and the viscous linear term in the Navier–Stokes equations. It is independent of the magnetic properties of the fluids. The magnetic Reynolds number represents the same type of ratio between non-linear and linear term in the induction equation for **B**. It is defined by

$$R_{\mathrm{m}} = \frac{UL}{\eta}. \tag{6}$$

Depending on the application, the magnetic Reynolds number, R_{m}, can vary tremendously. In astrophysical problems, R_{m} can be extremely high as a result of the dimensions of the objects studied. On the contrary, for most industrial flows involving liquid metal, R_{m} is very low, usually less than 10^{-2}. For low magnetic Reynolds number flows in the presence of an external magnetic field B^0, the non-linear MHD equations are simplified into the QS **approximation** in which the equation for the induced magnetic field does not need to be solved. In the QS approximation the Lorentz force is expressed in terms of the current density and the imposed magnetic field:

$$\mathbf{f}^L = \frac{1}{\rho}\mathbf{j} \times \mathbf{B}^0, \tag{7}$$

$$\mathbf{j} = \sigma(-\nabla\phi + \mathbf{u} \times \mathbf{B}^0), \tag{8}$$

where the electric potential ϕ is determined by imposing $\nabla\cdot\mathbf{j} = 0$. In this case, the interaction number N (also known as the Stuart number) characterizes the ratio between the Lorentz force and the non-linear convective term:

$$N \equiv \frac{\sigma(B^0)^2 L}{\rho U}, \tag{9}$$

where $\mathbf{B}^0$ is the magnitude of the applied external magnetic field.

2.2 LES methodology

An LES is a numerical experiment in which the large scales of the flow are simulated directly while the smallest scales of the flow are modelled. The practical motivations are fairly obvious. The largest scales of motion are the energy-containing scales. Their knowledge is often sufficient to predict most of the relevant properties of the flow. Moreover, their simulation is by far less costly than the full numerical simulation of turbulence. Indeed, typically more than 90% of the kinetic energy is associated to the largest scales of motion, which are described by about 0.1% of the dynamically active degrees of freedom. It is known however that turbulence is characterized by a wide spectrum of energy and, as a consequence, that the scale separation used in LES cannot be defined in terms of physical or phenomenological properties. On the contrary, it must be determined arbitrarily by the LES practitioner, depending mainly on the available computational resources. In practice, the scale separation in LES is achieved by applying an LES operator $\overline{\mathcal{G}}$ that splits the turbulent fields ϕ_i (which can be either u_i or b_i) into a large-scale part,

$$\overline{\phi}_i(\mathbf{x}) \equiv \overline{\mathcal{G}} \circ \phi_i, \tag{10}$$

which is solved numerically, and a small-scale remainder. The exact nature of $\overline{\mathcal{G}}$ will not be discussed here. It has been the subject of an extensive literature, especially in the applications of the LES approach to the Navier–Stokes problem. The important property of $\overline{\mathcal{G}}$ is to damp the phenomena that occur at scales smaller than a length scale $\overline{\Delta}$, while it leaves the largest scales almost unaffected. The choice of $\overline{\Delta}$ should correspond to the finest grid size achievable in a simulation considering the available computational resources. The equations for $\overline{\phi}_i$ are thus expected to be much easier to solve numerically than the original equations for ϕ_i. However, because of the non-linearities that appear in the hydrodynamic and MHD equations, the equations for $\overline{\phi}_i$ are not closed and require modelling efforts. The models are needed to represent the interactions between (i) the unresolved phenomena that occur at scales smaller than the grid resolution and which are represented by $\phi_i - \overline{\phi}_i$, and (ii) the large-scale phenomena represented by $\overline{\phi}_i$. For this reason, they are usually referred to as subgrid-scale models.

Beyond the above-mentioned practical motivations, the development of LES approaches have also been prompted by theoretical considerations. Indeed, large scales are usually very much dependent on the experimental conditions (boundary conditions, geometry, type of forcing, etc.) Their modelling is thus expected to be problem dependent. On the contrary, small scales in turbulence are supposed to have a more universal behaviour. The modelling of τ_{ij} is thus expected to be "almost problem independent" and could be better supported by theoretical approaches.

2.3 LES of MHD flows

Before presenting the application of LES to MHD flows, it is worth considering briefly the case of non-conductive fluids. For such fluids, the current $\mathbf{j}$ and the Lorentz force vanish. The application of the LES operator to the Navier–Stokes equations yields the following equations:

$$\partial_t \overline{u}_i = -\partial_i \overline{p} - \partial_j (\overline{u}_j \overline{u}_i) + \nu \nabla^2 \overline{u}_i - \partial_j \tau_{ij}, \tag{11}$$

where p represents the hydrodynamic pressure rescaled by ρ. The unknown term is given by

$$\tau_{ij}^u = \overline{u_i \, u_j} - \overline{u}_i \, \overline{u}_j. \tag{12}$$

This quantity, referred to as the subgrid-scale stress tensor, cannot be expressed in terms of the resolved velocity $\overline{u}_i$. In order to close the Eqs. (11), the tensor τ_{ij}^u is usually approximated by a model tensor (see § 2.4).

For flows characterized by a low R_m, the situation is not much more complex. Indeed, in many situations, the external magnetic field $\mathbf{B}^0$ can be assumed to be constant or, at least, to be weakly dependent on the position. In this case, the application of the operator $\overline{\mathcal{G}}$ to the evolution equations does not yield additional difficulty, since the Lorentz force is a linear function of the velocity:

$$\overline{\mathbf{f}^L} = \overline{\mathcal{G}} \circ \mathbf{f}^L = \frac{1}{\rho} \overline{\mathbf{j}} \times \mathbf{B}^0, \tag{13}$$

$$\overline{\mathbf{j}} = \overline{\mathcal{G}} \circ \mathbf{j} = \sigma(-\nabla\overline{\phi} + \overline{\mathbf{u}} \times \mathbf{B}^0). \tag{14}$$

Hence, the extension of the LES techniques for low-R_m flows is straightforward. The only term, which requires a modelling treatment is the same as that in the Navier–Stokes equation (12). The phenomenology of low-R_m flows may however be strongly affected by the external magnetic field, and it is not obvious that models performing well for the Navier–Stokes equations are valid for the QS approximation. In particular, bi-dimensionalisation effects observed in the presence of a strong external magnetic field may yield strongly anisotropic small scales. Their modelling in terms of an isotropic eddy viscosity is a priori questionable, though results presented in § 3.2 are very encouraging.

The application of LES to high magnetic Reynolds number flows, for which the complete non-linear MHD Eqs. (1)–(4) have to be solved, is slightly more complex. These equations are conveniently rewritten in terms of the reduced magnetic field $\mathbf{b} = \mathbf{B}/\sqrt{\rho\mu}$ as follows:

$$\partial_t u_i = -\partial_i p - \partial_j(u_j u_i - b_j b_i) + \nu\nabla^2 u_i, \tag{15}$$
$$\partial_t b_i = -\partial_j(u_j b_i - u_i b_j) + \eta\nabla^2 b_i, \tag{16}$$

where p now represents the sum of the hydrodynamic and magnetic pressures rescaled by ρ. The application of the LES operator to the non-linear MHD equations yield to the following structure of the LES equations:

$$\partial_t \overline{u}_i = -\partial_i \overline{p} - \partial_j(\overline{u}_j \overline{u}_i - \overline{b}_j \overline{b}_i) + \nu\nabla^2 \overline{u}_i - \partial_j \tau_{ij}^u, \tag{17}$$
$$\partial_t \overline{b}_i = -\partial_j(\overline{u}_j \overline{b}_i - \overline{u}_i \overline{b}_j) + \eta\nabla^2 \overline{b}_i - \partial_j \tau_{ij}^b, \tag{18}$$

where two unknown terms enter the LES equations and need to be modelled:

$$\tau_{ij}^u = (\overline{u_i u_j} - \overline{u}_i \overline{u}_j) - (\overline{b_i b_j} - \overline{b}_i \overline{b}_j), \tag{19}$$
$$\tau_{ij}^b = (\overline{u_j b_i} - \overline{u}_j \overline{b}_i) - (\overline{u_i b_j} - \overline{u}_i \overline{b}_j). \tag{20}$$

Hence, the extension of the LES technique to non-linear MHD does not generate additional technical difficulties. The major difference with the Navier–Stokes equations comes from the larger number of non-linearities, which yield extra unknown terms that must be modelled to account for the influence of the unresolved small scales on the large resolved scales.

2.4 LES models

A simple model consists in assuming a linear relation between the model tensor and the large-scale strain tensor $\overline{S}_{ij} = (\partial_i \overline{u}_j + \partial_j \overline{u}_i)/2$:

$$\tau_{ij}^{u,\mathrm{mod}} = -2\nu_e\, \overline{S}_{ij}. \tag{21}$$

In particular, if ν_e is constant, the Navier–Stokes equations for $\overline{u}_i$ are the same as for the original variable u_i with the change of viscosity $\nu \to \nu + \nu_e$. For this reason, the coefficient ν_e is traditionally referred to as the eddy (or effective or turbulent) viscosity. Similarly, an eddy magnetic diffusivity can be used to model the tensor τ_{ij}^b:

$$\tau_{ij}^{b,mod} = -2\, \eta_e\, \overline{S}_{ij}^b, \tag{22}$$

where the tensor $\overline{S}_{ij}^b = (\partial_i \overline{b}_j - \partial_j \overline{b}_i)/2$ is now the antisymmetric derivative of the resolved magnetic field.

In the application of LES to the Navier–Stokes problem, according to Kolmogorov's phenomenology, the eddy viscosity is supposed to scale like $\overline{\Delta}^{4/3}\, \overline{\epsilon}^{1/3}$, where $\overline{\epsilon}$ is the energy transfer rate. Depending on the way of estimating the energy transfer rate, different eddy viscosity models can be constructed. The most popular model has been derived in 1963 by Smagorinsky [8] who proposed the following eddy viscosity structure:

$$\nu_e = C_s\, \overline{\Delta}^2\, |\overline{S}|, \tag{23}$$

where $|\overline{S}| = \sqrt{2\overline{S}_{ij}\overline{S}_{ij}}$ and C_s is an arbitrary dimensionless parameter, usually referred to as the Smagorinsky constant. The direct extension of this model to the eddy magnetic diffusivity is straightforward:

$$\eta_e = D_s\, \overline{\Delta}^2\, |\overline{\mathbf{j}}|, \tag{24}$$

where it has been taken into account that the norm of the antisymmetric tensor $\overline{S}_{ij}^b$ is directly proportional to the current intensity thanks to relation (3).

Another, simpler model, has also been investigated. It is based on the Kolmogorov assumption that the energy transfer rate is scale independent in the inertial range of a turbulent flow. In this case, the eddy viscosity and the eddy magnetic diffusivity are modelled by:

$$\nu_e = C_k\, \overline{\Delta}^{4/3}, \qquad \eta_e = D_k\, \overline{\Delta}^{4/3}. \tag{25}$$

Contrary to C_s and D_s, the parameters C_k and D_k are *not* dimensionless. However, they are supposed to be scale invariant (i.e., independent of $\overline{\Delta}$), and this property is used to compute their value using the dynamic procedure presented in § 2.5. The models (23) and (24) will be referred to in the following as the Smagorinsky model (MS) while the models (25) will be referred to as the Kolmogorov model (MK). A mixed model (MM) in which the Kolmogorov eddy viscosity is used but no model for the eddy magnetic diffusivity ($\eta_e \approx 0$) is also used in some of the following examples. Finally, the case where no model is used ($\nu_e \approx \eta_e \approx 0$) is denoted by M0.

2.5 LES and dynamic procedure

The models τ_{ij}^{mod} that have been developed are often based on the phenomenological understanding of turbulent interactions between small unresolved scales and large resolved scales. The most popular models, like the MS, simply reduce to effective viscosity and magnetic diffusivity expressions derived using dimensional analysis. They usually perform fairly well, but have the drawback in that they contain unknown amplitude parameters, which have to be prescribed a priori and very often are adjusted by a trial and error approach. This difficulty has been overcome by the introduction of the dynamic procedure [9–11] in which a second operator $\widehat{\mathcal{G}}$ is introduced. In this presentation, the effect of the operator $\widehat{\mathcal{G}}$ is also considered as a convolution between a filter kernel and the turbulent fields, and the following notation will be used: $\widehat{\phi}_i = \widehat{\mathcal{G}} \circ \phi_i$. The dynamic procedure is then based on an identity, which relates the unknown tensor generated by the composition of $\overline{\mathcal{G}}$ and $\widehat{\mathcal{G}}$ to the original τ_{ij}. Indeed, if the operator $\widehat{\mathcal{G}} \circ \overline{\mathcal{G}}$ is applied to the MHD equations, unknown tensors, such as

$$T_{ij} = \widehat{\overline{\phi_i \phi_j}} - \widehat{\overline{\phi}}_i \, \widehat{\overline{\phi}}_j \ , \tag{26}$$

are generated. The interesting property of T_{ij} is that it is not independent of τ_{ij}, but is related to it through the Germano identity [10]:

$$T_{ij} = L_{ij} + \widehat{\tau}_{ij}, \tag{27}$$

where L_{ij} is known in terms of $\overline{\phi}_i$ and, consequently, does not require additional modelling:

$$L_{ij} = \widehat{\overline{\phi}_i \, \overline{\phi}_j} - \widehat{\overline{\phi}}_i \, \widehat{\overline{\phi}}_j. \tag{28}$$

The Eq. (27) is of course only valid for the exact and unknown tensors τ_{ij} and T_{ij}. As already mentioned, models for these tensors are usually based on dimensional analysis and contain unknown amplitude parameters. The operators $\overline{\mathcal{G}}$ and $\widehat{\mathcal{G}} \circ \overline{\mathcal{G}}$ can be chosen so that they define large-scale fields, which correspond to length scales in the same range. In this case, consistency in the modelling suggest to use the same type of models: $\tau_{ij}^{\text{mod}} = Cm_{ij}^{\tau}[\overline{v}_l]$ and $T_{ij}^{\text{mod}} = Cm_{ij}^{T}[\widehat{\overline{v}}_l]$. The difference $E_{ij} = L_{ij} + \widehat{Cm_{ij}^{\tau}} - Cm_{ij}^{T}$ between the right- and the left-hand sides of Eq. (27) can be considered as a measure of the performance of the model. The dynamic procedure uses this measure in order to prescribe the model parameter C by minimizing E_{ij}. When a homogeneous direction exists in the problem, the estimation for C is given by:

$$C \approx \frac{\langle L_{ij}M_{ij}\rangle_h}{\langle M_{ij}M_{ij}\rangle_h} , \qquad (29)$$

where $M_{ij} = \widehat{m^T_{ij}} - m^T_{ij}$ and the average $\langle \cdots \rangle_h$ is taken over the homogeneous direction(s). Obviously, this approach is restricted to special geometries with homogeneous direction(s). Complex geometries require an alternative treatment with a local definition of the parameter C [12].

3 Examples

Several techniques have been considered in order to assess the LES performances. When experimental results are available, the LES predictions are compared to the measurements of large-scale properties of the flow. However, realistic experimental set-ups correspond to complex geometries. In that case, the predictions of the LES could be unsatisfactory for at least two reasons: the models can perform poorly, or the numerical errors can affect the entire LES predictions. The purpose of this section is to give some examples for which the models provide a reasonable picture of the small-scale effects on the resolved flow. Hence, in order to avoid possible difficulty related to the contamination of LES predictions by numerical errors, the following technique has been adopted. First, very accurate high resolution DNS are produced for various parameter sets. These DNS are performed using a pseudospectral code. The geometry corresponds to a cubic or a rectangular box with periodic boundary conditions. The parameter sets (viscosity, magnetic diffusivity, lengths of the box, initial conditions) are chosen so that both the velocity and the magnetic field are well resolved. Second, the initial condition is filtered to a much lower resolution using a spectral cut-off, yielding the initial condition of the LES. The LES is thus performed with this lower resolution, for which a model is necessary, using exactly the same pseudospectral code. The results of the LES at various times are then compared to the filtered DNS. Such a comparison ensures that the evaluation properly reflects the performances without having to deal with possible influence of the numerical scheme.

3.1 Homogeneous turbulence at high magnetic Reynolds number

In the first example, a high resolution (512^3) DNS of decaying isotropic turbulence has been performed. The integration domain is a cube of linear extension 2π. The initial data set of the DNS has been cut-off-filtered from 512^3 to 64^3 Fourier modes, removing about 99.8% of the originally available information. The influence of the filtered scales on the remaining large scales of motion,

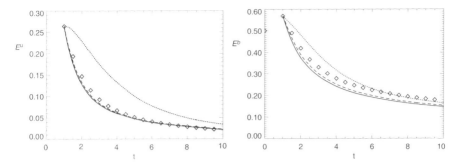

Fig. 1. Decay of the kinetic energy E^u (*left*) and magnetic energy E^b (*right*). Diamonds correspond to the reference DNS filtered to 64^3, the dotted line shows the result of the M0-LES without a model. The solid and broken lines represent the predictions of the LES used with model MK and model MS, respectively

which still contain 90% of the total energy, is accounted for through the models for τ_{ij}^u and τ_{ij}^b. The unknown parameters entering the definitions of the ν_e and η_e are systematically calibrated using the dynamic procedure.

A basic LES requirement is the reproduction of the temporal and spectral behaviour of the resolved kinetic (E^u) and magnetic (E^b) energies :

$$E^u = \frac{1}{V} \int d\mathbf{x} \frac{1}{2} \overline{u}_i(\mathbf{x}) \overline{u}_i(\mathbf{x}), \qquad E^b = \frac{1}{V} \int d\mathbf{x} \frac{1}{2} \overline{b}_i(\mathbf{x}) \overline{b}_i(\mathbf{x}). \qquad (30)$$

The time evolution of these quantities is shown in Fig. 1, where t is given, like for the reference DNS, in units of the large eddy turnover time. To evaluate the overall influence of the dynamic subgrid modelling, the dotted lines show the energy curves for a LES with no model M0. As expected, the M0-LES results strongly deviate from the filtered DNS data, since energy dissipation is mainly due to the high wavenumber modes, which have been cut off by the grid-filtering operation. A clear improvement is observed when the subgrid models MS and MK are applied. The evolution of E^u is well reproduced in both cases, showing that the main influence of the small-scale velocity field fluctuations on the kinetic energy is dissipative. The temporal development of the magnetic energy in the LES with MS and MK also satisfactorily fits the reference data, though one observes an offset between the LES results and the DNS.

The application of the virtually parameter-free dynamic subgrid models reproduces rather sensitive quantities like the angle-averaged energy spectra,

$$E^u = \int dk \, E_k^u(k), \qquad E^b = \int dk \, E_k^b(k), \qquad (31)$$

in good agreement with the DNS data. The kinetic energy spectrum E_k^u is shown in Fig. 2 at time $t = 6$, when E^K has decreased by a factor of about 6.5 . Both models, MS and MK, lead to spectra that follow the filtered DNS data up to the high wavenumber range of the LES system.

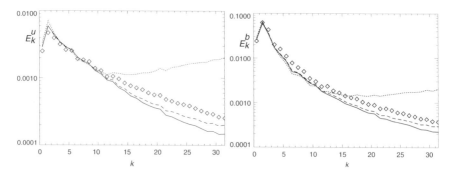

Fig. 2. Kinetic E_k^K and magnetic E_k^M energy spectra at time $t = 6$. Symbols are the same as in Fig. 1

The agreement of the LES results with the filtered DNS spectrum is evidently due to the subgrid models MS and MK, since the lack of such a model in the M0-LES causes a large accumulation of kinetic energy over nearly two-thirds of the LES wavenumber space, which is the consequence of the missing filtered-scale energy dissipation in combination with the direct spectral energy cascade. The same trends are observed for the angle-averaged magnetic energy spectrum E_k^b (Fig. 2). However, the LES show a spectral magnetic energy distribution that is too low across a wide range of scales when compared to the DNS spectrum. This observation suggests to neglect the dissipative effect of τ_{ij}^b. In the following example (§ 3.3), the MM obtained using the Kolmogorov scaling (25) for the eddy viscosity and no eddy magnetic diffusivity has been explicitly tested with a reasonable success.

3.2 Homogeneous turbulence at low magnetic Reynolds number

In this section, the LES methodology is illustrated in the context of low magnetic Reynolds number flows. As in § 3.1, a homogeneous flow is considered and analysed using a spectral code. Several DNS have been performed to generate accurate databases. These DNS runs all have a resolution consisting of 512^3 Fourier modes. The numerical experiments consist in describing the evolution of a freely decaying conductive flow that is subject to an externally applied magnetic field. This problem was first studied in 1976 by Schumann [13]. All the DNS start with the same initial condition obtained by time stepping a random phase velocity field until turbulent indicators like the skewness of the velocity derivatives reach "quasi constant" values. At that stage, the flow is characterized by the following quantities: $Re = uL/\nu = 380$ (integral Reynolds number), $R_\lambda = \sqrt{15/\epsilon \nu u^2} = 84.1$ (microscale Reynolds number), and the magnetic field is switched on. Three different cases are then considered corresponding to different intensities of the applied magnetic field. The corresponding values of the interaction parameter are: $N = 0$, $N = 1$, $N = 10$

(the case $N = 0$ is thus a non-magnetic case which serves as a comparison with the other two magnetic cases).

3.3 Evolution of global kinetic energy

The total kinetic energy contained in the flow illustrates very well the influence of the external magnetic field. In Fig. 3, the usual acceleration of decay corresponding to higher values of the interaction parameter is observed for the DNS runs.

In order to assess the LES method, some DNS snapshots of the flow field have been truncated from the 512^3 resolution to a 32^3 resolution (using a sharp Fourier cut-off). The initial condition for the LES runs has been obtained in the same way by truncating the DNS field at $t = t_0$. To further stress the efficiency of the dynamic procedure, some LES runs were also performed using the classical MS (with a true constant C_s). For homogeneous and isotropic turbulence, one can obtain the estimate $C_s \simeq 0.025$ [14], which is the value retained here.

Figure 4 represents the time evolution of the resolved kinetic energy density predicted by the LES (MS) and compared to the filtered DNS. On each plot, another curve representing a simulation on the 32^3 mesh without SGS modelling (M0) is added to stress the action of the subgrid-scale models. The case $B^{ext} = 0$ serves as a benchmark to check that for non-conductive flows, and our LES code behaves as expected. In both the cases $N = 1$ and $N = 10$, the LES performs remarkably well. In the case $N = 1$ the difference between LES and unresolved DNS is very clear. In the case $N = 10$ and for this

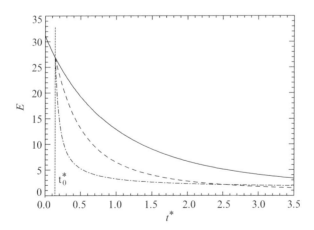

Fig. 3. Time history of the kinetic energy density E. The solid line represents the flow decay without applied magnetic field, the broken curve corresponds to the case $N = 1$, and the dash dot curve corresponds to the case $N = 10$. The dotted line represents the time t_0 at which the magnetic field is switched on (time has been normalized by the eddy turnover time at $t = t_0$)

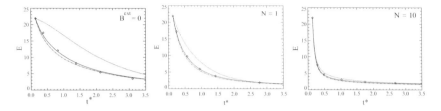

Fig. 4. Time history of the resolved kinetic energy density: LES vs. filtered DNS. The solid lines represent the predictions of the model MS (with dynamic procedure). The dash dot lines represent the classical Smagorinsky model (no dynamic procedure). The diamonds represent the filtered kinetic energy density obtained by truncating the DNS fields to a 32^3 resolution. The dotted lines represent the no-model case M0

diagnostic, the unresolved DNS does not depart significantly from the filtered DNS and LES (spectral diagnostics do however show a clear improvement of LES over unresolved DNS [15]). It is also noted that the LES with dynamic procedure performs consistently better than the LES with classical MS.

3.4 Anisotropy

It is well known that the action of the magnetic field in the QS approximation leads to a progressive suppression of spatial variations in the flow along the direction of the magnetic field. Figure 5 represent the contours of the kinetic energy density at three different times respectively for the filtered DNS and the LES with dynamic procedure (only the case $N = 10$ is shown because the effect is more pronounced for strong applied magnetic fields). As is obvious from the plots, the LES reproduces the filtered DNS structures very well.

Another form of anisotropy that appears in the presence of a magnetic field concerns the partition of energy between the components of the velocity. Let us assume that $\mathbf{B}^0 \parallel \mathbf{1}_z$ and denote $E_\parallel = \left\langle \frac{1}{2} u_z^2 \right\rangle$ and $E_\perp = \left\langle \frac{1}{4} (u_x^2 + u_y^2) \right\rangle$, where $< \cdots >$ stands for spatial averaging. It has been shown in [16] and [13] that, starting from an initial isotropic flow, one should have $E_\parallel / E_\perp \to 2$ for $t \to \infty$ (neglecting viscous and inertial effects). Here the situation is of course more complex since the simulations are performed at finite Reynolds numbers with both viscous and inertial effects present. It is thus interesting to examine how the LES simulations reproduce the energy partition anisotropy and this is illustrated in Fig. 6. Again, the LES with dynamic procedure reproduces the DNS behaviour very well.

3.5 Mixing layer turbulence at high magnetic Reynolds number

Although previous sections have been dedicated to homogeneous turbulence, most physical situations are characterized by inhomogeneous turbulent flows. Among them, the mixing layer represents one of the simplest configurations

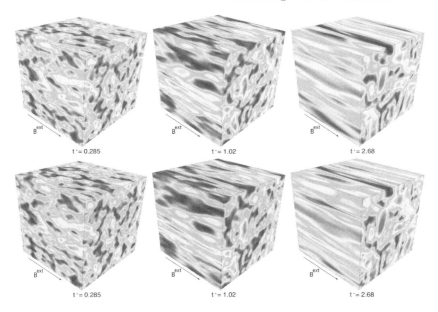

Fig. 5. Contours of the kinetic energy computed from the filtered DNS (*top*) and LES (*bottom*) with MS ($N = 10$ case). The different times at which the contours are calculated are indicated under the plots

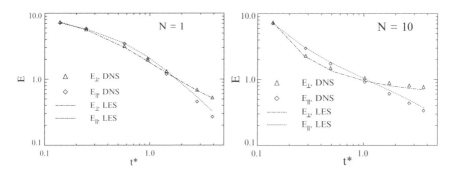

Fig. 6. Time history of the resolved kinetic energy density separated into parallel $E_{\parallel} = \left\langle \frac{1}{2} u_z^2 \right\rangle$ and perpendicular $E_{\perp} = \left\langle \frac{1}{4}(u_x^2 + u_y^2) \right\rangle$ contributions

[17]. It consists of two "distinct" homogeneous regions of different turbulent energy interacting through a layer of rapid transition. This configuration is an interesting test case, since the influence of inhomogeneity on the turbulence properties can be studied in detail without having to deal with the presence of a solid boundary. This particular situation may be observed in geophysics and astrophysics, where regions of different turbulent activities interact. It is thus at first a physical motivation that encourages us to study this case.

On the other hand, our motivation is also governed by practical computational aspects. Indeed, the mixing layer can be computed with the same pseudospectral code used in the preceding examples. In practice, the use of a discrete Fourier representation of the variables requires the flow to be periodic in each direction. This condition is satisfied also in the inhomogeneous direction if one considers a second mixing layer which performs the "reverse" transition compared to the first one. This also has the advantage that results gathered from both mixing layers can be averaged to improve the statistics.

Initializing homogeneous turbulence with a prescribed energy spectrum and preserving incompressibility is a standard procedure [18]. However, building an initial condition that has a prescribed inhomogeneous energy profile like in the mixing layer is not obvious in a pseudospectral code. Here, we have decided to impose in each plane perpendicular to the inhomogeneous direction (y by convention), the 2D spectral properties of the velocity and magnetic fields. In practice, this means that the amplitudes of the modes $u_i(k_x, y, k_z)$ and $b_i(k_x, y, k_z)$ are set at prescribed values that depend on $k_x^2 + k_z^2$ and y only. For values of y inside the homogeneous layers, these amplitudes basically correspond to isotropic turbulence with spectra chosen to reproduce the experimental set-up studied by Veeravali and Warhaft [19] used to study the shear-free mixing layer. Incompressibility is enforced using an iterative procedure that allows simultaneously to satisfy continuity and to match the desired spectra. As an illustration, the contour plot of the total (kinetic + magnetic) energy density after this initialization procedure is shown in Fig. 7. Once the procedure is finished, we let the flow decay freely.

The computational domain used in this example is a rectangular box of size $l_x = 2\pi$, $l_y = 4\pi$, and $l_z = 2\pi$ with periodic boundary conditions. The 4π-length is chosen along the inhomogeneity direction in order to ensure the existence of sufficiently large homogeneous layers. The kinematic viscosity and

Fig. 7. *Left*: some magnetic field lines at the initial time. *Right*: iso-surfaces of the electric current. In both figures, the coloured contours represent the total (i.e., kinetic + magnetic) energy density

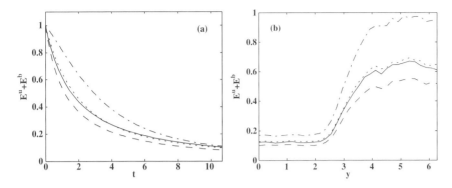

Fig. 8. (a) Time evolution of the total energy. (b) Profile of the total energy in the inhomogeneous direction. Legend: LES with model MK (*broken line*); model MM (*dotted line*); model M0 (*dash-dotted line*); Filtered DNS (*solid line*)

magnetic diffusivity are identical ($\nu = \eta = 1.5 \times 10^{-2}$) so that the magnetic Prandtl number is equal to one.

The reference DNS is performed using $256 \times 512 \times 256$ modes, while the initial LES fields are obtained by filtering down the DNS to $32 \times 64 \times 32$ modes.

Results are presented for the models MK, MM and M0. Figure 8a shows the decay of the volume-averaged total energy as a function of time and Fig. 8b shows a sample profile of the energy along the inhomogeneous direction (after the global energy has decayed by approximately a factor of 2). Both results indicate clearly that MM is the best performing model and that it significantly improves the agreement with DNS data.

4 Discussion

The purpose of this contribution is to introduce the LES methodology and to demonstrate its potential in modelling MHD turbulent flows, rather than advocating the use of a specific model. The models that have been presented are based on the very simple concepts of turbulent viscosity and turbulent magnetic diffusivity. They are systematically implemented, together with the dynamic procedure (see § 2.5), in order to avoid a lengthy discussion on the optimal value of the model parameters. The comparisons between filtered DNS and LES that have been presented here have to be considered as a proof of concept. There is certainly room for further developments and improvements of the small-scale models in MHD. Especially, the application of LES for MHD flows in complex geometries and in the presence of solid boundaries has almost not been developed so far. The major phenomenological differences, which exist between the classical boundary layer and the Hartmann layer in the presence of a magnetic field, certainly deserve further studies.

References

1. Frisch U (1995) Turbulence: The Legacy of A.N. Kolmogorov. Cambridge University Press, Cambridge
2. Rogallo RS, Moin P (1984) Numerical simulation of turbulent flows. Annu Rev Fluid Mech 16:99–137
3. Yoshizawa A (1987) Subgrid modeling for magnetohydrodynamic turbulent shear-flows. Phys Fluids 30(4):1089–1095
4. Zhou Y, Vahala G (1991) Aspects of subgrid modelling and large-eddy simulation of magnetohydrodynamic turbulence. J Plasma Phys 45:239–249
5. Theobald M, Fox P, Sofia S (1994) A subgrid-scale resistivity for magnetohydrodynamics. Phys Plasmas 1:3016–3032
6. Agullo O, Müller WC, Knaepen B, Carati D (2001) Large eddy simulation for decaying magnetohydrodynamics turbulence with dynamic subgrid modelling. Phys Plasmas 7:3502–3505
7. Müller WC, Carati C (2002) Dynamic gradient-diffusion subgrid models for incompressible magnetohydrodynamic turbulence. Phys Plasmas 9(3):824–834
8. Smagorinsky J (1963) General circulation experiments with the primitive equations: 1. the basic experiment. Month Weather Rev 91:99–164
9. Germano M, Piomelli U, Moin P, Cabot WH (1991) A dynamic subgrid-scale eddy-viscosity model. Phys Fluids A 3(7):1760–1765
10. Germano M (1992) Turbulence: the filtering approach. J Fluid Mech 238:325–336
11. Lilly DK (1992) A proposed modification of the germano subgrid-scale closure method. Phys Fluids 4:633–635
12. Ghosal S, Lund TS, Moin P, Akselvoll K (1995) A dynamic localization model for large-eddy simulation of turbulent flows. J Fluid Mech 286:229–255
13. Schumann U (1976) Numerical simulation of the transition from three- to two-dimensional turbulence under a uniform magnetic field. J Fluid Mech 74:31–58
14. Pope SB (2000) Turbulent Flows. Cambridge University Press, Cambridge
15. Knaepen B, Moin P (2004) Large-eddy simulation of conductive flows at low magnetic reynolds number. Phys Fluids 16:1255–1261
16. Moffatt HK (1967) On the suppression of turbulence by a uniform magnetic field. J Fluid Mech 28(3):571–592
17. Knaepen B, Debliquy O, Carati D (2004) DNS and LES of a shear-free mixing layer. J Fluid Mech 514:153–172
18. Rogallo RS (1981) Numerical experiments in homogeneous turbulence. NASA technical Memorandum 81315, NASA, Ames Research Center
19. Veeravalli S, Warhaft Z (1989) The shearless turbulence mixing layer. J Fluid Mech 207:191–229

The Growth of Magnetohydrodynamics in Latvia and Israel

Herman Branover

Ben-Gurion University of the Negev, Beersheba, Israel (hbranover@OKsatec.com)

1 Professor Igor Mikhailovich Kirko, Director of the Institute of Physics in Riga

The beginning of the rapid development of theoretical and applied magneto-hydrodynamics (MHD) during the end of the 1950s can be understood only by following the activities of a single talented, creative, and dedicated individual – a man who was appointed executive director of the newly established Institute of Physics at the Latvian Academy of Sciences. This was an unusual appointment because the person we are referring to, Professor Igor Mikhailovich Kirko, had just celebrated his 30th birthday. To the best of my knowledge, Kirko was the first to direct a scientific institution that made a broad experimental investigation on different phenomena of magnetohydrodynamics.

In order to prepare this chapter, I wrote to Professor Kirko asking him to describe how he established his institute. I had last seen Kirko in the early 1970s and was excited to renew my relationship with him. Probably over 90 years old now, Kirko is still very active and creative. In his response to my letter, he wrote me recollections detailing how he managed one of the most important institutions for fundamental research and technological application of numerous processes based on the results of profound theoretical studies. He also informed me that he is preparing a book on MHD. This chapter quotes freely from Kirko's recollections that he sent me, with his permission [1].

Naturally, Professor Kirko wanted to hire the best people for his new institute. He was fortunate to receive guidance on how to build his staff from a number of distinguished physicists, among them Professor Abraham Yoffe, the director of the Leningrad Physical Technical Institute of the Soviet Academy. Kirko recalls:

"... Yoffe advised me that an academic laboratory does not need more than ten researchers. In addition, there should be five expert technicians giving support to the rest of the team. For optimum efficiency, there should be one

S. Molokov et al. (eds.), Magnetohydrodynamics – Historical Evolution and Trends, 263–274. © 2007 Springer.

or more overlaps in the fields of expertise of the team members, so that in case of health or other problems, if one person cannot come to work, another team member can replace him. The technical personnel of each team would be well familiar with all the technical problems that the team deals with. That way deadlines could be met."

"I was very nervous," remembers Kirko, "about which main field of concentration to choose for the newly established Institute of Physics of the Latvian Academy. One day, one of my assistants showed me a new journal containing a picture of an electromagnetic induction pump working without valves and pistons. We were so excited by the picture in the journal and by the brief description of the new type of asynchronous pump that we decided on the spot to make a simple experiment to verify the device. We put some mercury into a bottle and inserted the bottle into the clearance of the stator of an asynchronous motor.

"What we saw seemed miraculous. The mercury in the bottle started rotating, while the shape of the free surface turned into that of a parabolic mirror. This—together with other considerations—convinced me that my new institute should start working in this direction. In those times, science planning was strictly centralized. Therefore, I had to report on our experiment and my desire for the institute to conduct MHD research to the Academic Secretary of the Department of Physics and Astronomy of the Academy of Sciences, Lev Artsimovich. Artsimovich's own work in thermo-nuclear fusion was highly classified. He approved my request and arranged for additional financing to add ten more scientific positions in my institute. Previously, I had been allotted only thirty positions for the entire Institute of Physics. Receiving ten additional positions was considered a great success for a young and inexperienced director. Now I had a vaguely defined area of scientific exploration."

Professor Kirko organized several teams of young researchers who had just completed their doctoral studies and had not found jobs yet. Professor Kirko writes:

"I came to the conclusion that I had to establish several teams working in parallel and studying the very few existing reports in parallel. The next step was to find talented people. Naturally, I contacted the physics department of the University of Latvia and the Polytechnic Institute. My efforts resulted in hiring several talented physicists, among them O. Lielausis, J. Lielpeter, A. Kalnin, and A. Mikelson. All of them had just finished their studies and, according to the regulations of that time had to accept the first position offered to them. I also hired several physicists who had finished their studies a few years earlier. Among them were I. Krumin and Y. Birzvalk."

Birzvalk, it is interesting to note, in addition to being well acquainted with contemporary physics, was deeply involved with classical literature. He was intrigued by the mystery surrounding Shakespeare. He analysed most of Shakespeare's works and was very active in promoting the idea that they were not written by one person.

"The problem with all these people", recalls Kirko, "was that none of them had studied hydrodynamics beyond the general introductory course. (I mean conventional hydrodynamics, let alone magnetohydrodynamics.)"

"I myself had graduated from the department of physics of the University of Moscow. There, too, I had not been exposed to hydrodynamics beyond some notions about the Bernoulli equation, and the transition from laminar to turbulent flow determined by a certain value of the Reynolds number. To resolve this severe problem, I went to the Polytechnical Institute of Leningrad to seek the advice of Professor Lev Loitsiansky, the foremost expert in hydrodynamics in the Soviet Union. He gave his full attention to my problem and offered advice and help on a permanent basis for the future."

"Concerning my immediate manpower problem, Loitsiansky suggested, 'I have a doctoral student, Herman Branover, who is equally knowledgeable in hydrodynamics and modern physics. If you give him a difficult problem in the physics of flows, he will come back in a few hours with a detailed solution. If you assume that the liquid is electroconductive and moves in an externally imposed magnetic field, he will come back in a few days to make a complete presentation, including theoretical studies and a well-developed program of one or more experiments. ... A few days ago he defended his candidate of science thesis, a wonderful piece of work, but our academic council rejected it by secret vote without any apparent reason. I don't know their real motivation. It could be caused by an anti-Semitic state of mind or by a desire to take revenge on me."

"What is your student doing now?"

"He said that he is going to abandon science and left for Riga to visit his family."

"Give me his address, and I'll try to talk with him."

"I found H. Branover in Riga, but he categorically declined my proposal to get engaged in MHD because he did not feel qualified. However, he agreed to take part in the weekly MHD seminars that I had initiated. ... At these seminars no one felt embarrassed. Neither the director nor the other colleagues tried to conceal their ignorance, and everyone studied hard. A few weeks later, Branover told me that he had searched several libraries in town in vain looking for substantial information about MHD. ... Ultimately, he agreed to accept the position I was offering him. Naturally, he concentrated on experimental studies fundamentally investigating the most general features of turbulence in MHD flows in ducts of different shape, especially of different cross section. He put forth special effort to construct a universal semi-empirical theory to enable the calculation of energy loss and heat transfer and mass in such flows.

"We started building primitive MHD mercury pumps at our small workshop."

Member of the Academy of Sciences Professor L. Artsimovich researched the phenomenon of the MHD dynamo. He studied how the structure of turbulence causes the spontaneous appearance of a magnetic field. An additional laboratory headed by Dr. A. Gailitis was created to work on these areas.

Professor Artsimovich also advised Professor Kirko to contact A. Komar and V. Glukhikh of the Electrosila plant in Leningrad. In Kirko's words:

"I left for Leningrad and told them openly about our first steps in the development of MHD conduction pumps by reproducing the well-known Hartmann's experiments. In response, I was told that all this was very interesting, but they had already created an induction pump supplied by three-phase mains at the laboratory in Leningrad. Our objectives, however, were more profound. We were studying a transition from laminar to turbulent flow and were able to estimate the drag coefficient and the effect of electromagnetic field on it. This aroused their interest, and they asked us to give them more details by secret mail."

"Why should we use secret mail? We are now preparing our first collection of articles on MHD for publication."

"Really? But there is a governmental list of secret topics, and MHD studies are not allowed to be published."

"But I have already obtained permission to publish."

"Well, your energy is outstanding!"

"This was the beginning of the friendship-rivalry of the two research teams in Riga and Leningrad."

Professor Kirko was as astute in building successful teams as he was in the financial and physical construction of his institute. He writes:

"Cooperation with Professor G. Shturman, an expert in electrical machinery, was started in a number of experimental and theoretical studies on the applicability of electrical machine theory to line induction pumps. In particular, we investigated so-called attenuation factors caused by reverse influence of liquid metal flow on the configuration of current lines and magnetic field distribution. Experimental studies of edge effects at the liquid metal inlet and outlet of the pump were of special interest. Here a strong influence of the magnetic Reynolds number and a dimensionless parameter determining the influence of the end effects were of considerable importance. J. Lielpeter constructed an annular electromagnetic channel, i.e., an induction pump without end effects, and its analog – a line pump with an inlet and outlet and an original system of the body suspension, which made it possible to establish the dynamic interaction between the working medium, i.e., flowing liquid metal, and the pump body very precisely. A rather precise method of estimating such pumps was developed on the basis of both theoretical data and numerous experiments. Our institute was among the first scientific institutions in the USSR working with lithium – a very dangerous metal."

By inventing financial tricks, Professor Kirko was able to legally pay expert mechanics a salary of 1,600 rubles:

"Paying special attention to our team of designers and mechanics, we managed to organize the necessary services thanks to our location in Salaspils, a small regional center outside Riga. This allowed us to spend considerable sums on wages for highly qualified workers attracting them by wages that sometimes exceeded salaries of senior researchers."

"We started to build up rapidly, to purchase equipment, to make a part of the necessary equipment by ourselves .... Several self-supporting enterprises were founded under the auspices of our Institute, such as a design bureau of magnetohydrodynamics in Riga that worked mainly according to our instructions, and a design bureau of MHD flow meters that developed novel types of these instruments.

"Solving the scientific problems was much harder than providing the material basis of our development. We committed ourselves to achieve very difficult goals. In those times, a director of an institution who failed or broke government contracts faced the prospect of grave penalties, and we all feared this."

Professor Kirko, now a Foreign Member of the Latvian Academy of Science, concludes his letter with a comment that accurate reporting forces me to quote: "In this respect, our institute was fortunate because of two talented scientists whose scientific work was especially intense and profound—Herman Branover and Olgert Lielausis. A brilliant scientific team was formed around them ...".

2 My personal debt to Professor Kirko

The Institute of Physics of the Latvian Academy of Sciences was fortunate to have a director who provided optimum conditions for his researchers and who took an active part in a great number of the research projects. As time went by, though, despite the ideal research conditions that Professor Kirko gave me, I felt that I could not continue living as a Soviet citizen working at the Institute. There were manifold reasons for this. Since 1964 I had been experiencing an awakening and began to believe in the Creator. The state enforced primitive atheism of Soviet universities suffocated me. The Communists crassly outlawed Jewish religious practice such as observing the Sabbath and keeping kosher. I felt that I was living a lie by being a Soviet scientist trying to secretly observe the laws of the Torah. I wrote a philosophic essay analysing my split personality. In this essay I disproved atheism and concluded that a return to belief and Jewish values was necessary. I arranged a meeting with members of the Israeli Embassy to give them my manuscript to take to Israel. Unfortunately, this meeting was observed.

I shall never forget how one bright March afternoon I was taken by the KGB to a most unpleasant interrogation. I feared that I would never see the sunlight again. Yet, the meeting seemed to go better than expected. They only wanted to know what was in the green notebook that I gave to a senior staff member of the Israeli Embassy. To my surprise and relief, they let me go and did not follow up.

According to rumor, Professor Kirko convinced the KGB that because of the importance of my research they should not to disrupt my work at the Institute by arresting me.

Kirko saved me a second time when I brought to him the letters of invitation I received from American scientists working in MHD asking me to come to their universities in the United States and give seminars on my work. To my shock, the chairwoman of the KGB department of the institute told me to write to each scientist who had invited me and tell him that I cannot come because my wife is ill. I did not write any such letters, but a few months later I started to receive letters from the same foreign researchers wishing my wife a speedy recovery. Someone in the secret service had answered the letters on my behalf!

In order not to complicate Professor Kirko any more, I resigned from the Institute at the beginning of 1971. After a long arduous struggle during which I was harassed and arrested several times by administrative arrest with no court proceedings or trial, I finally received permission to leave the Soviet Union. After the collapse of the USSR, I returned to visit Latvia in 1991, when I was elected a full foreign member of the Latvian Academy of Sciences.

3 Publications and conferences

In the mid-1960s the Institute of Physics started to publish its own journal, *Magnitnaya Gidrodinamika*, edited by the distinguished physicist Y. Birzvalk. The journal published the findings presented at the conferences that the Institute had initiated in order to discuss the scientific results and applications of MHD in different technologies. The first MHD conference took place in 1958. Since then, a conference was held in Latvia every 2 years, except from 1975 to 1996, when national economic hardship precluded such activity. During these years, eight international conferences were held in Israel.

During the economic malaise of the eighties and nineties the activities of the Institute stagnated to a certain degree. A number of researchers had left Riga, including Director Igor Kirko, who resettled in Perm, and myself, who immigrated to Beersheba, Israel. Despite these circumstances, the Institute of Physics team in Riga continued to obtain impressive results. Perhaps the most important achievement of this period was the construction and demonstration of a facility that presented the phenomenon of the MHD dynamo. In this volume there is a separate chapter describing the excellent work done by a special team under the guidance of Dr. A. Gailitis.

4 MHD in Beersheba

In Beersheba – free now to live openly as a believing, practicing Jew – I formed a new laboratory team using new equipment. About a dozen Ph.D. students wrote theses focusing mainly on spiral turbulence and the hazardous phenomenon of hurricane development. A detailed analysis of spiral turbulence

Fig. 1. The Etgar 3 mixer in Beersheba: A. El-Boher (*left*) with H. Branover

is presented in our book *Turbulence and Structures: Chaos, Fluctuations, and Helical Self-Organization in Nature and the Laboratory* [2].

The most important area of research that I have pursued in Beersheba is the continuation of my studies of spiral turbulence and further understanding the structure of turbulence when influenced by a strong magnetic field. It seems natural that turbulent fluctuation would be suppressed until it ultimately disappears. Measuring the pressure drop along the channel confirms this conclusion. To our great surprise, however, the first local measurement of turbulent velocity in a magnetic field showed not only that the turbulence does not disappear, but, to the contrary, it grows manifold. This led us to the conclusion that turbulence adjusts to the magnetic field and generates a new type of turbulence. This new type of turbulence is close to 2D in the plane perpendicular to the magnetic field. To be more precise, the turbulence becomes spiral. Such spiral turbulence leads to an inversed turbulent energy cascade. This means that turbulent energy in a magnetic field is not transferred into small-scale fluctuations when it is suppressed by viscosity, as is observed in regular turbulent flows. The energy instead is transferred into large-scale fluctuations having a very low level of turbulent energy dissipation. Such spiral turbulence with an inversed energy cascade is most important in the analysis of how hurricanes and other types of tropical atmospheric hazards develop.

The research team in Beersheba is now called the Joint Israeli-Russian Laboratory for Energy Research. It has closely collaborated with a number of Russian experts, particularly Professor S. Moiseev of the Institute of Space Research in Moscow. Until his death, Professor Moiseev came to Beersheba every year for 2–3 months and greatly contributed to the success of this laboratory.

Fig. 2. Part of the MHD family about to cut the cake

Fig. 3. Fifth Conference (Jerusalem, 1987). Participants J. Braun, F. Hussain, and Kopland (from *left* to *right*) in the talent competition, wearing berets, mimicking Herman Branover, chairman of the organizing committee

In addition to studying spiral turbulence, I continued my work creating a universal semi-empirical theory of MHD turbulence. Completed a few years ago, this work is referred to as Branover's Semi-Empirical Theory by energy turbulence researchers. This theory enables the calculation of pressure drop and other integral characteristics of flows in ducts of different cross-sectional shapes.

In parallel to the above, the team of the Joint Laboratory is intensifying the research of spiral turbulence in a multitude of applicative and theoretical areas. So far, our greatest success has been in regards to metallurgy and the role of turbulence with inversed energy cascade in the formation of hurricanes and other tropical catastrophic phenomena in the atmosphere. The leading contribution to the study of these two phenomena has been made

Fig. 4. Seventh Conference (Jerusalem, 1993). H. Branover, chairman of the organizing committee, briefing participants J. Braun and H.K. Moffatt during the talent contest

Fig. 5. Seventh Conference (Jerusalem, 1993). M. Petrick and R. Moreau with the organizing committee cake

Fig. 6. Seventh Conference (Jerusalem, 1993). M. Garnier (*left*), R. Moreau (*right*) and other French participants (B. Meneguzzi and F. Werkoff) in the talent contest

Fig. 7. A. Schendlin, the leader of the Russian MHD conversion center, with E. Matveeva

Fig. 8. A. Schendlin giving a bottle of Russian vodka as a prize to talent winner J. Meng

Fig. 9. Y. Unger, organizer of the talent contests, awarding the first place winner with a bottle of vodka.

by Professor A. Kapusta, Professor E. Golbraikh, Dr. S. Sukoriansky, and Dr. B. Mikhailovich. Until his untimely death in 2002, this group benefited from the invaluable leadership and initiative of Professor S. Moiseev. His

Fig. 10. P. Lykoudis performing his first-prize winning sirtaki (Greek dance)

influence is still felt in all areas of our work and will continue to be of pivotal importance in the coming years.

Another area of research involving a number of researchers of the Joint Laboratory is the development of a new concept of an MHD electrical power station, called the Etgar System. A completely designed system was developed jointly with Dr. M. Petrick and other researchers at the Argonne National Laboratory in Chicago. The performance and cost of such a system was analysed by Arthur D. Little, Inc. in Boston and found to be highly competitive with other systems.

Among other achievements attained by the Joint Laboratory, I wish to mention the development of MHD electrical power systems to be used in spaceships. We succeeded in achieving the lowest weight for an electric power-generating system.

5 The talented MHD family

One of the important activities of the Joint Laboratory is the organization of international conferences. Up until now, we have held eight such conferences, with registration of about 150 foreign scientists. People involved in MHD research sometimes feel lonely because it is still considered an esoteric field. Therefore, we try to make our conferences be a kind of family reunion. One way of creating this atmosphere is by running a talent competition at the gala dinner. We announce the talent show at the beginning of each conference. The participants are given full freedom to choose whether they want to dance, sing, or tell jokes. The idea is always received with such enthusiasm that we run the risk of participants investing more effort into their talent performance than their professional presentation. Sincere thanks go to my veteran Joint Laboratory colleague Yeshajahu Unger for the creative energy he puts into the talent shows that make us feel indeed like one big family.

References

1. Personal letter from Igor Kirko to Herman Branover, 17 June 2005
2. Branover H, Eidelman A, Golbraikh E, Moiseev S (1999) Turbulence and Structures: Chaos, Fluctuations, and Helical Self-organization in Nature and the Laboratory. Academic Press, San Diego

Some additional references of the most informative works on spiral turbulence and inverse energy cascades from among our hundreds of publications:

Branover H, Eidelman A, Nagorny M, Kireev M (1994) Magnetohydrodynamic simulation of quasi-two-dimensional geophysical turbulence. In Branover H, Unger Y (eds) Progress in Turbulent Research, AIAA 162:64–79

Branover H, Sukoriansky S, Talmage G, Greenspan E (1986) Turbulence and the feasibility of self-cooled liquid metal blankets for fusion reactors. Fusion Technol 10: 822-829

Branover H (1978) Magnetohydrodynamic Flow in Ducts. Wiley, New York

Branover H, Gershon P (1979) Experimental investigation of the origin of residual disturbances in turbulent MHD flows after laminarization. J Fluid Mech 94: 629–647

Velocity Measurement Techniques for Liquid Metal Flows

Sven Eckert, Andreas Cramer, and Gunter Gerbeth

Forschungszentrum Rossendorf, P.O.Box 510119, 01314 Dresden, Germany
(s.eckert@fz-rossendorf.de)

1 Introduction

Analysis and control of fluid flows, often subsidiary to industrial design issues, require measurements of the flow field. For classical transparent fluids such as water or gas a variety of well-developed techniques (laser Doppler and particle image velocimetry, Schlieren optics, interferometric techniques, etc.) have been established. In contrast, the situation regarding opaque liquids still lacks almost any commercial availability. Metallic and semiconductor melts often pose additional problems of high temperature and chemical aggressiveness, rendering any reliable determination of the flow field a challenging task. This review intends to summarise different approaches suitable for velocity measurements in liquid metal flows and to discuss perspectives, particularly in view of some recent developments (ultrasound, magnetic tomography). Focusing mainly on local velocity measurements, it is subsequently distinguished between invasive and non-invasive methods, leaving entirely aside the acquisition of temperature, pressure, and concentration, for which [1] may serve as a comprehensive reference.

2 Invasive techniques

In this context, invasiveness means insertion of a sensing unit into the medium under investigation, the consequence of which is twofold. We are not mainly concerned with probably adverse effects on the sensor owing to, e.g., high temperature or chemical aggressiveness, which ultimately boils down to a question of material science, rather than with the influence of the probe on the flow. This potential disturbance determines, besides their functional principles, the applicability of various types of anemometers to a considerable extent. On this note, different sensors are at first described and then discussed with particular attention to sensitivity.

S. Molokov et al. (eds.), Magnetohydrodynamics – Historical Evolution and Trends,
275–294. © 2007 *Springer.*

Velocity probes to be immersed into the fluid can be classified, according to the underlying physical effect, into force reaction, thermal, and conductive sensors. Note that neither this small list is downright complete nor is it possible to review all variants in each category due to the scope of this review. Following history, we start with the force reaction probes, because these had been the first employed in order to determine velocities in moving fluids.

2.1 Force reaction probes

These probes respond to the force exerted onto them by the flowing medium, which is in principle a pressure. Presumably, the best-known mechanical anemometer is the vane type used in weather stations in order to determine wind speed. It usually consists of a few hemispheres or cups attached to radial spokes. The rotation speed can be measured by a number of different mechanisms. Often a magnet, affixed to the shaft, traversing past a fixed coil induces a pulse for each revolution, or a digital shaft encoder is used. One may ask whether such rugged devices are of any benefit for magnetohydrodynamic (MHD) flow measurements. As far as integral stationary flow properties in certain configurations are a matter, the answer is certainly yes.

Recently, both Tallbäck et al. [2] and Taniguchi et al. [3] successfully measured angular velocities in an electromagnetically driven rotary liquid metal flow. Inserting vanes similar to the left one depicted in Fig. 1 having sizes of almost that of the container diameter, these authors determined an integral value that corresponds, e.g., to the flow rate in a pipe.

The commercial availability of such small impeller-based vanes, as shown on the right-hand side for less than $ 400 including a data station, might suggest to perform semi-local measurements also. Regarding the performance of moving mechanical parts-based semi-local sensors, it is instructive to have a look at similar devices. Szekely et al [4] made use of a linear arrangement consisting of a spring loaded rod onto the head of which a $\phi = 19$ mm stainless steel disc was fastened. Although the displacement of the rod owing to the drag exerted on the disc was sensitively recorded by a linear voltage differential

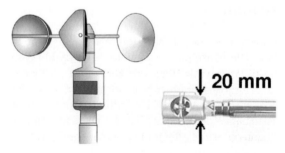

Fig. 1. Classical cup vane to determine wind speed (*left*) and miniaturised impeller-based vane probe (*right*)

transformer, the smallest velocity difference reported by the experimenters was 4 cm/s, and the smallest absolute value was 8 cm/s. For the suitability concerning the measurement of velocity fluctuations we quote the authors: "The inertia of the measuring system does not provide a great deal of insight into the structure of this turbulent flow." Force reaction probes comprising moving mechanical parts can be summarised to be restricted to time-averaged values at relatively high velocities and poor spatial resolution.

Another category of velocity probes makes use of directly measuring pressures, thus avoiding any moving mechanical part. The principle of operation of all these tubes is based on Bernoulli's law $p + \frac{\rho}{2}v^2 = p_0$, where p_0 denotes the total pressure, which is a constant, p the static pressure, and ρ and v the fluids density and velocity, respectively. In a stationary incompressible flow, the sum of the dynamic pressure $\frac{\rho}{2}v^2$ and p always results in the pressure within the resting fluid, which is that of the ambient atmosphere, plus the hydrostatic contribution $\rho g h$ of the fluid. Tube anemometers comprise basically a bend with one end directed in such a way that it faces the flow. As the kinetic energy is converted into potential one at the stagnation point, all tubes measure at least the total pressure p_0. Once the static pressure is known, the simpler Pitot tube allows the determination of the velocity according to Bernoulli's law. The static pressure p can only be determined accurately by measuring it in a manner such that the velocity pressure has no influence on the measurement at all. This is achieved by measuring it at right angle to the streamlines. The Prandtl tube sketched in Fig. 2 is an example of this, where p is determined through several static taps arranged circumferentially in the outer tube. A differential manometer thus allows for a direct measurement of the fluid velocity.

There exists not a huge number of research reports wherein Pitot and Prandtl tubes have been applied to liquid metal flows particularly addressing questions about sensitivity. From the few available it turns out that Pitot

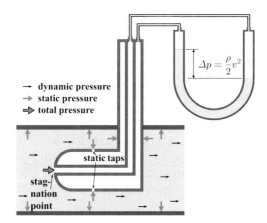

Fig. 2. Schematic diagram of a Prandtl tube

tubes are advantageous because they can be manufactured in smaller size. Typical outer diameters are in the range of a few millimetres. Moreau [5] points out that care has to be taken when magnetic fields are involved, a situation which is almost intrinsic to MHD experiments. Then, the stagnation pressure is not exactly equal to the fluid's loss of kinetic energy because of electromagnetic forces. It is estimated in [5] that this effect becomes significant for the smaller velocities of about 1 cm/s, which may be seen as the lower range of reliable operability of tube-based anemometers in liquid metals. Besides [5], for further reading see Branover et al. [6] and references therein.

It is obvious that this technique is not suitable for turbulence measurements if, e.g., a U-manometer is used as a pressure sensor owing to inertia of the fluid moving in the limbs. An attractive perspective of the method is offered by the availability of piezo-resistive pressure transducers. We successfully measured static pressure fluctuations in a 50 Hz AC electro-vortically driven flow through a ϕ =1 mm hole drilled into the chamber wall at sampling rates exceeding 1 kHz [7]. One may think about a Pitot tube consisting of a bent syringe and a piezo-transducer mounted at the other end. In [8], Pitot pressure surveys in a liquid metal atomization nozzle were reported, making use of a ϕ = 0.9 mm stainless steel tube. Operating also at the comparatively high rate of 1 kHz, the authors have been able to detect the transition from subsonic to supersonic flow regimes. It convincingly demonstrates the potential of tube-based anemometers.

With fibre flowmeters we return to the moving mechanical parts, but in a somewhat miniaturized variant and optical recording. In an early work, Griffiths and Nicol [9] mounted a single quartz fibre in a wall of a pipe in such a way that it protruded at a right angle to the flow direction. The deflection of the fibre tip was observed from the opposing side of the pipe by means of a travelling microscope. In an air experiment, velocities down to 10 cm/s were successfully recorded. Zhilin et al. [10] and Eckert et al. [11] constructed more complex sensors upon this principle, which were proven to work in liquid metals. A thin glass rod of several tens of μm in diameter was sealed into a thin-walled conical glass tube. The other end of this pointer was either blackened and brought into the light path where it led to absorption [10], or illuminated and observed by means of an endoscope [11]. Both techniques allowed for velocity resolution below 1 cm/s of two components. A remarkable feature of fibre sensors is the applicability to electro-vortical flows. Using the technique described in [11], Cramer et al. [7] determined the flow field in the comparable small volume of $2 \times 4 \times 2.5$ cm^3 throughflown by currents as high as several kiloamperes (kA) conveyed from a point source. Whereas the spatial resolution perpendicular to the sensor's axis was very high, the problem with the extended range of axial sensitivity was coped with by a mixed experimental–numerical approach. Based on bending moment theory, a numerical model of the probe was implemented predicting the integrated response of the sensor from a calculated flow field. These results were found to be in good agreement with the measurements.

2.2 Thermal anemometers

A wet finger in the air will detect the direction of the wind because a drop in temperature is felt on the surface facing the wind. Thermal anemometers act in a similar way, in that the passage of fluid takes heat away from a heated element at a rate dependent upon the velocity. This element, which is either a thin wire that can be made very short, or a metallic film on a quartz or ceramic substrate, is mounted at the end of a probe that can be inserted into the liquid under investigation. The mode of operation is either the change of electrical resistance at constant current, or the measurement of the current required to keep the resistance at a constant set point. Since the resistance will always be proportional to the temperature, the latter are frequently termed constant temperature anemometer. In practice, the resistance is measured or controlled by means of a Wheatstone bridge, one leg of which is the thermal probe.

Owing to their principle, the velocity readout of thermal probes seems instantaneous at first sight. However, even a very fine hot wire by itself cannot respond to changes in fluid velocity at frequencies above 500 Hz. By compensating for frequency lag with a non-linear amplifier this response can be increased to values exceeding 100 kHz. When compared with hot wires the cylindrical hot-film sensor, depicted in the upper right part of Fig. 3, has basically two advantages. A better frequency response is achieved because the sensitive part is distributed on the surface rather than on the entire cross section as with a wire. Secondly, the heat conduction to the supports (end losses) for a given length to diameter ratio are smaller due to the low thermal conductivity of the substrate material. A shorter sensing length thus can be used.

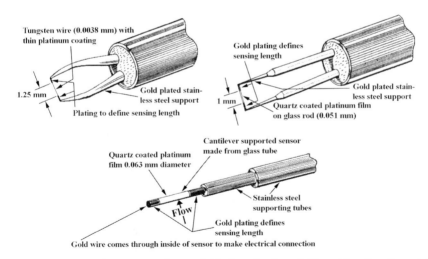

Fig. 3. Examples of commercially available hotwire (*top left*) and hotfilm (*top right* and *bottom*) probes. In particular, the lengths of the wire and the cantilever of the single-ended film sensor can be made very short

Hot-wire anemometers, employed mostly in gases by then, have been adopted for liquid metals in the 1960s. Sajben [12] reported on a system that was suitable in mercury at velocities from 1 to 12 cm/s. An improvement of this technique by Trakas et al. [13], also applied to a flow of mercury, revealed the capabilities of hot-film probes regarding velocity resolution. They were conditionally restricted upon principle to velocities of a few millimetre per second, which today may still be seen as the sensitivity threshold for an employment in liquid metals. Hot wire and film sensors are prone to fouling and deposition of debris and oxides, which change the transport properties of heat. Because they are thermal devices, it is important to compensate carefully for variations in ambient temperature and pressure. In particular, when applied in the low Prandtl number liquid metals, the high ratio between diffusive and convective heat transport leads to a significant decrease of resolution. All these drawbacks render the use of these thermal sensors in liquid metals somewhat inconvenient if not even tedious [14,15]. For a typical MHD application, Robinson and Larsson [16] is referred to, who determined the mean velocity field in a flow driven by a rotating magnetic field.

Quick response is one of the prerequisites for turbulence measurements. Further, this measuring task puts a severe restriction on the size of the sensor in order to resolve all scales of potentially significant vortices. Being not restricted to thermal probes, this means that the finite extension of the sensitive zone acts as a lowpass filter with respect to the time domain. Because thermal sensors fulfill both suppositions, they have been intensively used in the study of turbulence. To quote a few, Alemany et al. [17] is referred to, who investigated the influence of a DC magnetic field on the flow of mercury created by a moving grid with a hot-film sensor attached behind the grid. Recently, Petrović et al. [18] reported on the accuracy of turbulence measurements by hot wires. The article quotes a variety of modern studies devoted to hot wires and somehow addresses the question about their perspectives, in particular of such ones having up to 12 sensing wires.

2.3 Potential difference probes

Often these probes are also named the conductive anemometers. This may be due to the fact that the sensing wires of the probe are in electrical contact with the conducting medium. The basic principle consists in measuring a voltage drop $\Delta\phi$ induced by a magnetic field $\mathbf{B}$ across its wire spacing Δl according to Ohm's law: $\mathbf{j} = \sigma(E + \mathbf{u} \times \mathbf{B})$. In the absence of electric currents $\mathbf{j}$, the electric field strength E, expressed by a finite difference of the potential $E \approx \Delta\phi/\Delta l$ for sufficiently small sensor dimensions, is independent of the electrical conductivity σ, and is linearly related to the velocity $\mathbf{u}$. Determination of fluid velocities via this electromotive force (e.m.f.) dates back to Faraday [19]. He had tried vainly to measure the voltage induced across the river Thames by the motion of the water in the earth's magnetic field. Kolin [20] proposed to use a probe consisting of two wires, insulated except at the tip, with a separation of the wire tips in the order of a few thousandths of an inch. In an orthogonal arrangement with a homogeneous static measuring field, the potential difference induced between the wires should then be a direct measure of the corresponding velocity component.

The measuring magnetic field can be applied either globally over the entire melt volume, or locally confined to the wire tips. Again Kolin [21] was among the first who used also incorporated magnet probes. Equipped with a small electromagnet, the probe's sensitivity must have been poor. Ricou and Vives [22] reported on a feasible solution using rare-earth CORAMAG and ALNICO permanent magnets, the latter were even operable in aluminium melts. As any sensor immersed into the fluid poses an obstacle, the influence of which onto the flow increases with size, the question about how small a probe can be build becomes an important issue. From this point of view, the potential difference probes (PDP) using a globally applied field are seemingly advantageous because they essentially consist only of two wires. On the other hand, it is well known that static fields may damp the flow to be measured. Compared to a typical hot wire having $\phi = 1$ mm, the probes in [22] were several times larger. A globally applied field with the same strength as that acting in the incorporated probe in [22] certainly influences the flow significantly. Because the sensitivity of PDPs is determined by the product of field strength and wire spacing, we are concerned here with the usual compromise inherent to every measurement task. Without dismissing sensitivity, which was about 1 cm/s minimum velocity at a stated by the authors of [22] resolution of 1 mm/s, Weissenfluh [23] constructed PDPs having $\phi \approx 1$ mm. At this stage it may be summarized that PDPs compare well to hot wires regarding performance. The drawback that they are not suited in many configurations owing to the presence of electric currents (see [24]) is compensated by the ease of use.

Similar to thermal sensors, PDPs have been thoroughly employed for the measurement of mean velocities down to their resolution of around 1 mm/s [25, 26], and to determine turbulence characteristics of fast flows [27, 28]. An advantage of PDPs is the utilization in those cases where a strong magnetic field is intrinsic to the problem. This branch of investigations comprises basic research, e.g., two–dimensional (2D) turbulence [29,30], as well as applications in fusion technology [31]. In this context, one particular technique is worth noting. When the fluid flow becomes quasi-2D in a sufficiently high magnetic field, the electric potential does not vary considerably, neither in the core nor in the Hartmann layers. Davoust et al. [32] and Messadek and Moreau [33] acquired velocities non-invasively at the Hartmann wall by means of electrodes mounted in the wall. These measurements yielded to a good approximation the core velocity values in planes perpendicular to the applied field. Besides observing local fluid velocities in an ideally isothermal flow, adding at least a third electrode allows to account for temperature effects produced by thermo-electricity (Seebeck effect). It depends on the particular choice of geometry, the number of electrodes, and the materials thereof whether the fluid velocity is measurable without thermoelectric disturbances or the temperature can be measured in addition. Examples for such combined probes are to be found in [23, 31, 32].

Summarizing so far it becomes obvious that: (i) further miniaturization of incorporated magnet PDPs below 1 mm will lead to a serious decrease in sensitivity; (ii) the use of a globally applied field with the same strength of that locally acting in incorporated PDPs is often unacceptable because of the influence on the flow; (iii) any tip spacing $\Delta l > 1$ mm does hardly allow for turbulence measurements, according to Bolonev et al. [34], who determined experimentally the influence of Δl on the transfer function, which quantifies the above-mentioned spatial integration leading to a lowpass filtering in the time domain; (iv) a significant increase in sensitivity of e.m.f.-based measurements can consequently be achieved only by an as good as possible noise – and disturbance – free set-up of the electronic data acquisition system.

At least since the work of Remenieras and Hermant [35] tackling the problem of inductive transients and noise in e.m.f.-based velocity measurements, it became obvious that a fully differential-ended amplifier chain is mandatory despite of the low impedance source of the probe. Using state-of-the-art instrumentation with high impedance coupling between amplifier stages and meticulously avoiding systematic disturbances such as thermoelectricity, we have been able to extend the sensitivity of mean velocity measurements to 10^{-2} mm/s for a $\Delta l = 1$ mm probe. From the calibration curve in Fig. 4 it is seen that voltages less than 1 nV had to be acquired reliably. As sensitivity, resolution, and bandwidth are always a compromise, the performance of the measuring chain allowed the determination of velocity fluctuations in a flow driven by a rotating magnetic field commencing slightly above the threshold of linear stability. This corresponded to a mean velocity of a little less than 3 cm/s, over a wide range of the Taylor number. Figure 5 demonstrates that it

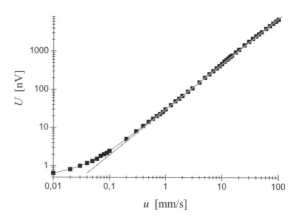

Fig. 4. Calibration curve of an incorporated magnet PDP using highly sensitive analog instrumentation. The probe response was non-linear below 1 mm/s

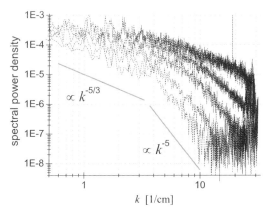

Fig. 5. Power spectra of velocity fluctuations in a flow driven by a rotating magnetic field from slightly above the linear stability threshold (steepest slope of inertial range) spanning a range of factor 15 in the governing parameter [36]

was always possible to resolve all scales of wavelengths. A detailed description of the experiment and the electronics can be found in Cramer et al. [36].

3 Non-invasive techniques

3.1 Ultrasonic methods

Ultrasonic methods are non-invasive, but not fully contactless. A continuous acoustic path from the ultrasonic transducer to the liquid under investigation is required for transmission of the ultrasonic wave into the flow region and for reception of the measuring signal.

Two common principles are known to apply ultrasound for measurements of fluid velocities: the ultrasonic Doppler and the transit time, also called the time-of-flight technique. The operating mode of ultrasonic flowmeters by the transit-time method is based on two sequential measurements: an ultrasonic pulse is sent between two transducers upstream and downstream through the liquid. The run-time difference between downstream acceleration and upstream deceleration delivers the averaged velocity. To obtain local information about the flow field, Johnson et al. [37] proposed a method to measure three-dimensional (3D) flows by transmitting and receiving ultrasonic beams along a multitude of lines. The arrangement is that each volume element is traversed by a set of lines having components in each direction for which flow components are to be reconstructed. Each propagation time measurement of the ultrasonic wave is an integral of a function of sound speed and fluid velocity along the particular line leading to a set of integral equations, which have to be inverted to obtain the unknown fluid velocity vector field.

A more promising way to measure local velocities is offered by the ultra-sound Doppler method, often called ultrasound Doppler velocimetry (UDV) or ultrasonic velocity profile (UVP) monitor. The origin of this technique can be retraced to the medical branch [38]. Owing to the pioneering work of Takeda [39, 40] it has also been established in physics and fluids engineering. The measuring principle is based on the pulsed echo technique. Ultrasound pulses of a few cycles are emitted from the transducer and travel along the measuring line. If such a pulse hits microparticles suspended in the liquid, a part of the ultrasonic energy is scattered. It can be received using a second transducer or by the same transducer working in the listening mode between two emissions. In the majority of cases the second variant is realized. The entire information of the velocity profile along the ultrasonic beam is con-tained in the echo. If the sound velocity of the liquid is known, the spatial position along the measuring line can be determined from the detected time delay between the burst emission and its reception. The movement of an ensemble of scattering particles inside the measuring volume will result in a small time shift of the signal structure between two consecutive bursts. The velocity is obtained from a correlation analysis between consecutive bursts. The measuring principle is sketched in Fig. 6. Owing to the Nyquist theo-rem, the product of measurable maximum velocity and penetration depth is limited by the sound velocity and the ultrasonic frequency. For a more detailed description of the basics of the measuring principle the reader is referred to Takeda [40].

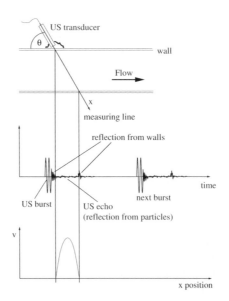

Fig. 6. UDV-measuring principle shown for a channel flow

We want to focus here on a few problems arising especially with the application of UDV to liquid metal flows, namely the load of the sensors due to high temperatures and the chemical aggressiveness of the melt, the transmission of the ultrasonic waves through container walls, the acoustic coupling between the transducer, the wall and the melt, as well as the allocation of suitable reflecting particles inside the liquid metal. The feasibility of velocity profile measurements in liquid metals using UDV has been demonstrated for the first time by Takeda [41], who measured velocity profiles in a T-tube filled with mercury at room temperature. Further successful applications have been published by Brito et al. [42] for liquid gallium and by Eckert and Gerbeth [43] for liquid sodium at a temperature of about 150°C. In many applications the ultrasonic transducer cannot be brought into direct contact with the liquid metal. The ultrasonic methods also allow measurements through the container wall as shown in Fig. 7 for the case of a channel flow. However, one has to take into account that any additional interface in the ultrasonic path diminishes the energy of the ultrasonic beam. One reason for such losses might be a mismatch between the acoustic impedances Z of the wall material and the liquid metal, which is rather pronounced for liquid sodium ($Z_{Na} = 2 \times 10^6$ Ns/m^3) flowing inside a channel of stainless steel ($Z_{SS} = 4.5 \times 10^7$ Ns/m^3). In this case, the transmission of a sufficient amount of ultrasonic energy through the wall can only be assured if the wall thickness meets almost exactly a multiple of half the wavelength of the ultrasonic wave in the wall material [43]. Another issue is the wetting at the inner wall. The occurrence of thin gas or oxide layers impedes the passover of the ultrasonic wave into the liquid metal. Brito et al. [42] performed UDV measurements in a vortex of liquid gallium confined in a cylindrical vessel made from different materials (polycarbonate,

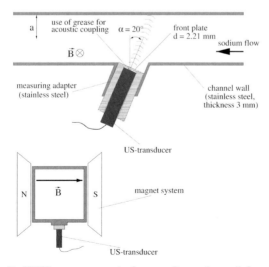

Fig. 7. UDV measurements for a sodium channel flow [43]

nylon, copper). The authors observed a continuous deterioration of the signal quality with progressing time of measurements. This phenomenon was related to oxide films developing at the inner cylinder wall. To prevent adherence of oxides at the wall the authors proposed a special coating with a cataphoretic film. Experiences gained with stainless steel and liquid sodium [43] confirm that oxide layers at the contact surface must be eliminated to guarantee a low-loss transmission of the ultrasonic wave.

The conventional piezoelectric transducers using lead zirconate titanate (PZT)-based materials are usually restricted to a temperature range below 200°C. Other piezoelectric materials with higher Curie temperatures like $GaPO_4$ or $LiNbO_3$ can work up to temperatures of 650°C or 900°C, respectively. Such sensors have already been used for fluid level detectors in liquid metal fast breeder reactors [44]. However, the piezoelectric coupling factor of the heat-resistant piezoelectric materials is by a factor of about 5 less than that for standard materials. This leads to a worse signal-to-noise ratio and, thus, results in a sensitivity, which is insufficient for UDV measurements. The application of acoustic wave guides as a buffer between the hot liquid and the piezoelectric elements is another approach to elude the temperature restriction of 200°C. Different types of acoustic waveguides, consisting in the simplest version of a solid cylinder of heat-resistant material, have already been applied to extend the working range of ultrasonic flowmeters towards higher temperatures [45, 46]. The structure of waveguides for Doppler shift measurements appears to be more sophisticated because a monomode propagation of the ultrasonic wave inside the waveguide is required. This results in a restriction for the thickness of the waveguide structures. Gelles [47] demonstrated the basic features of a fiber-acoustic waveguide consisting of a bundle of cylindrical fibres. Eckert et al. [48] presented a waveguide made of a stainless steel foil with a thickness of 0.125 mm as shown in Fig. 8. The thinner the waveguide structures, the higher the emission frequencies can be applied, and the lower the velocities can be measured. The operability of such steel waveguides has already been demonstrated in CuSn and aluminium at temperatures up to 750°C [48, 49].

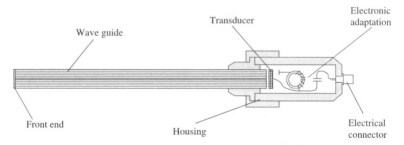

Fig. 8. Ultrasonic sensor with integrated acoustic waveguide for measurements in hot metallic melts

Doppler devices require the presence of scattering objects inside the fluid. Artificial or natural particles, gas bubbles, or fluctuations in density can serve for this purpose. There is a lack of quantitative studies in liquid metals focusing on the dependence of signal properties on parameters, such as concentration, morphology (e.g., size, shape) and acoustic properties of the suspended reflectors. The signal quality depends on the optimal particle concentration. Though low concentration does not disturb the propagation of the ultrasonic wave significantly, the sensitivity of the measurements deteriorates. On the other hand, high concentration improves the sensitivity but increases the attenuation and, in turn, limits the depth of the measurement. Scattering particles to be added to the flow should match the fluid density to avoid a slip between the fluid and particle motion and to guarantee homogeneous distribution in the entire fluid volume. Moreover, the particles need to be wetted by the liquid to avoid agglomeration effects. It is obviously favourable to work solely with natural impurities usually existent in metallic melts with a common, technical purity standard. Noble liquid metals, such as mercury, contain an insignificant amount of natural tracers, whereas, for instance, in liquid gallium or gallium alloys, a distinct oxidation cannot be avoided with reasonable effort. Here, the situation could arise that the UDV measurements might be complicated by too many tracers inside the measuring volume [42,50].

Another essential point of interest is the question with respect to the capability of the UDV technique for analysing turbulent velocity fluctuations. In the past, electromagnetic potential probes were used in MHD turbulence research to record local time series, and to calculate the frequency power spectrum [29]. Because of the statistical character of the measuring principle the UDV method is inferior regarding the time resolution of both measuring techniques. A number of US bursts have to be superposed in order to get a reliable velocity signal. Depending on the distinct experimental conditions this requirement typically leads to time resolution of between 10 and 100 ms. On the other hand, the UDV technique delivers the local velocity simultaneously at different locations along one measuring line. Usually, the turbulent energy $E(k)$ is derived from the frequency power spectrum $P(f)$ by employing Taylor's hypothesis. In many applications, for instance, the electromagnetic stirring in confined geometries, this assumption becomes questionable because a clearly dominating mean flow, which moves a frozen turbulent structure, does not exist. Regardless of the limitations in time resolution, the UDV method allows a direct calculation of the velocity structure functions, and therefore provides information about the scaling properties of the flow. Takeda [51,52] studied the transition from laminar flow to turbulent in a rotating Taylor–Couette system by measuring the spatiotemporal velocity field. To analyse the velocity structure quantitatively he applied spatial and temporal Fourier transform and orthogonal decomposition techniques. Related studies on thermal turbulence in mercury have recently been published by Mashiko et al. [53] and Tsuji et al. [54].

The amount of publications dealing with UDV measurements in liquid metal flows is still manageable. This measurement technique can provide valuable insight in miscellaneous flow situations, occurring for instance, during electromagnetic stirring [50], during solidification of metallic alloys [55], inside a mercury target for a spallation source [56] or in MHD two-phase flows [57].

3.2 Radioscopic techniques

Visible light cannot be used for flow visualization in metallic melts because penetration of macroscopic metallic layers requires photon energies of at least 10 keV. On the other hand, radioscopic techniques working with short-wavelength radiation, such as x-rays or nuclear radiation, have been employed for in situ investigations of kinetics and morphology of solid–liquid interfaces during solidification. Information about the flow pattern can also be obtained. Szekely [58] determined the turbulent diffusivity in liquid steel using radioactive tracers. For this reason he introduced a capsule containing radioactive gold into the centre of the bath. Samples of the steel were periodically taken out at certain positions and the radioactive content was measured. Stewart and Weinberg [59] introduced radioactive material into liquid tin to delineate the flow pattern. After a certain period of time, the system was rapidly quenched in order to freeze the tracer position. The tracer profile was taken as representative of the flow pattern. Obviously, these first realizations have to be considered as fairly crude and by no means non-invasive. An in situ monitoring of the tracer movement in the melt is necessary. Kakimoto et al. [60] report about a direct observation of the flow structure in molten silicon by x-ray radiography. The authors developed a multilayered tracer consisting of a small tungsten cylinder in the sensor. The tungsten was covered by layers of SiO_2 and carbon to adjust the density to that of silicon and to wet the tracer by the molten silicon. X-rays penetrating the silicon pool during the process were detected by an image intensifier. Because of the much larger absorption coefficient the momentary position of the singular tungsten particle can be followed by the visualization system allowing the reconstruction of the particle trajectory.

Another approach is the visualization of the density field as proposed by Koster et al. [61–64]. X-ray absorption within material depends on the mass attenuation coefficient, fluid density, and the material thickness in the direction of the penetrating radiation. If the density is altered by temperature, the method provides a temperature field visualization being related to the velocity field in natural convection. This radioscopic technique was tested with a natural convection benchmark study in liquid gallium [63]. The weak dependence of the density on temperature in metallic melts requires additional efforts, for instance, to carefully avoid beam scattering in the environment, to achieve excellent resolution of the radioscopic system. Koster et al. [63] published a highest resolution in detection of local density changes of 0.02%. A very recent development is the application of high frame-rate neutron radiography to investigate liquid metal two-phase flows. Saito et al. [65, 66] performed experiments using the JRR-3M nuclear research reactor providing high neutron fluxes. Gold–cadmium particles were added to a lead–bismuth melt, and 2D velocity fields were reconstructed using particle tracking velocimetry.

3.3 Flow tomography from measurements of the induced magnetic field

Magnetoencephalography (MEG) is well established in the medical branch as a convenient method to study the brain function and diseases, such as epilepsy [67]. Low electric currents flowing inside the neurons generate magnetic fields which can be measured outside the body, thus providing a remarkably accurate representation of the local brain activity. Is it conceivable to use a similar principle for flow measurements? To answer this question, let us consider an electrically conducting liquid flowing within a certain volume. By imposing an external magnetic field, such an unknown flow field will generate a distribution of induced currents inside the liquid and thereby an induced magnetic field. The latter is present inside, as well as outside of the melt volume. The structure of the induced field obviously contains information about the flow. A reliable interpretation of this information would provide a fully contactless method to determine 3D velocity fields. The strength of the applied fields must be sufficiently weak, so that the flow to be measured is not influenced. However, this measuring principle can also be applied in cases where stronger magnetic fields are already present in the process under consideration, for instance, in continuous casting with an electromagnetic brake or in single crystal growth processes.

The first attempt to utilize this principle for flow measurements was undertaken by Köhler et al. [68]. The authors applied a few local sensors to detect the flow velocity of liquid steel in the mould in close vicinity to the sensor position. The sensors consisting of permanent magnets and highly sensitive magnetic field detectors were positioned close to the wall of the mould. Because of difficulties regarding the sensor calibration, the velocity information was obtained by correlating the output signal of two adjacent sensors. The question is obvious whether a complete reconstruction of the velocity field can be realized using a sufficient number of magnetic field sensors around the fluid volume to be measured. Stefani et al. [69] showed that the sole measurement of the induced magnetic field, even using numerous sensors, cannot deliver a unique solution of the problem as long as the electrical potential at the surface of the fluid volume is not taken into account. The determination of the electric potential requires a set of electrodes at the fluid surface implicating that the principle thus becomes less attractive for hot and aggressive fluids or for facilities where the fluid boundary is not accessible owing to technological reasons. The problem can be solved by subsequent application of various external magnetic fields to the same flow field [70]. The imposition of two orthogonal magnetic fields represents a certain minimum configuration for a fully contactless flow tomography. An experimental demonstration of a contactless inductive flow tomography (CIFT) has been reported by Stefani et al. [71]. The scheme of this experiment is shown in Fig. 9.

A cylindrical vessel with an aspect ratio close to 1 contains about 4.5 l of the eutectic alloy GaInSn. A propeller forces a flow inside the vessel up to maximum velocities of $1\,\mathrm{m/s}$, which corresponds to a magnetic Reynolds number $R_\mathrm{m} \approx 0.4$. Two pairs of Helmholtz coils consecutively produce axial and transverse magnetic fields. The induced magnetic fields are measured by 49 Hall sensors at different positions around the vessel. The main problem of the method is that the values of the induced magnetic fields are some orders of magnitude lower than the applied field. The authors let the propeller rotate in both directions, resulting either in an upward or in a downward pumping with different flow structures. Whereas the downward pumping produces both a main poloidal roll and a toroidal motion, the latter one is, to a large extent, inhibited by guiding blades for the upwards pumping. The CIFT technique was able to discriminate between those different flow patterns [71]. By comparing these measurements with the UDV technique it was further shown that not only the structure, but also the range of the velocity scale was correctly reproduced, see the right part of Fig. 9.

A particular advantage of CIFT is the transient resolution of the full 3D flow structure in steps of several seconds. Hence, slowly changing flow fields in various processes can be traced in time. Further developments of this measuring principle will use also AC magnetic fields to improve the depth resolution of the determined velocity field.

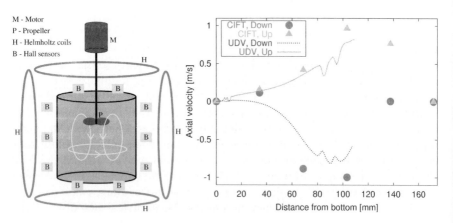

Fig. 9. Scheme of the CIFT demonstration experiment (*left*), and comparison of CIFT and UDV velocity measurements (*right*) for the axial velocities along the central vertical axis of the cylinder (UDV measurements are only shown up to the propeller position, whereafter they become unreliable) [71]

References

1. Brusey BW, Brussiere JF, Dubois M, Moreau A (eds) (1999) Advanced Sensors for Metal Processing. Canadian Institute of Mining, Metallurgy and Petroleum, Montreal
2. Tallbäck GR, Lavers JD, Beitelman LS (2003) Simulation and measurement of EMS induced fluid flow in billet/bloom casting systems. In: Asai S, Fautrelle Y, Gillon P (eds) Proceedings of the 4th International Symposium on Electromagnetic Processing of Materials, Lyon, France, pp 154–159
3. Taniguchi S, Maitake K, Okubo M, Ando T, Ueno K (2003) Rotary stirring of liquid metal without free surface deformation by combination of rotational and vertical traveling magnetic fields. ibid, pp 339–343
4. Szekely J, Chang CW, Ryan RE (1977) The measurement and prediction of the melt velocities in a turbulent, electromagnetically driven recirculating low melting alloy system. Metal Trans 8B:333–338
5. Moreau R (1978) Local and instantaneous measurements in liquid metal MHD. Proc Dynamic Flow Conf, pp 65–79
6. Branover H, Gelfgat YM, Tsinober AB, Shtern AB, Shcherbinin EV (1966) The application of Pitot and Prandtl tubes in magnetohydrodynamic experiments. Magnetohydrodynamics 2:55–58
7. Cramer A, Gerbeth G, Terhoeven P, Krätzschmar A (2004) Fluid velocity measurements in electro–vortical flows. Mat and Manufact Processes 19:665–678
8. Mates SP, Settles GS (1995) A flow visualization study of the gas dynamics of liquid metal atomization nozzles. In: Proceedings of the International Conference on Powder Metallurgy and Particulate Materials, Seattle, USA
9. Griffiths RT, Nicol AA (1965) A fibre flowmeter suitable for very low flow rates. J Sci Instrum 42:797–799
10. Zhilin VG, Zvyagin KV, Ivochkin YP, Oksman AA (1989) Diagnostics of liquid metal flows using fibre-optic velocity sensor. In: Lielpeteris M, Moreau R (eds) Liquid Metal Magnetohydrodynamics, Kluwer Academic, Dordrecht, 373–379
11. Eckert S, Gerbeth G, Witke W (2000) A new mechano–optical technique to measure local velocities in opaque fluids. Flow Meas Instrum 11:71–78
12. Sajben M (1965) Hot wire anemometer in liquid mercury. Rev Sci Instrum 36:945–953
13. Trakas C, Tabeling P, Chabrerie JP (1983) Low–velocity calibration of hot–film sensors in mercury. J Phys E: Sci Instrum 16:568–570
14. Argyropoulos SA (2000) Measuring velocity in high–temperature liquid metals: a review. Skand J Metallurgy 30:273–285
15. Reed CB, Picologlou BF, Dauzvardis PV, Bailey JL (1986) Techniques for measurement of velocity in liquid–metal MHD flows. Fusion Technol 10:813–821
16. Robinson T, Larsson K (1973) An experimental investigation of a magnetically driven rotating liquid-metal flow. J Fluid Mech 60:641–664
17. Alemany A, Moreau R, Sulem PL, Frisch U (1979) Influence of an external magnetic field on homogeneous turbulence. J de Méchanique 18:277–313
18. Petrović DV, Vukoslavčević PV, Wallace JM (2003) The accuracy of turbulent velocity component measurements by multi–sensor hot wire probes: a new approach to an old problem. Exp Fluids 34:130–139
19. Faraday M (1832) Experimental researches in electricity – second series (Bakerian lecture). Phil Trans Roy Soc 175:197–244

20. Kolin A (1943) Electromagnetic method for the determination of velocity distribution in fluid flow. Phys Rev 63:218–219

21. Kolin A (1944) Electromagnetic velometry. I. A method for the determination of fluid velocity distribution in space and time. J Appl Phys 15:150–164

22. Ricou R, Vives C (1982) Local velocity and mass transfer measurements in molten metals using an incorporated probe. Int J Heat Mass Transfer 25:1579–1588

23. Weissenfluh T (1985) Probes for local velocity and temperature measurements in liquid metal flow. Int J Heat Mass Transfer 28:1563–1574

24. Tsinober A, Kit E, Teitel M (1987) On the relevance of the potential–difference method for turbulence measurements. J Fluid Mech 175:447–461

25. Gelfgat YM, Gelfgat AY (2004) Experimental and numerical study of rotating magnetic field driven flow in cylindrical enclosures with different aspect ratios. Magnetohydrodynamics 40:147–160

26. Barz RU, Gerbeth G, Wunderwald U, Buhrig E, Gelfgat YM (1997) Modelling of the isothermal melt flow due to rotating magnetic fields in crystal growth. J Cryst Growth 180:410–421

27. Grossman LM, Charwat AF (1952) The measurement of turbulent velocity fluctuations by the method of magnetic induction. Rev Sci Instrum 23:741–747

28. Bojarevičs A, Bojarevičs V, Gelfgat YM, Pericleous K (1999) Liquid metal turbulent flow dynamics in a cylindrical container with free surface: experiment and numerical analysis. Magnetohydrodynamics 35:258–277

29. Kolesnikov YB, Tsinober AB (1972) Two–dimensional flow behind a cylinder. Magnetohydrodynamics 8:300–307

30. Eckert S, Gerbeth G, Witke W, Langenbrunner H (2001) MHD turbulence measurements in a sodium channel flow exposed to a transverse magnetic field. Int J Heat Fluid Flow 22:358–364

31. Burr U, Barleon L, Müller U, Tsinober AB (2000) Turbulent transport of momentum and heat in magnetoydrodynamic rectangular duct flow with strong sidewall jets. J Fluid Mech 406:247–279

32. Davoust L, Cowley MD, Moreau R, Bolcato R (1999) Buoyancy–driven convection with a uniform magnetic field. Part 2. Experimental investigation. J Fluid Mech 400:59–90

33. Messadek K, Moreau R (2002) An experimental investigation of MHD quasi two-dimensional turbulent shear flows. J Fluid Mech 456:137–159

34. Bolonev N, Charenko A, Eidelmann A (1976) About the correction of turbulence spectra measured using conductivity anemometers. Ing Phys J 2:243–247 (in Russian)

35. Remenieras G, Hermant C (1954) Mesure électromagnétique des vitesses dans les liquides. Houille Blanche 9:732–746

36. Cramer A, Varshney K, Gundrum T, Gerbeth G (2006) Experimental study on the sensitivity and accuracy of electric potential local flow measurements. Flow Meas Instrum 17:1–11

37. Johnson SA, Greenleaf JF, Tanaka M, Flandro G (1977) Reconstructing three-dimensional temperature and fluid velocity vector fields from acoustic transmission measurements. ISA Trans 16:3–15

38. Atkinson P (1976) A fundamental interpretation of ultrasonic Doppler velocimeters. Ultrasound Med Biol 2:107–111

39. Takeda Y (1986) Velocity profile measurement by ultrasound Doppler shift method. Int J Heat Fluid Flow 7:313–318

40. Takeda Y (1991) Development of an ultrasound velocity profile monitor. Nucl Eng Design 126:277–284
41. Takeda Y (1987) Measurement of velocity profile of mercury flow by ultrasound Doppler shift method. Nucl Technol 79:120–124
42. Brito D, Nataf H-C, Cardin P, Aubert J, Masson JP (2001) Ultrasonic Doppler velocimetry in liquid gallium. Exp Fluids 31:653–663
43. Eckert S, Gerbeth G (2002) Velocity measurements in liquid sodium by means of ultrasound Doppler velocimetry. Exp Fluids 32:542–546
44. Boehmer LS, Smith RW (1976) Ultrasonic instrument for continuous measurement of sodium levels in fast breeder reactors. IEEE Trans Nucl Sci 23:359–362
45. Liu Y, Lynnworth LC, Zimmerman MA (1998) Buffer waveguides for flow measurement in hot fluids. Ultrasonics 36:305–315
46. Jen C-K, Legoux J-G, Parent L (2000) Experimental evaluation of clad metallic buffer rods for high temperature ultrasonic measurements. NDT&E In 33:145–153
47. Gelles IL (1966) Optical-fiber ultrasonic delay lines. J Acoust Soc Am 39:1111–1119
48. Eckert S, Gerbeth G, Melnikov VI (2003) Velocity measurements at high temperatures by ultrasound Doppler velocimetry using an acoustic wave guide. Exp Fluids 35:381–388
49. Eckert S, Gerbeth G, Gundrum T, Stefani F (2005) Velocity measurements in metallic melts. In: Proceedings of 2005 ASME FED Summer Meeting, FEDSM2005-77089
50. Cramer A, Zhang C, Eckert S (2004) Local flow structures in liquid metals measured by ultrasonic Doppler velocimetry. Flow Meas Instrum 15:145–153
51. Takeda Y (1999) Quasi-periodic state and transition to turbulence in a rotating Couette system. J Fluid Mech 389: 81–99
52. Takeda Y, Fischer WE, Sakakibara J (1993) Measurement of energy spectral density of a flow in a rotating Couette system. Phys Rev Lett 70:3569–3571
53. Mashiko T, Tsuji Y, Mizuno T, Sano M (2004) Instantaneous measurement of velocity fields in developed thermal turbulence in mercury. Phys Rev E 69:036306
54. Tsuji Y, Mizuno T, Mashiko T, Sano M (2005) Mean wind in convective turbulence of mercury. Phys Rev Lett 94:034501
55. Eckert S, Willers B, Gerbeth G (2005) Measurements of the bulk velocity during solidification of metallic alloys. Metall Mater Trans A 36:267–270
56. Takeda Y, Kikura H, Bauer G (1998) Flow measurement in a SINQ mockup target using mercury. In: Proceedings of 1998 ASME FED Summer Meeting, FEDSM98-5057
57. Zhang C, Eckert S, Gerbeth G (2005) Experimental study of a single bubble motion in a liquid metal column exposed to a DC magnetic field. Int J Multiphase Flow 31:824–842
58. Szekely J (1964) Experimental study of the rate of metal mixing in an open-hearth furnace. Journal ISIJ 202:505–508
59. Stewart MJ, Weinberg F (1972) Fluid flow in liquid metals. Experimental observations. J Cryst Growth 12:228–238
60. Kakimoto K, Eguchi M, Watanabe H, Hibiya T (1988) Direct observation by X-ray radiography of convection of molten silicon in the Czochralski growth method. J Cryst Growth 88:365–370

61. Campbell TA, Koster JN (1994) Visualization of liquid/solid interface morphologies in gallium subject to natural convection. J Cryst Growth 140:414–425

62. Campbell TA, Koster JN (1995) Radioscopic visualization of Indium Antimonide growth by the vertical Bridgman-Stockbarger technique. J Cryst Growth 147:408–410

63. Koster JN, Seidel T, Derebail R (1997) A radioscopic technique to study convective fluid dynamics in opaque liquid metals. J Fluid Mech 343:29–41

64. Derebail R, Koster JN (1998) Visualization study of melting and solidification in convecting hypoeutectic Ga-In alloy. Int J Heat Mass Transfer 41:2537–2548

65. Saito Y, Mishima K, Tobita Y, Suzuki T, Matsubayashi M (2005) Measurements of liquid-metal two-phase flow by using neutron radiography and electrical conductivity probe. Exp Therm Fluid Sci 29:323–330

66. Saito Y, Mishima K, Tobita Y, Suzuki T, Matsubayashi M, Lim IC, Cha JE (2005) Application of high frame-rate neutron radiography to liquid-metal two-phase flow research. Nucl Instrum Meth Phys Res A 542:168–174

67. Hämäläinen M, Hari R, Ilmoniemi RJ, Knuutila J, Lounasmaa OV (1993) Magnetoencephalography – theory, instrumentation, and applications to noninvasive studies of the working human brain. Rev Mod Phys 65:413–497

68. Köhler KU, Andrzejewski P, Julius E, Haubrich H (1994) Measurements of steel flow in the mould. In: Asai S (ed) Proceedings of International Symposium on Electromagnetic Processing of Materials, Nagoya, Japan, pp 344–349

69. Stefani F, Gerbeth G (1999) Velocity reconstruction in conducting fluids from magnetic field and electric potential measurements. Inverse Problems 15:771–786

70. Stefani F, Gerbeth G (2000) A contactless method for velocity reconstruction in electrically conducting fluids. Meas Sci Technol 11:758–765

71. Stefani F, Gundrum T, Gerbeth G (2004) Contactless inductive flow tomography. Phys Rev E 70:056306

Flow Control and Propulsion in Poor Conductors

Tom Weier, Victor Shatrov, and Gunter Gerbeth

Forschungszentrum Rossendorf, P.O. Box 510119, D-01314 Dresden, Germany
(T.Weier,V.Shatrov,G.Gerbeth)@fz-rossendorf.de

1 Introduction

The possibility to act on a fluid flow in a contactless way, offered by magnetohydrodynamics (MHD), stimulated the imagination of aerodynamists and naval engineers relatively early.

Ritchie [1] appears to be the first using electromagnetic forces to pump electrolytes in 1832. Figure 1 shows two of his apparatuses. Basically, the horizontal magnetic field component near the pole of a permanent magnet (N) interacts with the mainly vertical electric field between two ring electrodes (w, w') to set the dilute acid in the angular gap (AB) into rotational motion.

In the 1950s, a multitude of aerospace applications of MHD flow control techniques has been envisioned using the fact that at high enough speeds air gets ionised by the action of shock waves and frictional heating, and thus becomes a conductor. Such high-speed conditions are typical for re-entry problems. Resler and Sears [2] and Busemann [3] proposed, among others, to use magnetic fields to control heat transfer, to decelerate or to accelerate vehicles, and to prevent flow separation. Although enthusiasm for the practical application of these ideas waned later on, the topic is now again under investigation in connection with scramjets, e.g., [4,5], heat transfer mitigation [6] and electromagnetic braking for re-entry vehicles [7]. An overview of magnetoaerodynamic (MAD) research can be found in the proceedings of a recent von Kármán institute lecture series [8]. MAD deals with compressible fluids at hypersonic speeds, which represents quite a complex problem on its own, and will not be discussed here. Henceforth, the discussion is limited to incompressible fluids, mainly seawater. Due to the focus of our own work, the survey may be somewhat biased towards the boundary layer control (BLC).

S. Molokov et al. (eds.), Magnetohydrodynamics – Historical Evolution and Trends,
295–312. © 2007 *Springer.*

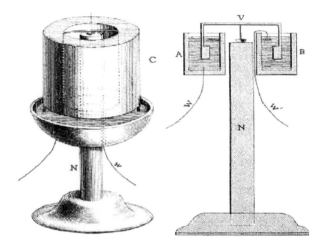

Fig. 1. Ritchie's electromagnetically rotated electrolyte columns [1]

2 Electromagnetic propulsion

Electromagnetic propulsion in seawater has been proposed by Rice [9] already in 1961. According to Friauf [10] and Way [11] the idea had attracted the attention of several inventors at that time. The main reason for this attraction has been the seemingly elegant operating principle using no moving parts. Proposed applications include the silent propulsion of naval submarines [12], the use in high-speed cargo submarines [11], and the propulsion of future high-speed surface ships without the danger of cavitation [13].

Conventionally (see, e.g., [11,14]), the electromagnetic propulsion methods are subdivided into four groups, as shown schematically in Fig. 2. If both the electric and magnetic fields are imposed, the propulsion scheme is termed "conductive" (Fig. 2a, c); if only an alternating magnetic field is applied, the method is referred to as "inductive" (Fig. 2b, d). The internal flow systems (Fig. 2a, b) use a duct with an electromagnetic pump, while the fields penetrate into the surrounding sea for external systems (Fig. 2c, d).

The arrangement of flush-mounted electrodes and magnets suggested by Rice [9], and shown in Fig. 3, belongs to category c, the external conductive propulsion. In 1966, Way [11] built a model submarine named EMS–1 with an electromagnetic thruster of the external conductive type at the University of California, Santa Barbara. The model was approximately 3 m long and had a displacement of approximately 400 kg. A dipole electromagnet provided a magnetic induction at the hull of 0.015 T. Powered by lead-acid batteries, the submarine reached a maximum velocity of approximately 0.8 knots (0.4 m/s).

This experiment even arrested the attention of mass media at that time. Nevertheless, in addition to the principal possibility to propel a marine vessel by the electromagnetic forces, some fundamental problems inherent in electromagnetic propulsion in seawater, already noted by Friauf [10] and Phillips [15],

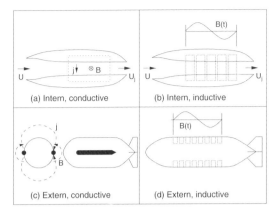

Fig. 2. Classification of electromagnetic propulsion methods according to Way [11]

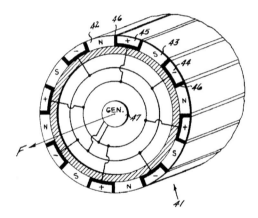

Fig. 3. Arrangement of electrodes (45) and permanent magnets (42) proposed by Rice [9] to propel a cylindrical body in seawater

were now demonstrated in practice. Especially, the unfavourable ratio of power input to available thrust was striking. After a period of active research, US activities in electromagnetic propulsion declined, apparently towards the end of the 1960s.

The reason for the efficiency deficit is easily explained. Regardless of the electromagnetic propulsion method, the Lorentz force density **f** producing the thrust is due to a current density **j** and a magnetic induction **B**, namely

$$\mathbf{f} = \mathbf{j} \times \mathbf{B}. \tag{1}$$

Since in seawater applications the magnetic Reynolds number is small, the induced magnetic fields can be neglected and **B** becomes the applied field only. The current density is given by the Ohm's law

$$\mathbf{j} = \sigma(\mathbf{E} + \mathbf{U} \times \mathbf{B}). \tag{2}$$

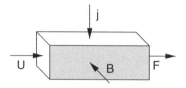

Fig. 4. Operating principle of an internal conductive propulsor

Here, $\mathbf{E}$ denotes the electric field strength, $\mathbf{U}$ the flow field, and σ the electric conductivity of the medium, respectively. Owing to its simplicity, internal conductive propulsion (type a) as sketched in Fig. 4 is chosen to illustrate the main performance criteria of electromagnetic propulsion in seawater following Thibault [16].

Of primary interest are the necessary power input to achieve a specific thrust (energy balance) and the total thrust available to propel the vessel at a specific velocity (momentum balance). The thrust per unit volume equals the Lorentz force density with the absolute value $f = jB$ ($f = |\mathbf{f}|, \ldots$) in the case sketched in Fig. 4, assuming uniform and orthogonal electromagnetic and flow fields. The ideal electrical to mechanical efficiency η is the ratio of propulsive power or thrust per unit volume $p_T = jBU$ to the total power supplied per unit volume $p_E = jE$, where U denotes the flow velocity and E the electric field strength. This gives:

$$\eta = \frac{p_T}{p_E} = \frac{UB}{E} = \frac{1}{\phi}. \tag{3}$$

Thus, it appears that the ideal efficiency is the inverse of the load factor ϕ

$$\phi = \frac{E}{UB}, \tag{4}$$

giving the ratio of applied E to the induced UB electric field (see, e.g., [17]).

Expressing the current density in Eq. (1) by Ohm's law (2) and taking into account that the induced electric field acts against the applied one, the total electromagnetic thrust F can be expressed as follows:

$$F = \sigma U B^2 (\phi - 1) V. \tag{5}$$

Here V denotes the volume of the duct. Thus, for maximum ideal efficiency ($\eta = \phi = 1$), the attainable thrust is zero. However, to propel a vessel, the usually non-zero total hydrodynamic drag D has to be balanced by the thrust. In a rough estimate, the total drag in turbulent flows is proportional to the square of the flow velocity: $D = kU^2$. Taking into account the relation $D = F$ and Eqs. (3), (5), it follows that the ideal electric efficiency is:

$$\eta = \frac{1}{1 + \frac{kU}{\sigma V B^2}}. \tag{6}$$

To maximise the efficiency at a given velocity U, the product VB^2 should therefore be chosen as large as possible. This has been realised quite early. Doragh, aware of the need for for high magnetic fields, suggested to use superconducting magnets already in 1963 [18].

Similar observations can be made for outer conductive propulsion methods (see, e.g., [11]), and for inductive methods (see, e.g., [15]). The main conclusion is always the need to deploy the highest possible magnetic induction in the largest available volume.

Inductive arrangements have been investigated, for example, by Phillips as early as 1962 [15] and later by Khonichev and Yakovlev [19], where the latter paper was the first one treating the coupled electromagnetic and hydrodynamic parts of the problem. Generally, the practical applicability of the inductive approach is limited by the fact that superconducting magnets providing an alternating magnetic field are not easily available. Therefore, the maximum magnetic field strength and consequently the efficiency are quite limited. However, Saji et al. [20] proposed and demonstrated experimentally an ingenious approach to the problem consisting in rotating the magnet including the cryostat. Unfortunately, the effort required in a real application might countervail the advantages gained. Intensive theoretical studies of the inductive approach have been performed by Yakovlev and co-workers in the late 1970s and early 1980s [21–23].

Saji and co-workers built two model ships, SEMD-1 and ST-500, with superconducting magnets with racetrack coils in the 1970s. In both cases, the propulsors were of the external conductive type. SEMD-1 [24] tested in 1976 [25] had a 0.6 m long and relatively bulky magnet with a maximum induction of $\sim$1 T mounted below the vessels hull. A maximum efficiency of 0.1% has been determined from tests in a tub, where the model was at rest and mounted to a force balance. In 1979, a second model ST-500 has been built and operated [25], now with a magnet of 2 T maximum induction and smoothly integrated into the hull of the vessel. In towing tank tests a maximum speed of 0.6 m/s has been reached with a total thrust of 15 N.

While thrusters of the external conductive type have been further investigated mainly numerically, e.g., [26–29], research concentrated on internal conductive propulsion in the 1980s and 1990s. This type has been favoured, because it allows for large thruster volumes with relatively homogeneous electromagnetic field distributions [16]. The most noticed achievement in this field has without doubt been the successful sea trial of the YAMATO-1 in 1992 [30], a 30 m long ship with 185 t displacement propelled by two electromagnetic thrusters with a mean induction of 4 T delivering 8,000 N of thrust each. YAMATO-1 reached top speed of 6.6 knots and a maximum electrical efficiency of 1.4% [30]. Considerably higher efficiencies have been reported for land-based experiments at Naval Undersea Warfare Center (NUWC) and Argonne National Laboratory (ANL) in the USA, where electromagnetic thrusters have been integrated in closed seawater loops. Meng et al. [31] found a maximum efficiency of $\sim$2.7% for a magnetic induction of 3.3 T and a load factor of $\phi \approx 20$. While the efficiency could be further increased to nearly 10% for a 6 T magnet at NUWC [32], as high values as 38% are reported by Meng et al. [32] for experiments with a 6 T magnet of 1 m bore diameter at ANL. However, this impressive device had a weight of more than 173 t.

For the simple crossed field arrangement sketched in Fig. 4 and used in a modified form for the thrusters of YAMATO-1 and in the experiments at NUWC and ANL, superconducting dipole magnets are necessary. These kind of magnets require massive structural enforcement to withstand the large magnetic forces, resulting in heavyweight constructions and relatively low maximum magnetic field strength.

For the same bore diameter, superconducting solenoids allow for lower weight and higher maximum field strength than dipole magnets. Recent thruster developments concentrated therefore on the use of solenoids. However, the axial magnetic field requires a special arrangement of electrodes and baffles forming the so called "helical thruster" (Fig. 5) as proposed and demonstrated by Bashkatov [33] as well as by Tada [34] in 1991.

However, the increase of electrical efficiency due to the higher magnetic field strength is accompanied by an increase of hydrodynamic losses in the thruster introduced by the baffle and other guiding plates. In 1995, Lin and Gilbert [12] used a 12 T helical thruster in a closed seawater loop and measured nearly 20% electrical efficiency. In 1998, Chinese researchers operated a 3.5 m long model ship HEMS-1 with 1 t displacement in a seawater pool [35,36]. Equipped with a helical thruster with 5 T induction, a maximum speed of 0.68 m/s has been measured. In 1999, a helical thruster based on a 14 T solenoid has been run by a Chinese–Japanese group in a closed seawater loop [13, 37]. Ideal efficiencies exceeding 60% have been found, while the maximum efficiency including all losses is 13% for a load factor of 2.6 [37].

Meanwhile, worldwide activities in MHD-propulsion decreased, magnetic inductions and bore diameters of currently affordable superconducting magnets allow only for thruster efficiencies far below that of competing propulsion methods. This penalty currently outweighs all envisaged advantages. However, driven by the simplicity of the approach and the fascination it exerts on contemporary art, MHD-propulsion made its way into the classroom after all [38].

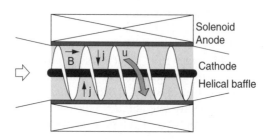

Fig. 5. Sketch of a helical thruster

3 Electromagnetic flow control

3.1 Drag reduction

Fluid dynamic drag can be of different origin. Electromagnetic flow control in seawater has mainly concentrated on the skin friction drag and to a lesser extent on the form drag. Wave drag has been dealt with by Petit [39] but while the demonstration experiment used a dilute acid, the focus of his work was on shock wave cancellation in hypersonic flight [40, 41]. The following discussion is limited to skin friction and form drag.

3.1.1 Transition delay

In 1961, Gailitis and Lielausis [42] proposed to use a stripwise arrangement of electrodes and magnets as sketched in Fig. 6 to delay the transition of a laminar boundary layer. Except for the plane geometry, electric and magnetic field sources are similar to the propulsion system patented by Rice [9] in the same year, see Fig. 3. However, the idea of Gailitis and Lielausis was not to propel the plate by the electromagnetic forces, but to compensate for the viscous losses of the near wall flow. For the electrode–magnet–arrangement in Fig. 6 the ratio of the electromagnetic to frictional forces, i.e., the Hartmann number Z, can be written as

$$Z = \frac{1}{8\pi} \frac{j_0 M_0 a^2}{\rho \nu U_\infty}. \tag{7}$$

Here M_0 denotes the magnetisation of the permanent magnets, j_0 the applied current density, a the width of the electrodes, ρ the fluids density, ν its kinematic viscosity, and U_∞ the outer flow velocity, respectively. For $Z = 1$, the growth of the boundary layer can be inhibited. Assuming the Lorentz force density distribution to be uniform in the spanwise direction and exponentially decaying with the distance y from the wall, an exponential distribution of the wall parallel velocity component u, namely

$$\frac{u}{U_\infty} = 1 - e^{-\frac{\pi}{a}y}, \tag{8}$$

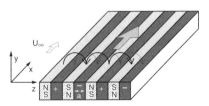

Fig. 6. Stripwise geometry of electrodes and magnets for a streamwise wall-parallel force as proposed by Gailitis and Lielausis [42]

follows as a solution of the boundary layer as well as the Navier–Stokes equations. The exponential boundary layer profile is similar to that of the asymptotic suction boundary layer and should therefore posses similar stability characteristics, i.e., a critical Reynolds number based on the displacement thickness of about $Re_{\delta_1 \mathrm{crit}} = 4.7 \times 10^4$ compared to only $Re_{\delta_1 \mathrm{crit}} = 520$ for the Blasius boundary layer [43]. The development of an exponential profile for a force distribution $f \sim \exp(-\pi y/a)$ and $Z = 1$ has been shown numerically by Tsinober and Shtern [44] by solving the boundary layer equations. Experimental work at that time has been limited to qualitative observations of flow cases considerably different from zero-pressure-gradient boundary layers [45]. Meanwhile, also experimental evidence is available. Figure 7 presents Laser–Doppler measurements showing convincingly the transformation of a Blasius boundary layer profile with $Re_{\delta_1} \approx 290$ to an exponential one [46].

Transition delay promises a huge potential for skin friction drag savings. Comparing typical laminar to turbulent skin friction drag, these savings may even be large enough to offset very low electrical efficiencies η (3). However, as is well known, linear stability of the asymptotic boundary layer profile alone is not a sufficient condition for the transition delay. In practice, many additional effects, e.g., receptivity to the disturbance environment and the influence of the real force distribution, have to be taken into account. So far, these aspects have only partially been addressed. The evolution of the boundary layer and the stability of the accompanying profiles has been studied by Zhilyaev et al. [47] and later by Albrecht et al. [48].

In 1962, Phillips [15] estimated power requirements for boundary layer stabilisation with induced fields and found them far exceeding possible savings.

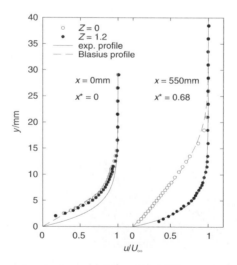

Fig. 7. Development of an exponential boundary layer profile under the influence of the Lorentz force

To the knowledge of the authors, for stabilisation purposes, Lorentz forces have up to now only be considered for modifications of the mean flow profile. Techniques acting on the disturbances (wave cancellation) offer potentially higher efficiencies, but are coupled to sophisticated sensor–actuator systems.

3.1.2 Turbulent boundary layers

Since transition control is practically limited to length Reynolds numbers $Re_x < 4 \times 10^7$ [49], techniques for skin friction reduction in turbulent boundary layers (TBL) are desirable in many cases. Though Shtern [50] discussed already in 1970 the possibility to limit the growth of a TBL by a streamwise Lorentz force, it was only at the beginning of the 1990s that electromagnetic control of TBLs became of increasing interest. While Meng [51] followed the ideas of [9] and the work on BLC in the 1960s reviewed by Tsinober [52] and Lielausis et al. [45], Nosenchuck and Brown developed a different approach based on wall normal forces in 1993 [53]. Especially the experiments by Nosenchuck and Brown [53], who coined the term electromagnetic turbulence control (EMTC) [54], were very well received at that time, e.g., [55], and sparked further research.

Mainly three different force configurations have been investigated in order to control TBLs: wall parallel streamwise (Fig. 6), wall parallel spanwise (Fig. 8, left), and nominally wall normal forces (Fig. 8, right).

Wall parallel forces in the streamwise direction have been applied, e.g., in the experiments of Henoch and Stace [56] and Weier et al. [57] as well as in the numerical analysis of Crawford and Karniadakis [58]. This force configuration increases instead of reducing wall shear stress, because the acceleration of the near wall fluid leads to a higher slope of the mean velocity profile in the streamwise direction. However, the momentum gain due to the Lorentz force surpasses the friction drag rise. While mean velocity and skin friction are increased near the wall, their fluctuating components are damped for higher momentum input [56,57]. Shtern's [50] concept of a TBL of constant thickness by means of streamwise forces has been experimentally verified in [57].

Nosenchuck and Brown [54], O'Sullivan and Biringen [59], Thibault and Rossi [60], and others used nominally wall normal, time-dependent forces. Nosenchuck and co-workers reported several successful experiments with a

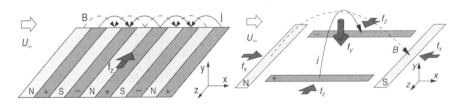

Fig. 8. Magnet/electrode arrangements for spanwise (*left*) and wall–normal forces (*right*)

multitude of electromagnetic actuators ("tiles") generating turbulent skin friction reductions of more than 90% [54], 55% [61], and a total drag decrease of more than 50% [62]. The physical mechanism behind this drag reduction is supposed to be a global reorganisation of the boundary layer into rotational periodic structures, cf. [63] and the sketches and flow visualisations in [64]. However, other groups were unable to reproduce these results [65]. As pointed out by Rossi and Thibault [66], the real force distribution produced by the electromagnetic tiles is quite complex and may play a crucial role in the experiments.

Time-dependent wall parallel forces in spanwise direction have been investigated numerically, among others, by Berger et al. [67], and Du et al. [65] and experimentally by Pang and Choi [68], and Breuer et al. [69]. Drag reductions ranging from 10% for the directly measured mean drag coefficient [69] to 40% for the local skin friction [68] have been found, indicating that this type of forcing is indeed able to reduce the skin friction drag of turbulent flows. The drag reduction mechanism is supposed to be similar to that suggested for spanwise oscillating walls [68]. Nevertheless, the energy balance of the approach is not favourable.

3.1.3 Form drag of bluff bodies

Compared to skin friction reduction, the use of Lorentz forces to control flow separation received less attention. Probably, the first experimental demonstration of separation prevention, as well as provocation, on the half-cylinder has been given by Crausse and Cachon [70].

Selected flow visualisations from their paper are reproduced in Fig. 9. Note that not all field sources are inside the body. Although Crausse and Cachon did not perform any force measurements, it is obvious from Fig. 9 that a

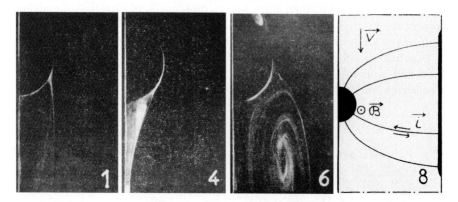

Fig. 9. Control of flow separating from the half-cylinder by the electromagnetic forces. (From [70].) Unforced flow (1), force downstream (4), force upstream (6), and the field configuration (8)

Lorentz force directed downstream reduces the size of the separation bubble behind the half-cylinder considerably, thereby reducing form drag as well. Similar experiments, but with an interchanged role of electric and magnetic fields and an additional conductivity gradient, where performed in 1961 by Lielausis [45, 71].

Successful electromagnetic control of the flow around a circular cylinder has been reported by Petit [39] for electrodes embedded in the cylinder and an externally applied magnetic field. A circular cylinder equipped with electrodes and permanent magnets generating a wall parallel force in the streamwise direction was used in the experiments and numerical calculations of Weier et al. [72]. Similar configurations have later been investigated by Kim and Lee [73], Posdziech and Grundmann [74], and Chen and Aubry [75]. While skin friction drag is increased by this force configuration, form drag is strongly reduced for an initially separated flow at Reynolds numbers Re of the order of 100. For stronger forcing the increase in skin friction drag dominates the form drag decrease. The total drag on the cylinder under these conditions is, however, negative due to the electromagnetically generated thrust.

Shatrov and Yakovlev [27] studied numerically the flow around a sphere with mainly wall parallel Lorentz force for Re up to 1,000. For increasing interaction parameter, i.e., the ratio of the electromagnetic to inertial forces, the size of the separation region is first reduced. Later, separation is suppressed completely resulting in a strong decrease of form drag. Skin friction drag is increased, as is always the case for streamwise wall parallel forces acting downstream. At sufficiently high interaction parameters, the sphere is driven upstream by the Lorentz forces. In a subsequent paper [76], Shatrov and Yakovlev extended the investigated Reynolds number range up to 10^5 treating the problem of a steady and axially averaged flow. For large Reynolds number, the total drag on the sphere was reduced four times, and despite the moderate electrical efficiency of $\eta \approx 40\%$, the total energy consumption was reduced as well. Note that this "moderate" electrical efficiency is high compared to what has been reached in MHD propulsion experiments, since low load factors at still sufficient momentum input, i.e., a strong magnetic field, is easier to realise numerically.

Very recently, Shatrov and Gerbeth [77] have shown that it is possible to reduce the drag on a sphere by three orders of magnitude using an optimised field distribution. Since a high load factor was assumed, efficiency of this drag reduction is nevertheless quite small.

3.2 Lift and manoeuvrability

Besides the drag reduction, there are other goals in flow control. A prominent one is the prevention of separation in order to generate a certain lift used for manoeuvring or stabilisation of marine vessels. For these applications, energetical efficiency is not always a primary goal. The emphasis is rather on the viability of a specific lift increase compared to the uncontrolled case.

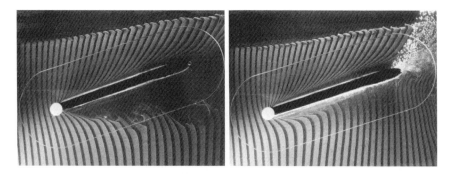

Fig. 10. Separated flow on the suction side of an inclined flat plate (*left*). Reattached flow due to a wall parallel streamwise Lorentz (*right*) [57]

Separation suppression at the suction side of inclined hydrofoils is, as in the case of bluff bodies, easily achieved by a streamwise wall parallel force acting in the flow direction. The flow visualisations in Fig. 10 demonstrate separation control in case of an 18° inclined flat plate at a chord length Reynolds number of $Re = 1.2 \times 10^4$. Electrodes and magnets are distributed practically along the whole chord length, leading to a uniform acceleration of the boundary layer flow along the plate.

The reattached flow shown in the right part of Fig. 10, while on one hand reducing the drag, on the other hand also re-establishes the lift of the plate. Figure 11 demonstrates this with measurements of the lift coefficient C_L on a PTL IV hydrofoil. At a fixed angle of attack of 17°, the suction side flow is already separated at the low chord length Reynolds numbers $3.4 \ldots 5.8 \times 10^4$. The Lorentz force influence is characterised by an electromagnetic momentum coefficient c_μ defined in analogy to the one used in separation control by blowing [80]. c_μ links the total electromagnetic momentum input to the dynamic pressure and has been shown by Weier et al. [78] to collapse separation control data of different experiments and enables a direct comparison to alternative control methods [81]. Two control regimes can be distinguished in Fig. 11: at small momentum coefficients, the boundary layer is gradually reattached to the foil's surface, a process leading to a steep increase in lift. Above a certain momentum coefficient $c_{\mu r}$, necessary for complete reattachment, further lift increase can be observed which is weaker and is proportional to the square root of c_μ. These two regimes have been observed earlier in separation control by blowing, see the right part of Fig. 11, and termed BLC and "circulation control" by Poisson-Quinton [80].

Scale up of the experimental results reveal that power requirements for the original design based on conventional permanent magnets may prevent its application at sea going vessels.

In analogy to oscillatory suction and blowing [82], time-periodic Lorentz forces can be used to excite the separated flow. This indeed reduces the

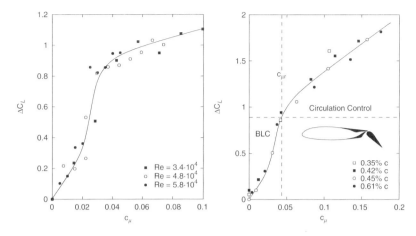

Fig. 11. Lift increase versus momentum coefficient for an electromagnetically controlled hydrofoil at 17° (*left*) [78] and for steady blowing over a 45° inclined flap on a NACA 23015 according to [79] and [80] (*right*)

momentum input necessary to recover the lift of the attached flow by more than an order of magnitude [81]. However, increasing the maximum lift requires momentum coefficients comparable to those necessary with steady electromagnetic forces. Despite this, electromagnetic forces have proven to be a flexible tool to study periodic excitation of separated flows. It has been shown, for example, that using different excitation wave forms, increases in the efficiency of 70% at constant efficient momentum input are possible [81].

4 Conclusions

Obviously the crux of the electromagnetic flow control, as well as the propulsion in poorly conducting fluids, is the efficiency limited by the achievable magnetic field strength. Among others, Busemann [3] realised similar fundamental problems concerning MAD already in 1961: *"Practically we are only at one tenth of the conductivity and one tenth of the magnetic field strength ..."*. But at the same time he raised hopes that these unfavourable conditions might be overcome: *"The worst enemy of rigid mathematical proofs is the designing engineer, who accomplishes the 'impossible' by simply violating the assumptions of the proof."* For an energetically efficient electromagnetic propulsion or drag reduction, this engineer is yet to come. On the other hand, already today the electromagnetic force control might be of use for applications where the energetic balance is not the primary goal as described above for lift production.

Until this time arrives, practical applicability of electromagnetic forces for both propulsion as well as flow control, depends on the progress in readily available sources for high magnetic fields. At the same time dealing numerically with low load factors enforces the solution of coupled flow and electric fields as demonstrated by Shatrov and Gerbeth [83], a problem not unfamiliar to traditional MHD but gladly avoided in the majority of present papers on EMTC. In most of the works up to now, the electromagnetic flow control was applied using a few simple force configurations only. A key potential for optimisation lies in the tailoring of the electromagnetic forces, both in the spatial as well as in the time domain. An example for the spatial optimisation potential was recently given in [77]. The time optimisation of the acting Lorentz forces may eventually result in a reactive solution [49].

Regardless of the efficiency, Lorentz forces have attractive features to offer for basic research on flow control mechanisms. They are a unique possibility for an easily controllable momentum source of unlimited bandwidth and great flexibility.

The main problem for applications, namely efficiency, is different if one changes the point of interest to areas other than ship building. Current densities are an inherent feature of electrochemical processes, leaving only the skilful placements of magnetic sources to control momentum and thereby mass transfer, space–time–yield, etc. However, this is a topic on its own and is reviewed by Alemany and Chopart [84] in this volume.

References

1. Ritchie W (1832) Experimental researches in voltaic electricity and electromagnetism. Phil Trans Roy Soc London 122:279–298
2. Resler EL Jr, Sears WR (1958) The prospects for magneto–aerodynamics. J Aero Sci 25:235–245, 258
3. Busemann A (1961) Is aerodynamics breaking an ionic barrier? NASA-TM-X-56147
4. Fraishtadt V, Kuranov A, Sheikin E (1998) Use of MHD systems in hypersonic aircraft. Tech Phys 43:1309–1313
5. Macheret S, Shneider M, Miles R (2002) Magnetohydrodynamic control of hypersonic flows and scramjet inlets using electron beam ionization. AIAA J 40:74–81
6. Poggie J, Gaitonde D (2002) Magnetic control of flow past a blunt body: Numerical validation and exploration. Phys Fluids 14:1720–1731
7. Moses R (2005) Regenerative aerobraking. Space Technology and Applications International Forum, Albuquerque
8. Chazot O, Zuber M (2003) Introduction to Magneto–Fluid–Dynamics for Aerospace Applications. VKI LS 2004–01, Rhode-Saint-Genèse
9. Rice W (1961) Propulsion system. US Patent 2,997,013
10. Friauf J (1961) Electromagnetic ship propulsion. ASNE J:139–142
11. Way S (1967) Electromagnetic propulsion for cargo submarines. AIAA–paper 1967–0363
12. Lin T, Gilbert J (1995) Studies of helical magnetohydrodynamic seawater flow in fields up to twelve teslas. J Prop Power 11:1349–1355

13. Nishigaki K, Sha C, Takeda M, Peng Y, Zhou K, Yang D, Suyama AH, Qing Q, Yan L, Kiyoshi T, Wada H (2000) Elementary study on superconducting electromagnetic ships with helical insulating wall. Cryogenics 40:353–359
14. Convert D (1995) Propulsion Magnetohydrodynamique en Eau de Mer. Ph.D. thesis, Universite Joseph Fourier, Grenoble
15. Phillips O (1962) The prospects for magnetohydrodynamic ship propulsion. J Ship Res 5:43–51
16. Thibault J (1994) Status of MHD ship propulsion. In: 2nd International Conference on Energy Transfer in MHD Flows, Aussois
17. Sutton G, Sherman A (1965) Engineering Magnetohydrodynamics. McGraw-Hill, New York
18. Doss E, Geyer H (1990) The need for superconducting magnets for MHD seawater propulsion. In: Proceedings of the 25th Intersociety Energy Conversion Engineering Conference, Reno
19. Khonichev V, Yakovlev V (1978) Motion of a sphere by a variable magnetic dipole in an infinite conductive fluid, produced by a variable magnetic dipole located within the sphere. J Appl Mech Techn Phys 19:760–765
20. Saji Y, Iwata A, Sato M, Kita H (1992) Fundamental studies of a superconducting electro-magnetic ship thruster to be driven by an alternating magnetic field. Adv Cryog Eng 37:463–471
21. Khonichev V, Yakovlev V (1980) Motion of a plane plate of finite width in a viscous conductive liquid, produced by electromagnetic forces. J Appl Mech Techn Phys 21:77–84
22. Shatrov V, Yakovlev V (1981) Change in the hydrodynamic drag of a sphere set in motion by electrodynamic forces. J Appl Mech Techn Phys 22:817–823
23. Yakovlev V (1980) Theory of an induction MHD propeller with a free field. J Appl Mech Techn Phys 21:376–384
24. Saji Y, Kitano M, Iwata A (1978) Basic study of superconducting electromagnetic thrust device for propulsion in seawater. Adv Cryog Eng 23:159–169
25. Iwata A, Saji Y, Sato S (1980) Construction of model ship ST–500 with superconducting electromagnetic thrust system. In: Rizutto C (ed) Proceedings of the 8th International Cryogenic Engineering Conference, Genova
26. Khonichev V, Yakovlev V (1980) Theory of a free-field conduction propulsion unit. J Appl Mech Techn Phys 21:666–673
27. Shatrov V, Yakovlev V (1985) Hydrodynamic drag of a ball containing a conduction-type source of electromagnetic fields. J Appl Mech Techn Phys 26:19–24
28. Pohjavirta A, Kettunen L (1991) Feasibility study of an electromagnetic thruster for ship propulsion. IEEE Trans Mag 27:3735–3742
29. Convert D, Thibault JP (1995) External MHD propulsion. Magnetohydrodynamics 31:290–297
30. Motora S, Takezawa S (1994) Development of MHD ship propulsion and results of sea trials of an experimental ship YAMATAO–1. In: 2nd International Conference on Energy Transfer in MHD Flows, Aussois
31. Meng J, Henoch C, Hrubes J (1994) Seawater electromagnetohydrodynamics: A new frontier. Magnetohydrodynamics 30:401–418
32. Meng J, Hendricks P, Hrubes J, Henoch C (1995) Experimental studies of a seawater superconducting electromagnetic thruster: A continuing quest for higher magnetohydrodynamic propulsion efficiency. Magnetohydrodynamics 31:279–289

33. Bashkatov V (1991) Reactive forces in magneto-hydrodynamics and their application for MHD-jet propulsive ocean ships. In: International Symposium on Superconducting Magnetohydrodynamic Ship Propulsion, Kobe
34. Tada E (1992) Propulsive analysis for high efficient superconducting EMT–powered ships. In: Tani J, Takagi T (eds) Electromagnetic Forces and Applications. Elsevier, Amsterdam
35. Yan L, Sha C, Zhou K, Peng Y, Yang A, Qin J (2000) Progress of the MHD ship propulsion project in China. IEEE Trans Appl Superconductivity 10:951–954
36. Yan L, Wang Z, Xue C, Gao Z, Zhao B (2000) Development of the superconducting magnet system for HEMS-1 MHD model ship. IEEE Trans Appl Superconductivity 10:955–958
37. Yan L, Sha C, Peng Y, Zhou K, Yang A, Qing Q, Nishigaki K, Takeda M, Suyama D, Kiyoshi T, Wada H (2002) Results from a 14 T superconducting MHD propulsion experiment. AIAA-paper 2002-2172
38. Font G, Dudley S (2004) Magnetohydrodynamic propulsion for the classroom. Phys Teach 42:410–415
39. Petit JP (1983) Is supersonic flight, without shock wave, possible? In: Proceedings of the 8th International Conference on MHD Electrical Power Generation, Moscow
40. Lebrun B, Petit J (1989) Shock wave annihilation by MHD action in supersonic flows: Quasi-one dimensional steady analysis and thermal blockage. Eur J Mech B Fluids 8:163–178
41. Lebrun B, Petit J (1989) Shock wave annihilation by MHD action in supersonic flows: Two-dimensional steady non-isentropic analysis, Anti-shock criterion, and shock tube simulations for isentropic flows. Eur J Mech B Fluids 8:307–326
42. Gailitis A, Lielausis O (1961) On a possibility to reduce the hydrodynamic resistance of a plate in an electrolyte. Appl Magnetohydrodynamics, Rep Phys Inst 12:143–146
43. Drazin P, Reid W (1981) Hydrodynamic Stability. Cambridge University Press, Cambridge
44. Tsinober AB, Shtern AG (1967) On the possibility to increase the stability of the flow in the boundary layer by means of crossed electric and magnetic fields. Magnetohydrodynamics 3:103–105
45. Lielausis O, Gailitis A, Dukure R (1991) Boundary layer control by means of electromagnetic forces. In: Proceedings of the International Conference on Energy Transfer in Magnetohydrodynamic Flows, Cadarache
46. Weier T, Albrecht T, Mutschke G, Gerbeth G (2004) Seawater flow transition and separation control. In: International Workshop on Flow Control by Tailored Magnetic Fields, Dresden
47. Zhilyaev M, Khmel T, Yakovlev V (1991) Boundary-layer stability in magneto-hydrodynamic streamlining of a plate with an internal source of electromagnetic fields. Magnetohydrodynamics 27:184–189
48. Albrecht T, Grundmann R, Mutschke G, Gerbeth G (2005) Numerical investigation of transition control in low conductive fluids. In: Joint 15th Riga and 6th PAMIR International Conference Fundamental and Applied MHD, Rigas Jurmala
49. Gad-el Hak M (2000) Flow Control: Passive, Active, and Reactive Flow Management. Cambridge University Press, Cambridge
50. Shtern A (1970) Feasibility of modifying the boundary layer by crossed electric and magnetic fields. Magnetohydrodynamics 6:407–411

51. Meng J (1993) Magnetohydrdynamic boundary layer control system. US Patent 5,273,465
52. Tsinober A (1990) MHD flow drag reduction. In: Bushnell D, Hefner J (eds) Viscous Drag Reduction in Boundary Layers. AIAA, Washington, DC
53. Nosenchuck D, Brown G (1993) Control of turbulent wall shear stress using arrays of TFM tiles. Bull Am Phys Soc 12:2197
54. Nosenchuck D, Brown G (1993) Discrete spatial control of wall shear stress in a turbulent boundary layer. In: So R, Speziale C, Launder B (eds) Near–Wall Turbulent Flows. Elsevier, Amsterdam
55. Moin P, Bewley T (1994) Feedback control of turbulence. Appl Mech Rev 47:S3–S13
56. Henoch C, Stace J (1995) Experimental investigation of a salt water turbulent boundary layer modified by an applied streamwise magnetohydrodynamic body force. Phys Fluids 7:1371–1383
57. Weier T, Fey U, Gerbeth G, Mutschke G, Lielausis O, Platacis E (2001) Boundary layer control by means of wall parallel Lorentz forces. Magnetohydrodynamics 37:177–186
58. Crawford CH, Karniadakis GE (1997) Reynolds stress analysis of EMHD–controlled wall turbulence: Part I. Streamwise forcing. Phys Fluids 9:788–806
59. O'Sullivan P, Biringen S (1998) Direct numerical simulations of low Reynolds number turbulent channel flow with EMHD control. Phys Fluids 10:1169–1181
60. Thibault JP, Rossi L (2003) Electromagnetic flow control: characteristic numbers and flow regimes of a wall-normal actuator. J Phys D: Appl Phys 36:2559–2568
61. Nosenchuck D, Brown G, Culver H, Eng T, Huang I (1995) Spatial and temporal characteristics of boundary layers controlled with the Lorentz force. In: 12th Australian Fluid Mechanics Conference, Sydney
62. Nosenchuck D (1996) Boundary layer control using the Lorentz force on an axisymmetric body. Bull Am Phys Soc 41:1719
63. Nosenchuck D (1994) Electromagnetic turbulent boundary-layer control. Bull Am Phys Soc 39:1938
64. Nosenchuck D, Brown G (1995) Multiple electromagnetic tiles for boundary layer control. US Patent 5,437,421
65. Du Y, Symeonidis V, Karniadakis G (2002) Drag reduction in wall–bounded turbulence via a transverse travelling wave. J Fluid Mech 457:1–34
66. Rossi L, Thibault JP (2002) Investigation of wall normal electromagnetic actuator for seawater flow control. J Turb 3:005
67. Berger TW, Kim J, Lee C, Lim J (2000) Turbulent boundary layer control utilizing the Lorentz force. Phys Fluids 12:631–649
68. Pang J, Choi KS (2004) Turbulent drag reduction by Lorentz force oscillation. Phys Fluids 16:L35–L38
69. Breuer K, Park J, Henoch C (2004) Actuation and control of a turbulent channel flow using Lorentz forces. Phys Fluids 16:897–907
70. Crausse É, Cachon P (1954) Actions électromagnétiques sur les liquides en mouvement, notamment dans la couche limite d' obstacles immergés. Comptes rendus hebdomadaires des séances de l' Académie des Sciences 238:2488–2490
71. Lielausis O (1961) Effect of electromagnetic forces on the flow of liquid metals and electrolytes. Ph.D. thesis, Academy of Sciences of the Latvian SSR, Institute of Physics, Riga

72. Weier T, Gerbeth G, Mutschke G, Platacis E, Lielausis O (1998) Experiments on cylinder wake stabilization in an electrolyte solution by means of electromagnetic forces localized on the cylinder surface. Exp Thermal Fluid Sci 16:84–91

73. Kim S, Lee C (2000) Investigation of the flow around a circular cylinder under the influence of an electromagnetic force. Exp Fluids 28:252–260

74. Posdziech O, Grundmann R (2001) Electromagnetic control of seawater flow around circular cylinders. Eur J Mech B Fluids 20:255–274

75. Chen Z, Aubry N (2005) Active control of cylinder wake. Comm Nonlin Sci Num Sim 10:205–216

76. Shatrov V, Yakovlev V (1990) The possibility of reducing hydrodynamic resistance through magnetohydrodynamic streaming of a sphere. Magnetohydrodynamics 26:114–119

77. Shatrov V, Gerbeth G (2005) Electromagnetic flow control leading to a strong drag reduction of a sphere. Fluid Dyn Res 36:153–173

78. Weier T, Gerbeth G, Mutschke G, Lielausis O, Lammers G (2003) Control of flow separation using electromagnetic forces. Flow Turb Comb 71:5–17

79. Schwier W (1943) Blasversuche zur Auftriebssteigerung am Profil 23015 mit verschiedenen Klappenformen. FB 1865, Zentrale f wiss Berichtswesen, Berlin–Adlershof

80. Poisson-Quinton P (1956) Einige physikalische Betrachtungen über das Ausblasen an Tragflügeln. Jahrbuch der WGL:29–51

81. Weier T, Gerbeth G (2004) Control of separated flows by time periodic Lorentz forces. Eur J Mech B Fluids 23:835–849

82. Greenblatt D, Wygnanski I (2000) The control of flow separation by periodic excitation. Prog Aero Sci 36:487–545

83. Shatrov V, Gerbeth G (2005) On magnetohydrodynamic drag reduction and its efficiency. In: Joint 15th Riga and 6th PAMIR International Conference on Fundamental and Applied MHD, Rigas Jurmala

84. Alemany A, Chopart JP (2006) An outline of magnetoelectrochemistry. This volume

Part IV

Electromagnetic Processing of Materials

Overview of Electromagnetic Processing of Materials

Shigeo Asai

Department of Materials, Physics and Energy Engineering, Graduate School of
Engineering, Nagoya University, Furo-cho, Chikusa-ku, Nagoya, 464-8603, Japan
(c42538a@cc.nagoya-u.ac.jp)

Summary. History of electromagnetic processing of materials (EPM) is described
and several functions utilized in EPM are reviewed. Main activities of EPM are
summarized with the view on mass production and applications of high magnetic
fields related to nanotechnology. Future trends and prospects of EPM are discussed.

1 Introduction

In a metal industry, electric energy has been used as heat energy for an
extended period of time because of cleanliness, high controllability, and high
energy density. Technologies using electric energy have been developed rather
early and went ahead without sufficient background of scientific understanding. Good examples are electromagnetic levitation and electromagnetic mixing, which were invented very early, in 1923 and 1932, respectively. To bridge
the gap between the technology and the scientific understanding, Magnetohydrodynamics (MHD) which had been established by Alfvén in 1942, was first
introduced at the IUTAM conference entitled "The Application of Magnetohydrodynamics to Metallurgy", held in Cambridge, England in 1982 [1]. Before
the conference, a research laboratory, MADYLAM aiming at the applications
of MHD, has been established in CNRS in Grenoble, France. Encouraged
by the Cambridge symposium, the Iron and Steel Institute of Japan (ISIJ)
inaugurated the Committee of Electromagnetic Metallurgy in 1985. The new
research activity, which began in the iron and steel industry, has grown to hold
in 1994 the first International Symposium on Electromagnetic Processing of
Materials (EPM) in Nagoya, Japan. The term EPM, which has been established by combining the two channels, metallurgy and MHD, has formally
been used for the first time at this symposium. Since the first international
symposium, it has been held every 3 years in France and Japan alternatively.

Hitherto, the activities of EPM have been devoted to the economical
aspect relating to mass production and nanotechnology aspect of high-quality

S. Molokov et al. (eds.), Magnetohydrodynamics – Historical Evolution and Trends,
315–327. © 2007 *Springer.*

materials. Furthermore, EPM activity is spreading into a new area, solving environment problems.

2 Functions of electromagnetism applied to materials processing

Several functions making use of the Lorentz force are applicable to the materials processing as follows. The function of *shape controlling* is based on the magnetic pressure given as $P_m = B^2/2\mu$. The function of *fluid driving* is induced by imposing a direct electric current and a magnetic field, $\mathbf{F} = \mathbf{J} \times \mathbf{B}$, or by imposing a traveling magnetic field. The function of *flow suppressing* appears when applying a direct magnetic field to moving molten metal, based on the principle of $\mathbf{F} = \sigma(\mathbf{v} \times \mathbf{B}) \times \mathbf{B}$. The function of *levitating* appears when gravity force balances the electromagnetic one, $\mathbf{J} \times \mathbf{B} = \mathbf{g}$. When the electromagnetic force is much larger than both gravity and the adhesion force due to surface tension, $|\mathbf{J} \times \mathbf{B}| > \max\{|\rho\mathbf{g}|, 6\sigma/a^2\}$, the function of *splashing* takes place. The Joule heat, $q = |\mathbf{J}|^2/\sigma$, provides the function of *heat generating*.

Regarding magnetization force given as $(\chi/\mu)(\mathbf{B} \cdot \nabla)\mathbf{B}$ and $\mathbf{M} \times \mathbf{B}$, there are two kinds of forces available in materials processing. One is the force pulling ferromagnetic and paramagnetic materials to a magnet and repulsing diamagnetic ones. The other is the force rotating materials to a magnetic field direction as a compass rotates to the north direction on Earth.

Figure 1 reveals an overview of the electromagnetic processing of materials as a tree. The roots indicate the academic background supporting this engineering field as follows:

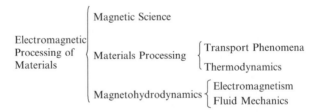

That is, EPM is based on magnetic science, materials processing, and MHD, where the functions of electromagnetism are utilized for processing of materials, including the electrically conductive and non-conductive substances. The branches predict functions of electromagnetism and the leaves in each branch show processes and technologies related to the corresponding function as described in the above. Furthermore, the development of superconducting magnetic technologies has made helium-free superconducting magnets available, and this promises to open new fields for practical industrial applications.

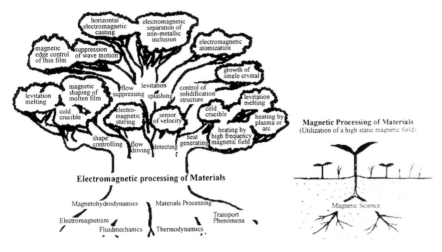

Fig. 1. A tree of electromagnetic processing of materials

3 EPM in mass productions

The practical application of EPM has begun in the field of economic mass productions in steel and aluminium industries and a substantially large part of EPM activities are concentrated in this field at present. Typical examples are electromagnetic stirring and electromagnetic braking in a mold of a continuous casting process of steel. Traveling alternating and static magnetic fields are used there, respectively. Stirring has contributed to the improvement in the surface quality of slabs, while braking has lead to the improvement in the inner quality. Electromagnetic casting (EMC) process invented by Getselev [2] in 1966 is the most significant and successful example in aluminium industry, where a fixed alternating magnetic field with kilohertz frequency is applied to make use of the function of the shape controlling. Surface defects of cast metals usually appearing near the surface have been eliminated by non-contact between the mold and metal. In 1986, Vives et al. [3] proposed a CREM process, in which an alternating magnetic field with commercial frequency was imposed from the outside of a mold. The experimental result reported by them was appreciated due to nice surface quality, similar to that in products of EMC. Being stimulated by this experimental result, several fundamental research works, aiming to apply Vives's result to a continuous casting process for steels, started at the Committee of EPM organized by ISIJ. They were taken over to the International Project involving Japan, France, and Sweden during the period of 1995–2000. Within this Project, an intermittent alternating magnetic field has been developed for casting, and applied from the outside of a mold. The researchers faced two crucial problems, namely the magnetic decay in a copper mold, and the mold deformation due to thermal stress in the mold. The experimental results have proved that the smooth surface is achievable even in steel casts by the new casting. Another success story is a casting process of bimetallic slabs [4], where the function of flow suppressing in a static magnetic field is used. For melting titanium scraps, a large-scale cold crucible with about 500 kg melting capacity has been developed utilizing the functions of levitation and heat generation. The process combining a cold crucible with a precise casting technology has been developed for titanium-alloy products, such as turbochargers and golf clubs [5]. The direct induction skull melting [6] is another promising technology for melting materials with rather low electrical conductivity, such as silicon and ceramics.

4 Applications of high magnetic fields in EPM

4.1 Classification of functions associated with high magnetic fields

Owing to the development of superconductive technologies, which made high magnetic fields available within a rather large space even in conventional-scale laboratories, the technologies relating to crystal orientation, structure alignment, and spin chemistry have been introduced in the field of EPM. Table 1 shows the classification of functions accompanied by a high magnetic field in EPM. The high magnetic field enables not only to enhance various functions based on the Lorentz force, but also to induce several functions based on the magnetization force. The crystal orientation and the structure alignment in non-magnetic materials are typical examples of the use of the magnetization force.

The possibility of mass transport and mass rotation due to the magnetization force has been studied for several processes, such as solidification [7–10], electro-deposition [11], vapour-deposition [12–14], and solid-phase reaction [15]. It is now recognized that the application of a high magnetic field is surely useful and promising method in EPM.

Table 1. Utilization of a High Static Magnetic Field in EPM

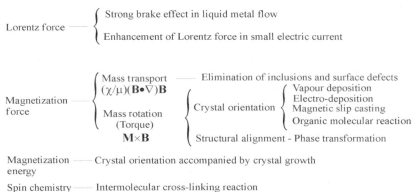

4.2 Qualitative evaluation of phase transformation

4.2.1 Principle

Magnetic susceptibility of a mixture with two components is given by the equation

$$\chi_m = f_1\chi_1 + f_2\chi_2, \tag{1}$$

where f_1 and f_2 are fractions of the two components, respectively. In addition, equation

$$f_1 + f_2 = 1 \tag{2}$$

obviously holds.

Once the magnetic susceptibility χ_m is measured, the fractions of components in the mixture can be derived from Eqs. (1) and (2) to give:

$$f_1 = \frac{\chi_m - \chi_2}{\chi_1 - \chi_2}, \qquad f_2 = \frac{\chi_m - \chi_1}{\chi_2 - \chi_1}. \tag{3}$$

Here, the magnetic susceptibility can be obtained by the use of Gouy method [16,17] which is based on the measurement of the magnetization force F_z, namely:

$$\chi_m = \frac{2L\mu_0}{m_s(B_L^2 - B_0^2)} F_z, \tag{4}$$

where L and m_s are the length and the mass of the specimen, respectively, μ_0 is the magnetic permeability, and B_L and B_0 are magnetic flux densities at the top and the bottom of the specimen, respectively. The magnetization force F_z can be obtained from the difference between the weights of a specimen measured with and without magnetic field.

When we apply the principle to evaluate a phase fraction change during a phase transformation, we have to measure temperature of the specimen, together with the magnetization force. Then, we need to evaluate the values of χ_1 and χ_2 appearing in Eq. (3) beforehand, since the magnetic susceptibility is a function of temperature.

4.2.2 Measuring solid fraction during solidification

We can obtain the relationship between the magnetic susceptibility and temperature measured during the solidification of an alloy as shown in Fig. 2. It is found that the magnetic susceptibilities of both solid and liquid phases can be expressed as a linear function of the temperature around the melting point with good approximation. That is, the magnetic susceptibilities in the single solid and liquid phases are given by equations

$$\chi_{ml} = C_{l1}T + C_{l2}, \tag{5}$$

$$\chi_{ms} = C_{s1}T + C_{s2}. \tag{6}$$

By substituting χ_{ml} and χ_{ms} evaluated from Eqs. (5) and (6) into Eq. (3), the relation between the solid fraction and temperature during the

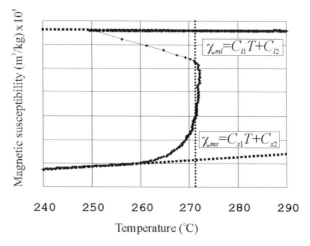

Fig. 2. Calculation of the solid fraction

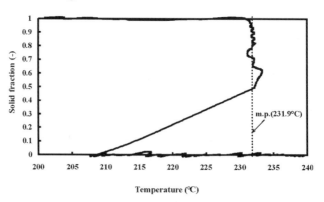

Fig. 3. The relation between temperature and solid fraction for zinc (cooling)

solidification of zinc is obtained as shown in Fig. 3. It can be noticed that the solid phase of about 50 mass% has precipitated at the point where the recalescence finishes and the temperature starts rising up to the melting point.

The method developed here can be applied to in situ measurement of various phase transformations in solid, liquid, and gas phases, and will promise better and deeper understanding of phase transformations and reactions in the near future.

4.3 Crystal orientation in high magnetic field

4.3.1 Theory of crystal texture control

Recently, it has been found that crystal orientation in materials can be controlled by the imposition of high magnetic fields. This principle can be applied not only to magnetic materials, but also to non-magnetic materials with asymmetric unit cells [16–26].

When a non-magnetic substance is magnetized in a magnetic field, the energy for magnetization of the substance is given by the equation

$$U = -\int_0^{B/\mu_0} M dB_{in}, \tag{7}$$

where M is the magnetization, B and B_{in} are the imposed magnetic flux density and the magnetic flux density in the substance, respectively, and μ_0 is the permeability in vacuum ($4\pi \times 10^{-7}$ H/m). The principle of control of the crystal orientation using magnetic field is that magnetic torque rotates crystals to take stable crystal orientation so as to decrease the magnetization energy.

Let us consider the crystal structure with magnetic anisotropy, that is, the magnetic susceptibility is different in each crystal direction. The value of the magnetization energy is given by the equation

$$U = -\frac{\chi}{2\mu_0 (1 + N\chi)^2} B^2, \tag{8}$$

which has been derived from Eq. (7). This determines the preferred crystal direction depending on the magnetic susceptibility of each crystal axis and the crystal shape. In the above N is the demagnetization factor. Let χ_c and $\chi_{a,b}$ represent the c-axis and the a- or b-axis of the magnetic susceptibility, respectively. When $\chi_c > \chi_{a,b}$, i.e. $U_c < U_{a,b}$, the c-axes of crystals is the preferred one in parallel to the direction of the magnetic field. In contrast, when $\chi_c < \chi_{a,b}$, i.e., $U_c > U_{a,b}$, the a- or b-axis of crystals is the preferred one in parallel to the magnetic field. That is, the c-axis of crystals aligns to all of the directions in the plane perpendicular to the imposed magnetic field.

Four necessary conditions have to be satisfied for the crystal orientation under the imposition of a magnetic field. Firstly a unit crystal cell of materials to be oriented should have magnetic anisotropy. The second is that the magnetization energy provided by the magnetic field should be higher than the thermal energy to cause thermal perturbation. The third condition is that the materials should be in the weak constraint medium, in which a particle composed by the materials can rotate by such a feeble magnetization force. The fourth is that each particle composed by a single crystal should be dispersed in the medium.

4.3.2 Vapour-deposition process [13]

A crucible filled with target material of bismuth with 5 nine purity was put into a vacuum chamber set in the bore of a superconducting magnet generating a magnetic field of 12 T at the maximum intensity, and a glass plate as a substrate was set perpendicular to the magnetic field direction at the position with the maximum magnetic flux density in the bore. After the degree of

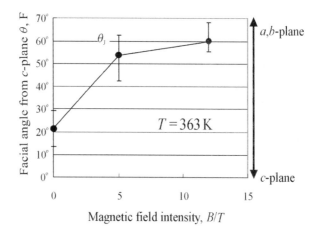

Fig. 4. The relation between magnetic intensity and facial angle from the c-plane in bismuth films

vacuum in the chamber reached a value of 5×10^{-3} Pa, bismuth in the crucible was heated up to 1,073 K by an electric heater.

Figure 4 shows the relation between the magnetic field intensity and the facial angle θ_F, the definition of which is given in the reference [27]. The rotation to the a, b-plane increases with the increase in the magnetic field intensity. This result agrees with the theoretical prediction based on Eq. (8).

4.3.3 Electro-deposition process

A copper substrate as cathode and a zinc plate as anode were set in a vessel as an electrolytic cell. The magnetic field of 12 T was imposed perpendicular to the cathode substrate plane. The detail of the experimental condition is given in [11]. Figure 5 shows the relations between the orientation index and the imposed magnetic flux density in the electrodeposits obtained at $J = 700$ A/m^2. The higher the magnetic field, the more the c-plane orientation is elicited. This result agrees with the theoretical derivation based on the magnetization energy given in Eq. (8).

4.3.4 Applications of magnetic fields in slip casting process

A novel process where a high magnetic field is imposed during slip casting was proposed to fabricate crystal-orientated ceramics [28, 29]. Here another novel process is proposed, in which a specimen is rotated during the slip casting under a high magnetic field. Figure 6 shows schematically the functions of the magnetic field and rotation of crucible. Regarding the substance whose magnetic susceptibility in the a- or b-axis is higher than that in the c-axis, $\chi_c < \chi_{a,b}$, one-directional crystal orientation can not be obtained in a slip

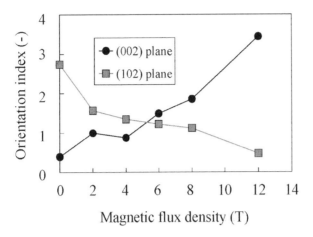

Fig. 5. Relations between magnetic flux density and orientation index of zinc electrodeposits obtained at $J = 700$

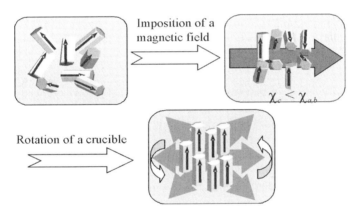

Fig. 6. Schematic view of the functions of magnetic field and rotation of a crucible under a magnetic field

casting under a high magnetic field, because the free choice of crystal orientation exists both in the a- and b-axes. When the magnetic field is imposed on the suspension, the c-axis of particles can align in various directions in the plane perpendicular to the magnetic field direction. When a rotating magnetic field is imposed on a fixed specimen, the c-axis of particles will be perpendicular to the plane in which the magnetic field is rotating. From the viewpoint of relative motion, the condition whereby the specimen is fixed and the magnetic field rotates is equivalent to the case where the specimen rotates in a fixed magnetic field. It is suggested that in this configuration the c-axis of particles will align to the direction of gravity. The usefulness of the newly proposed process has been confirmed in fabrication of Si_3N_4 ceramics. In order

Fig. 7. SEM micrographs of specimens made of α-Si_3N_4 powder with β-Si_3N_4 seeds: (a), (b) without magnetic field; (c), (d) with magnetic field of 10 T under rotation of crucible

to examine the effect of rotation, green samples have been prepared by rotating under the magnetic field of 10 T. Moreover, for the sake of comparison, another sample has also been made without the magnetic field. After drying, the green samples were embedded in powder bed of 60 wt % Si_3N_4 + 40 wt % BN set in a graphite crucible and maintained at 1800°C for 1.5 h in N_2 without a magnetic field. Figure 7 shows the scanning electron microscope (SEM) micrograph of the polished surfaces of specimen. It can be seen in Figs. 7a, b that β-Si_3N_4 rod grains appear randomly distributed in the specimen, which has been prepared without the exposure to the magnetic field. In the case of the specimen prepared with the rotation under the magnetic field, a highly textured material has been obtained as shown in Figs. 7c, d.

4.3.5 Crystal orientation in metal solidification

A zinc film (10×28 mm^2) prepared by dipping a steel plate into a molten zinc bath was set into a stainless steel pipe, which was inserted from the upper part of a magnet bore. The sample plane was set up at the position with the maximum magnetic flux density in the direction either parallel or perpendicular to the magnetic field. A thermocouple was inserted through the midair part of the stainless steel pipe and the temperature of the sample was measured by the thermocouple connected with the plane. The crucible was filled with argon gas to prevent the oxidation of the sample and the

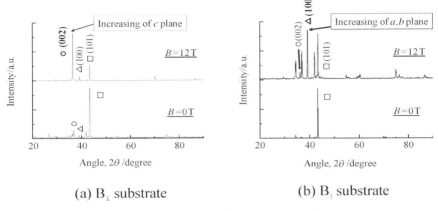

Fig. 8. X-ray diffraction patterns of zinc

temperature was kept in liquid and solid zone of zinc for 3 min. Then the furnace was cooled down.

The diffraction pattern of x-rays on the samples, which were exposed to the magnetic field in perpendicular and parallel directions is shown in Fig. 8a and b, respectively. In Fig. 8a, the peak (101) plane was detected stronger in the sample obtained with no magnetic field imposed. On the other hand, when the magnetic field of 12 T was imposed, the peak (101) decreased, and the peak (002) corresponding to the c-plane appeared stronger. The diffraction pattern of x-rays on the sample to which the magnetic field was imposed in parallel is shown in Fig. 8b. The peak (101) detected stronger in the sample obtained with no magnetic field, the same as the previous result shown in Fig. 8a. However, when the magnetic field of 12 T was imposed, the peak (101) decreased, and the peak (100) corresponding to the a, b-plane appeared stronger. That is, regardless of the direction of the imposed magnetic field, the zinc crystals aligned in the direction predicted by the magnetization energy mentioned before.

5 Future prospects for EPM

Concerning the future prospects for EPM, both the research trends and industrial applications should be mentioned. One of the most crucial aspects for the former is to find new functions, which are implied by the imposition of electric and magnetic fields. Stirring of molten metals, suppressing liquid metal motion, heating metals, levitating metals, and separating non-conductive materials in a conducting medium, etc., are well-known functions based on the Lorentz force. Aligning crystal orientation and transporting materials are the other well-known functions based on the magnetization force. Recently, the introduction of high magnetic fields into EPM has provided several new functions, such as generating micro-eddy motion, the so-called micro-MHD effect [30], and patterning of non-conductive particles [31], which were found in the interaction between the electro-deposition reactions and the Lorentz force. On the other hand, enhancing crystal orientation in sintering process [32, 33], shifting solid–solid phase transformation [34], crystal orientation during its transformation [35], and self-assembling and patterning of particles [36] are new functions, which have been explained using the magnetization energy or the polarization effect.

Hitherto, industrial applications have concentrated on applications of functions related to the Lorentz force. The functions of the magnetization force induced by high magnetic fields, have scarcely been seen in practical applications. The functions based on the Lorentz force, such as stirring of molten metals induced by traveling or alternating magnetic fields, suppressing molten metal motion by static magnetic fields, and heating metals by high frequency magnetic fields, have been used in metal industries for a long time. In 1990s, cold crucible technology based on the functions of levitating and heating, which was invented in 1920, was redeveloped for a great demand of chemically reactive metals, and for metals with high meting point, such as titanium and silicon. In this technology, scaling up and the development of ejecting method of a molten metal are main topics for the 2000s. The cold crucible with the capacity of up to 500 kg has been developed in Japan [37], but the full success story of the ejecting method has not been seen yet. The technology of soft-contacting solidification, whereby an alternating magnetic field is imposed from outside of continuous casting mold to reduce the pressure between the mold and a molten metal, and to provide the reduction of cooling rate, was developed in a steel industry in the middle of 1990s. It is going to be applied to continuous casting of metals with low latent heat per volume, such as aluminium and magnesium. The unstable solidification will be reasonably prevented by the surface heating effect induced by high-frequency magnetic field.

The activities in EPM are also spreading into a new area of solving environmental problems.

References

1. Moffatt HK, Proctor MRE (eds) (1982) Proceedings of the Symposium of the International Union of Theoretical and Applied Mechanics, Cambridge, UK, Sept 1982. The Metals Society
2. Getselev ZN (1969) US Patent 3467166
3. Vives Ch, Forest B, Riquest JP (1986) French Patent No 841470
4. Takeuchi E, Tanaka H, Kajioka H (1994) In: International Symposium on Materials, Oct 1994, ISIJ, 364
5. Demukai N, Ichiyanagi S, Shibata T (1992) Committee of Electromagnetic Processing of Materials, ISIJ No 9–4
6. Kaneko K, Gillon P, Landaud D (2003) In: Proceedings of the 4th International Symposium on Electromagnetic Processing of Materials, Lyon, 250–259

7. Morikawa H, Sassa K, Asai S (1998) Mater Trans 39:814–818
8. Yasuda H, Tokieda K, Ohnaka I (2000) Mater Trans 41:1005–1021
9. Legrand BA, Chateigner D, Perrier R, Tournier R (1997) J Magn Magn Mater 173:20–28
10. Noudem JG, Beille J, Bourgault D, Chateigner D, Tournier R (1996) Physica C 264:325–330
11. Taniguchi T, Sassa K, Asai S (2000) Mater Trans 41:981–984
12. Mitani S, Bai HL, Wang ZJ, Fujimori H, Motokawa M (2000) In: Proceedings of the 3rd International Symposium on Electromagnetic Processing of Materials, Japan, ISIJ, pp 630–634
13. Tahashi M, Sassa K, Hirabayashi I, Asai S (2000) Mater Trans 41:985–990
14. Awaji S, Watanabe K, Ma Y, Motokawa M (2001) Physica B 294-295:482–485
15. Ito M, Sassa K, Doyama M, Yamada S, Asai S (2000) TANSO 191:37–41
16. Iguchi Y (1991) Experimental Chemical Course 9. Maruzen Tokyo, pp 439–450
17. Suzuki N (1974) Metal Data Book. Japan Metal Institute, Maruzen Tokyo, pp 10, 18
18. Lu Y, Nagata A, Watanabe K, Nojima T, Sugawara K, Hanada S, Kamada S (2003) Physica C 392:453–457
19. He SS, Zhang YDD, Zhao X, Zuo L, He JCC, Watanabe K, Zhang T, Nishijima G, Esling C (2003) Adv Eng Mater 5:579–583
20. Chen P, Maeda H, Watanabe K, Motokawa M (2000) Physica C 337:160–164
21. Chen P, Maeda H, Watanabe K, Motokawa M, Kitaguchi H, Kumakura H (1999) Physica C 324:172–176
22. Chen P, Maeda H, Kakimoto K, Zhang PX, Watanabe K, Motokawa M (1999) Physica C 320:96–100
23. Zimmerman MH, Faber KT, Fuller ER Jr (1997) J Am Ceram Soc 80:2725–2729
24. Farrel E, Chandrasekhar BS, DeGuire MR, Fang MM, Kogan VG, Clem JR, Finnemore DK (1987) Phys Rev B 36:4025–4027
25. Ferreira M, Maple MB, Zhou H, Hake RR, Lee BW, Seaman CL, Kuric MV, Guertin RP (1988) Appl Phys A 7:105–110
26. Paulik W, Faber KT, Fuller ER Jr (1994) J Am Ceram Soc 77:454–458
27. Tahashi M, Ishihara M, Sassa K, Asai S (2003) Mater Trans 44:285–289
28. Inoue K, Sassa K, Yokogawa Y, Sakka Y, Okido M, Asai S (2003) Mater Trans 44:1133–1137
29. Sakka Y, Suzuki TS, Tanabe N, Asai S, Kitazawa K (2002) Jpn J Appl Phys 41:1416–1418
30. Aoyagi R, Asanuma M (1999) In: Proceedings of the 3rd Meeting, International Symposium on New Magneto-science'99, JST, 1999, Omiya, Japan pp 22–42
31. Yamada T, Asai S (2001) J Jpn Inst Metals 65:910–915
32. Murakami Y, Sassa K, Asai S (2006) J Am Ceram Soc (submitted)
33. Sassa K, Murakami Y, Asai S (2004) In: Proceedings of Asia-Euro Workshop on Electromagnetic Processing of Materials, Sept 2004, Northeastern University, Shenyang, China, pp 69–75
34. Xu Y, Ohtsuka H, Wada H (2000) J Magn Soc Jpn 24:655–658
35. Yasuda H, Ohnaka I, Yamamoto Y, Tokieda K, Kishio K (2003) Mater Trans 44:2207–2212
36. Takayama T, Ikezoe Y, Uetake H, Hirota N, Kitazawa K (2003) Jpn J Appl Phys 27:299–302
37. Ninagawa S, Nagao M, Kusamichi T, Nakagawa K, Fukumoto H (2000) KOBELCO Tech Rev No 23, Apr 2000:19–22

Applications of High Magnetic Fields in Materials Processing

Hideyuki Yasuda

Department of Adaptive Machine Systems, Graduate School of Engineering, Osaka University, Suita, Osaka 565-0871, Japan (yasuda@ams.eng.osaka-u.ac.jp)

1 High magnetic fields in solidification processing

Magnetic fields are often used in materials processing, especially in solidification processing for metallic alloys and semiconductors. For example, both static and alternating magnetic fields have been used extensively to control melt flow and solidified structures in the continuous casting of steels [1, 2]. Static magnetic fields are used as electromagnetic brakes and alternating magnetic fields are used as electromagnetic stirrers. Alternating magnetic fields with rather high frequencies are used to hold melt pools and to achieve soft contacts with molds. It is well known that electromagnetic processing can significantly improve the quality of products.

The effect of static magnetic fields on macrosegregation has first been reported in [3], based on observations of semiconductors grown by the horizontal Bridgman technique. The magnetic field was found to suppress the temperature fluctuations in the melt and the grown crystal did not exhibit compositional striations. Static magnetic fields were also used for Czochralski growth of semiconductor crystals InSb [4]. The imposed magnetic field eliminated the macrosegregation, which was associated with unsteady state flow during crystal growth. In the solidification and crystal growth processes, both conventional electromagnets and permanent magnets are used with magnetic field intensities typically less than 2 T. It was recognized that conventional magnetic field intensities are sufficiently high to reduce convection.

It has been reported that a sufficient amount of suppression of convection during the Bridgman technique resulted in diffusion-controlled segregation [5]. The imposed magnetic field of 3 T is required to obtain diffusion-controlled growth. The growth of single crystals in the presence of a magnetic field has been carried out to improve crystal qualities [6–8]. Pioneering studies prove that macrosegregation can be modified by high magnetic fields. In particular, the exposure of high magnetic fields is found to be a powerful tool for suppressing the macroscopic melt flow sufficiently, and for achieving diffusion-controlled growth.

S. Molokov et al. (eds.), Magnetohydrodynamics – Historical Evolution and Trends, 329–344. © 2007 Springer.

In addition to the modification of the macrosegregation, high magnetic fields were used to control solidified structures in multiphase solidification. Pb–Bi and Sn–Cd peritectic alloys were unidirectionally solidified under a magnetic field of 10 T [9]. The diffusion-controlled solute transfer resulted in a periodic structure in which the two constituent phases alternatively grew perpendicular to the growth direction. An alternative use for high magnetic fields is to achieve an intrinsic growth mode and to control solidified structures.

Recently, interest has focused on the use of high magnetic fields to suppress microscopic flow and to modify microstructures. The Hartmann number,

$$Ha = Ba\sqrt{\sigma/\eta}, \tag{1}$$

is often used as a measure of the influence of static magnetic fields on the fluid flow. Here, σ and η are the electrical conductivity and viscosity, respectively, B is the magnetic flux density, and a is the length scale of the system. With increasing magnetic fields, the influence of the magnetic field is obvious, even for smaller systems. It is expected that a high magnetic field can be used to control microscopic fluid flow. Thus, the micro magnetohydrodynamic (μ-MHD) effect is recognized as a tool for controlling the microstructure of materials.

In the 1990s, cryocooled superconducting magnets were developed and became popular. The advantages of such magnets are their high magnetic fields, their ability to run long-term and continuously, and their room-temperature bore. Typically, a magnetic field of 10 T can be imposed in a room-temperature bore of size 10 cm. In a relatively large bore at room temperature, the superposition of alternating magnetic fields can easily be achieved. Thus, the magnetic fields are designed for various purposes in materials processing, leading to a new area of study of electromagnetic processing under high static magnetic fields. For example, a cold crucible was inserted into a superconducting magnet [10, 11]. Titanium, which is one of the most reactive metals, was statically melted in a cold crucible. Studies first proved the potential of the simultaneous application of alternating and static magnetic fields for handling melts. The levitation technique has also been developed [12–14]. The design of functional magnetic fields should be noted as an important technique in the electromagnetic processing of materials (EPM).

Under high magnetic fields, not only MHD but also thermodynamic effects are important. The magnetic energy, E_m, is given by the following equation:

$$E_m = -\mathbf{M} \cdot \mathbf{H}. \tag{2}$$

In general, the magnetic energy is extremely small, even for ferromagnetic materials under a conventional magnetic field. However, the magnetic energy influences the phase transformation in the solid state when a high magnetic field is imposed [15, 16]. The results proved that a high magnetic field enables the development of novel material processing through MHD and the thermodynamic effect.

This paragraph presents examples of material processing under high magnetic fields. Firstly, the modification of a solidified structure by the μ-MHD effect is demonstrated. Secondly, a novel levitation technique as well as its application is presented. Thirdly, the thermodynamic effect of the high magnetic field is briefly explained.

2 Reduction of the melt flow in the microscopic region

2.1 Monotectic solidification

This paragraph describes the control of a solidified structure of monotectic alloy by a static magnetic field and the fabrication of porous media using the solidified structure. The high magnetic field has been used to suppress the microscopic melt flow and motion of immiscible droplets, typically of 10 μm diameter.

A monotectic reaction is defined by the simultaneous production of a solid phase (S) and a liquid phase (L1) from a liquid phase (L), L → S + L1. The reaction is essentially the same as the eutectic reaction defined by L → S1 + S2, except that in the monotectic reaction one of the products is liquid. In these reactions, coupled growth can often occur, in which two constituent phases cooperatively grow by exchanging solute in the vicinity of the solidifying front [17]. The coupled growth results in a lamellar structure or a rod structure.

It is well known that the aligned rod structure can be produced for monotectic alloys in monotectic compositions by unidirectional solidification as well as for eutectic alloys [18–20]. Compared to eutectic structures [17], very specific growth conditions are required for developing the aligned structure in monotectic alloys. For example, a regular structure has rarely been produced in hypermonotectic compositions. The liquid–liquid interface can promote melt flow due to the temperature and concentration dependence of the interfacial energy between the two liquid phases. The liquid phase produced through the monotectic reaction moves easily in the melt due to convection and density differences. Consequently, the inhomogeneous dispersion of the minor phase particles, called "gravity segregation", is often induced during conventional solidification.

In spite of the difficulties associated with solidification, there are some attractive features of the aligned structure of monotectic alloys. One is that the shape of the minor phases is truly cylindrical because they are liquid when major phases solidify. The other feature is that minor phase rods of the same diameter are regularly aligned with each other, because the major and minor phases grow cooperatively. It is a sort of self-organization process. If the fibrous minor phase is removed from the matrix, porous media in which deep pores with the same diameter are regularly aligned in the matrix, can be fabricated [18–21]. Thus, it is of interest to control the monotectic solidification.

2.2 Effect of high magnetic fields on monotectic solidification

Hypermonotectic Al-10 at % In alloys have been unidirectionally solidified under magnetic fields of 10 T [22]. As is shown in Fig. 1, the imposition of a static magnetic field during unidirectional solidification successfully achieved an aligned rodlike structure, even for the hypermonotectic composition (10 at % In). The three-dimensional (3D) image has been reconstructed from computerized tomography using a synchrotron radiation facility [23]. The continuous In rods with diameters of 10–20 μm are regularly aligned parallel to each other.

The formation of the aligned structure is explained by considering the melt flow in the vicinity of the solidifying front. For the hypermonotectic alloys, the In droplet nucleates on the solidifying front of the Al phase. The In liquid droplets can be pushed by the solidifying front [24–27]. Figure 2 is a schematic illustration of an In droplet at the solidifying front. In the models [24–27], the interfacial energy difference causes a repulsive force, and the melt flow into the gap between the droplet and the solidifying front causes a drag force. The drag force increases with increasing diameter of the droplet. When the diameter exceeds a certain value, the droplet is engulfed by the front. The sequence of the nucleation process, the pushing and the engulfment result in a random distribution of the In droplets in the matrix.

The imposition of a static magnetic field during monotectic solidification influences the melt flow around the droplets and the movement of droplets.

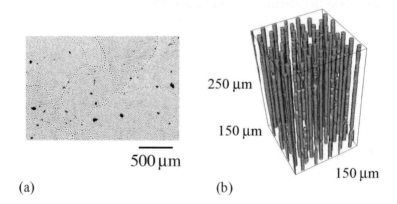

250 μm

150 μm

500 μm

150 μm

(a)   (b)

Fig. 1. (a) Transverse section of Al-10 at % In monotectic alloys solidified at 10 T and (b) 3D image obtained by micro x-ray CT. The black and gray are Al and In, respectively. In the 3D image, the Al phase has been removed

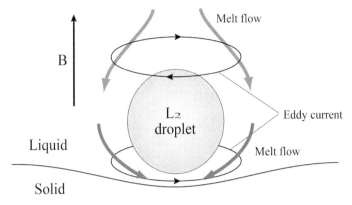

Fig. 2. Schematic illustration of an L_2 droplet at the solidifying front under a static magnetic field

The melt flow around the droplet is similar to the melt flow of a rising particle in a metallic melt [28–30]. The analogy is used to explain the pushing/engulfment of the droplets at the solidifying front. The melt flows above and below the sphere have a horizontal component and the eddy current is induced. The Lorentz force caused by the imposed magnetic field and the eddy current break the horizontal flow. As a result, the static magnetic field enhances the drag force. The MHD analysis [28–30] and the experimental result [31] indicate that the influence of the magnetic field becomes significant in the case when the Hartmann number is sufficiently larger than unity.

In unidirectional solidification, the Hartmann number of the In droplet (10μm diameter) is estimated to be roughly equal to 5 at a magnetic field of 10 T. This suggests that engulfment is enhanced by the magnetic field. The local flow is also reduced, as well as the flow around the droplet. Therefore, the solute transfer is relatively well controlled by the diffusion and consequently the coupled growth between the Al-rich solid and In-rich liquid occurs even at hypermonotectic compositions. According to the analysis, high magnetic fields of the order of 10 T are required to control the microstructure of the multiphase solidification, since the microstructure is typically of the order of 10^{-6} m.

Electrochemical dissolution successfully removes the In rods from the matrix [22]. Figure 3 shows the porous Al produced using the monotectic alloys solidified under the high magnetic field. Deep pores whose depths were more than 500 μm were produced by monotectic solidification under a magnetic field and electrochemical dissolution. The study implies that a high magnetic field is a powerful tool for controlling the microstructure and contributes to the fabrication of functional materials. MHD is expected to integrate high magnetic fields into materials processing.

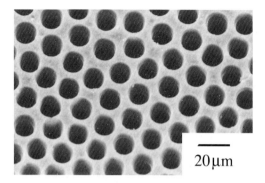

Fig. 3. Porous aluminum fabricated by the selective dissolution of the In phase from the Al–In alloy solidified under a magnetic field

3 Superposition of magnetic fields

3.1 Electromagnetic vibrations

From the viewpoint of materials processing, alternating magnetic fields have active roles on melts, while static magnetic fields have passive ones. Furthermore, the effect of alternating magnetic fields depends on the frequency. For example, relatively low frequencies are effective for stirring and high frequencies are effective for inducing magnetic pressures. Thus, the superposition of various magnetic fields is a promising method for improving the quality of products.

In the continuous casting of steel, it has been found that an alternating magnetic field imposed from the outside of a mould improves the surface quality of steel [1, 2]. The hydrostatic pressure between a mould and molten metal is reduced due to the magnetic pressure and the stabilization of the meniscus of the molten steel. In addition, an intermittent alternating magnetic field has also been imposed to introduce the synchronized oscillation of the meniscus [2]. The superposition of alternating magnetic fields was found to improve the surface quality of cast steel.

As mentioned above, a large room-temperature bore gives rise to the superposition of an electric current and alternating magnetic fields in a high static magnetic field. Compression waves in melts have useful functions such as degassing, the acceleration of the reaction rate, the refinement of solidified structures, and the dispersion of immiscible substances. Mechanical methods, such as the electrostrictive and the magnetostrictive techniques are restricted in metallurgical processes because of the contamination of devices in high-temperature melts. The generation of compression waves in melts by the application of high-frequency electromagnetic fields has been proposed [32]. To intensify the compression waves, an alternating magnetic field (60 Hz) was superimposed on a high static magnetic field [33]. The observed pressure was close to atmospheric one.

 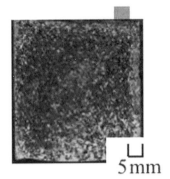

Without vibration   **With vibration**

Fig. 4. The effect of the electromagnetic vibration on the solidified structure. The imposition of the electromagnetic vibration during the solidification significantly contributed to the refinement. (After Professor Iwai, Nagoya University.)

Simultaneously, imposing an alternating current and a static magnetic field has been found to refine the grain size of solidified alloys [34, 35]. Recently, a novel method has been proposed [36, 37]. An alternating current was locally imposed during solidification. Vibrations occurred near the electrodes and propagated in the melt. Figure 4 shows the solidified structure. The vibrations increased the number of seed crystals and consequently a refined structure was obtained in the whole of the casting. The refinement of the crystal grains in the solidified structure will be beneficial for improvement of the mechanical properties.

3.2 Containerless process by magnetic fields

The electromagnetic levitator is the most popular method for processing metallic melts using the containerless method [38–40]. The method has been widely used to investigate solidification from undercooled melts [41] and to measure thermophysical properties, even the metastable states [38]. The electromagnetic force due to an alternating magnetic field is given by

$$\mathbf{F} = -\frac{1}{2\mu}\nabla(\mathbf{B} \cdot \mathbf{B}) + \frac{1}{\mu}(\mathbf{B} \cdot \nabla)\mathbf{B}. \tag{3}$$

Here $\mathbf{B}$ is the magnetic flux density and μ is the magnetic permeability. The force of gravity balances the electromagnetic force given by the first term in Eq. (3). The second term is the rotational term, which causes electromagnetic stirring. Even when the first term is dominant at high frequencies, the second term remains nonzero. Therefore, electromagnetic levitation intrinsically induces convection and oscillations in melts.

Besides the electromagnetic force, the magnetization force, which originates in the interaction between the magnetization and the external magnetic field, can be used to levitate melts in high static magnetic fields. The magnetization force is given by Eq. (4):

$$\mathbf{F} = -\frac{1}{2\mu}\chi\nabla \cdot \mathbf{B}^2. \tag{4}$$

Here, χ is the magnetic susceptibility. Diamagnetic materials were levitated by the magnetization force [42–44]. Since magnetization is a body force, a pseudo-microgravity condition can be achieved. Thus, the levitation method by the static magnetic field can be used to avoid violent vibrations and strong convection. In addition to diamagnetic materials, paramagnetic ones have also been levitated using magneto-Archimedes levitation [45]. In this method, the susceptibility of materials to levitation is relatively negative with respect to the susceptibility of the atmosphere. Although the method can be used for various materials, conventional superconducting magnets cannot levitate most of the melts used in metallurgical processes. Thus, it is still desirable to develop another method that is able to levitate melts at high temperatures.

Recently, an electromagnetic levitation method, which simultaneously imposes alternating and static magnetic fields, has been developed [14]. A theoretical approach has also been developed for fluid flow in melts levitated using alternating and static magnetic fields [46]. The melt flow can be suppressed by imposing a static magnetic field. Figure 5 is a schematic illustration of an electromagnetic levitator. A cryogen-free superconducting magnet

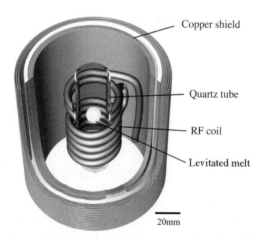

Fig. 5. Schematic illustration of the electromagnetic levitation apparatus using alternating and static magnetic fields

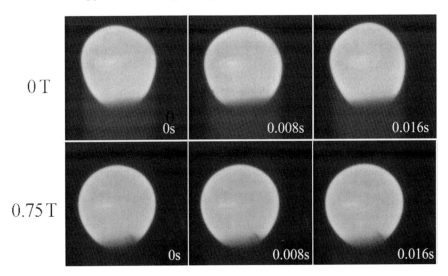

Fig. 6. Levitated copper melts at 0 T and 0.75 T

imposes a static magnetic field. An RF generator (frequency: 200 kHz, power: 20 kW) has been connected to the levitation coil. Figure 6 shows levitated copper melts [14]. Oscillations and convection were only observed for magnetic fields below 0.5 T. Only rotation, for which the axis is parallel to the static magnetic field, was observed in magnetic fields exceeding 1 T.

The motion of levitated melts can be classified into five categories: oscillation, convection, rotation (rotation axis perpendicular to the static magnetic field), rotation (rotation axis parallel to the static magnetic field), and movement of the center of gravity. According to analysis of the electromagnetic force, all modes except rotation (rotation axis parallel to the static magnetic field) can be suppressed by the static magnetic field. The experimental results clearly indicate that the simultaneous imposition of alternating and static magnetic fields achieve a stable levitation in which metallic melts are levitated without melt flow or oscillation.

3.3 Application of electromagnetic levitation

Refined grains, such as equiaxed grains, are preferable for most castings. The morphology of solidified structures has been investigated by the conventional levitation method. In the Ni–Cu system, the morphological transition from equiaxed to columnar grains occurs at a lower critical undercooling, and the other transition from columnar to equiaxed grains occurs at a higher critical undercooling [41]. The fragmentation of the dendrites during the period following the recalescence on the basis of the model has been evaluated [47,48]. The estimated morphology agrees qualitatively with the experimental results.

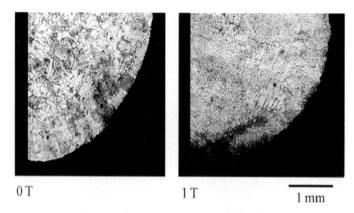

0 T   1 T   1 mm

Fig. 7. Microstructure of middle carbon steels solidified at 0 T and 1 T using the
levitation method using alternating and static magnetic fields

However, the authors also point out that the lower critical undercooling estimated by the model is in relatively poor agreement with the experimental
result [47]. The melt flow may also result in a poor agreement. However, the
melt flow cannot be modified by the conventional method. The novel levitation
method using the simultaneous application of alternating and static magnetic
fields achieves levitation without convection. Thus, it is of interest to examine
the effect of the melt flow on the microstructure from the undercooled melt.

Figure 7 shows the solidified structures of middle carbon steel using the
levitation method [49]. The equiaxed grains are obtained when a static magnetic field is not imposed during solidification. In contrast, dendrites grow into
the center of the specimen and columnar grains are obtained when a static
magnetic field of 1 T is imposed. The transition from equiaxed to columnar
grains by imposing a static magnetic field has also been observed for Cu–Ag
and Fe–Ni alloys [49]. Experiments using the novel levitation method show
that the transition from equiaxed to columnar grains by reducing the melt
flow is a universal phenomenon.

A possible mechanism for equiaxed grain formation is the fragmentation
of dendrite arms in the present case [50, 51]. Primary dendrite arms are frag-
mented due to the instability of the cylindrical shape [47, 48]. The surface
tension drives the shape change from a cylindrical to a spherical shape to
minimize the interfacial energy in the system. Since the shape change of the
dendrite arms is controlled by the solute diffusion around the arms, the solute
transfer can significantly affect the fragmentation. This transition induced
by the magnetic field indicates that the melt flow significantly enhances the
solute transfer and consequently causes fragmentation of the dendrite arms.
The experimental results first proved that solute transfer due to the melt flow
in the mushy zone contributed dominantly to fragmentation during solidification at the lower undercooling region.

Another application of the containerless processes is the measurement of the thermophysical properties of melts at high temperatures. For example, the oscillation drop method has been widely used for measuring the surface tension of various melts [39, 40, 52, 53]. This is evaluated from the Rayleigh frequency of the melt oscillation due to the surface tension. For measurements under terrestrial conditions, the melt becomes aspherical due to the electromagnetic and gravitational forces. The deviation from a spherical shape results in the split of the fundamental oscillation frequencies. For the $l = 1$ modes [39], the oscillations of the $m = 0$, $|m| = 1$ and $|m| = 2$ modes exist. In cases where the observed frequencies are identified, the surface tension can be evaluated by the corrected equation [40]. However, it is not always easy to perform mode identification [54].

As mentioned in § 3.2, static magnetic fields suppress the melt flow in levitated melt. Recently, damping of the oscillation by a static magnetic field has been observed [55]. The shape of the levitated melt has been traced as shown in Fig. 8. Traces of the melt show that the electromagnetic force agitated several modes of the oscillation at 0 T. At 0.6 T, the vertical length was almost constant and the horizontal one changed periodically. The shape change suggests that the oscillation of the $l = 2$, $|m| = 2$ mode remained, and that the other modes were preferentially suppressed. An advantage of the damping is that mode identification is not needed in the oscillation. Thus, the levitation method may simplify the evaluation of the surface tension. It is expected that the levitation method using alternating and static magnetic fields will become a useful new technique for the measurement of surface tension.

Melt levitated without melt flow is also beneficial for the measurement of other physical properties. For example, the volume of the melt can be precisely measured because the shape of the melt does not fluctuate. Consequently, the density of the melt is evaluated. In addition, other thermophysical properties such as the thermal conductivity, viscosity, and solute diffusivity may

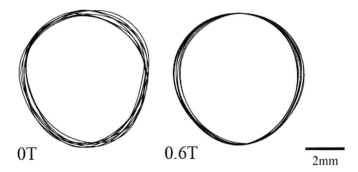

0T   0.6T   2mm

Fig. 8. Traces of levitated copper melts

be measured. The superposition of static magnetic fields on electromagnetic levitation is expected to have various applications.

4 Thermodynamic effect on the microstructure evolution

The development of crystallographically aligned or textured microstructures improves material properties. It has been recognized that the use of high magnetic fields can be a powerful tool for achieving crystallographically aligned microstructures (referred to as aligned structures) even for paramagnetic and diamagnetic materials. For example, the aligned structure of the Y–Ba–Cu–O superconductor has been fabricated using the anisotropy of magnetic susceptibility [56–59]. Aligned microstructures have also been obtained by sintering for some ceramics [60–63]. In these processes, the crystalline particles rotate in a preferential direction according to the magnetic anisotropy energy. The alignment is essentially explained by considering the rotation in the fluids.

Recently, crystallographical alignment and textured structures have been observed for ferromagnetic materials [64–67] and for paramagnetic ones [68,69] during annealing under a high magnetic field. For example, textured structures have been observed in Zn bicrystals during annealing in a high magnetic field [68]. The results suggest that the high magnetic field influences the thermodynamic equilibrium in the solid state, even for paramagnetic materials. From the viewpoint of materials processing, the thermodynamics, including the magnetic effect, need to be considered carefully, in addition to MHD.

References

1. Fujii T (2000) State of art of electromagnetic processing in Japanese iron and steel industry. In: Proceedings of the 3rd International Symposium on Electromagnetic Processing of Materials (EPM2000), Nagoya, Japan, pp 14–19; Takeuchi E, Miyazawa K (2000) Electromagnetic casting technology of steel. ibid, pp 20–27
2. Asai S (2000) Recent development and prospect of electromagnetic processing of materials. Sci Technol Adv Mater 1:191–200 (see also references therein)
3. Utech HP, Flemings MC (1966) Elimination of solute banding in indium antimonide crystals by growth in a magnetic field. J Appl Phys 37:2021
4. Witt AF, Herman CJ, Gatos HC (1970) Czochralski-type crystal growth in transverse magnetic fields. J Mater Sci 5:822
5. Matthiesen VH, Wargo MJ, Motakef S, Carlson DJ, Nakos JS, Witt WF (1987) Dopant segregation during vertical Bridgman-Stockbarger growth with melt stabilization by strong axial magnetic fields. J Cryst Growth 85:557–560
6. Becla P, Han J-C, Motakef S (1992) Application of strong vertical magnetic-fields to growth of II-VI-pseudo-binary alloys – HgMnTe. J Cryst Growth 121:394–398

7. Watring DA, Lehoczky SL (1996) Magneto-hydrodynamic damping of convection during vertical Bridgman-Stockbarger growth of HgCdTe. J Cryst Growth 167:478

8. Park YJ, Min S-K, Hahn S-H, Yoon J-K (1995) Application of an axial magnetic-field to vertical gradient freeze GsAs single-crystal growth. J Cryst Growth 154:10

9. Yasuda H, Tokieda K, Ohnaka I (2000) Effect of magnetic field on periodic structure formation in Pb-Bi and Sn-Cd peritectic alloys. Mater Trans 41:1005–1012

10. Bonvalot M, Courtois P, Gillon P, Tournier R (1995) Magnetic-levitation stabilized by eddy currents. J Magn Magn Mater 151:283

11. Gillon P (2000) Uses of intense d.c. magnetic fields in materials processing. Mater Sci Eng A287:146

12. Yasuda H, Ohnaka I, Ninomiya Y, Ishii R, Kishio K (2001) Levitation of metallic melt by simultaneous imposition of static and alternating magnetic fields for materials processing. In: Proceedings of the 5th International Symposium on Magnetic Suspension Technology, Turin, p 185

13. Ninomiya Y, Yasuda H, Ohnaka I, Ishii R, Fujita S, Kishio K (2003) Stable levitation of the metallic melt by simultaneous imposition of alternative and high static magnetic field. Trans Mater Res Soc Jpn 28:283

14. Yasuda H, Ohnaka I, Ninomiya Y, Ishii R, Fujita S, Kishio K (2004) Levitation of metallic melt by using the simultaneous imposition of the alternating and the static magnetic fields. J Cryst Growth 260:475

15. Yasuda H, Nakahira A, Ohnaka I, Yamamoto Y, Kishio K (2003) Formation of crystallographically aligned grains during coarsening in a magnetic field. Mater Trans 43:2555–2562

16. Yasuda H, Ohnaka I, Kawakami O, Ueno K, Kishio K (2003) Effect of magnetic field on solidification in Cu- Pb monotectic alloys. ISIJ Int 43:942–949

17. Jackson KA, Hunt JD (1966) Lamellar and rod eutectic growth. Trans Metall Soc AIME 236:1129

18. Grugel RN, Hellawell A (1981) Alloy solidification in systems containing a liquid miscibility gap. Metall Trans A12:669

19. Kamio A, Kumai S, Tezuka H (1991) Solidification structure of monotectic alloys. Mater Sci Eng A146:105

20. Dhindaw BK, Stefanescu DM, Singh AK, Curreri PA (1988) Directional solidification of Cu-Pb and Bi-Ga monotectic alloys under normal gravity and during parabolic flight. Metall Trans A19:2839

21. Angers LM, Grugel RN, Hellawell A, Draper CW (1982) Selective etching and laser melting studies of monotectic composite structures. In: Lemkey FD, Cline HE, McLean M (eds) In Situ Composites IV. Elsevier Science New York, p 205

22. Yasuda H, Ohnaka I, Fujimoto S, Sugiyama A, Hayashi Y, Yamamoto M, Tsuchiyama A, Nakano T, Uesugi K, Kishio K (2004) Fabrication of porous aluminum with deep pores by using Al-In monotectic solidification and electro-chemical etching. Mater Lett 58:911–915

23. Uesugi K, Suzuki Y, Yagi N, Tsuchiyama A, Nakano T (2001) Development of high spatial resolution x-ray CT system at BL47XU in SPring-8. Nucl Instr Meth Phys Res A467-468:853

24. Uhlmann DR, Chrlmers B, Jackson KA (1964) Interaction between particles and a solid–liquid interface. J Appl Phys 35:2986

25. Chernov AA, Temkin DE, Mel'nikova AM (1977) Growth kinetics and capture of impurities during gas-phase crystallization. Sov Phys Crystallogr 22:656

26. Pötschke J, Rogge V (1989) On the behavior of foreign particles at an advancing solid liquid interface. J Cryst Growth 94:726

27. Stefanescu DM, Dhindaw BK, Kasar AS, Moitra A (1988) Behavior of ceramic particles at the solid-liquid metal interface in metal matrix composites. Metall Trans 19A:2847

28. Chester W (1961) The effect of a magnetic field on the flow of a conducting fluid past a body of revolution. J Fluid Mech 10:459

29. Chester W, Moore DW (1961) The effect of a magnetic field on the flow of a conducting fluid past a circular disk. J Fluid Mech 10:466

30. Ueno K, Yasuda H (2003) MHD oseenlet and flow field rising with a sphere under strong vertical magnetic field. Magnetohydrodynamics 39:547–554

31. Yasuda H, Ohnaka I, Kawakami O, Ueno K, Kishio K (2003) Effect of magnetic field on solidification in Cu-Pb monotectic alloys. ISIJ Int 43:942

32. Amano S, Iwai K, Asai S (1997) Non-contact generation of compression waves in a liquid metal by imposing a high frequency electromagnetic field. ISIJ Int 37:962–966

33. Iwai K, Wang Q, Momiyama T, Asai S (2000) Generation of compression waves in a liquid metal by the simultaneous imposition of DC and AC magnetic fields. In: Proceedings of the 3rd International Symposium on Electromagnetic Processing of Materials (EPM2000), Nagoya, Japan, ISIJ, p 61

34. Vives C (1996) Effects of forced electromagnetic vibrations during the solidification of aluminum alloys. 2. Solidification in the presence of colinear variable and stationary magnetic fields. Metall Trans 27B:457

35. Radjai A, Miwa K, Nishino T (1998) An investigation of the effects caused by electromagnetic vibrations in a hypereutectic Al-Si alloy melt. Metall Trans 29A:1477

36. Takagi T, Iwai K, Asai S (2003) Solidified structure of Al alloys by a local imposition of an electromagnetic oscillating force. ISIJ Int 43:842–848

37. Sugiura K, Iwai K (2004) Effect of operating parameters of an electromagnetic refining process on the solidified structure. ISIJ Int 44:1410–1415

38. Egry I (1999) Structure and properties of levitated liquid metals. J Non-Cryst Solid 250–252:63

39. Rayleigh L (1879) On the capillary phenomena of jets. Proc R Soc Lond 29:71

40. Cummings DL, Blackburn DA (1991) Oscillations of magnetically levitated aspherical droplets. J Fluid Mech 224:395–416

41. Norman AF, Eckler K, Zambon A, Gatner F, Moir SA, Ramous E, Herlach DM, Greer AL (1998) Application of microstructure-selection maps to droplet solidification: a case study of the Ni-Cu system. Acta Mater 46:3355

42. Beaugnon E, Tournier R (1991) Levitation of organic materials. Nature 349:470

43. Beaugnon E, Fabregue D, Billy D, Nappa J, Tournier R (2001) Dynamics of magnetically levitated droplets. Physica B 294–295:715

44. Motokawa M, Hamai M, Sato T, Mogi I, Awaji S, Watanabe K, Kitamura N, Makihara M (2001) Magnetic levitation experiments in Tohoku university. Physica B 294-295:279

45. Ikezoe Y, Hirota N, Nakagawa J, Kitazawa K (1998) Making water levitate. Nature 393:749–750

46. Shatrov V, Priede J, Gerbeth G (2003) Three-dimensional linear stability analysis of the flow in a liquid spherical droplet driven by an alternating magnetic field. Phys Fluids 15:668

47. Schwarz M, Karma A, Eckler K, Herlach DM (1994) Physical-mechanism of grain-refinement in solidification of undercooled melts. Phys Rev Lett 73:1380

48. Schwarz M, Karma A, Eckler K, Herlach DM (1994) Physical-mechanism of grain-refinement in solidification of undercooled melts. Phys Rev Lett 73:2940

49. Yasuda H, Ohnaka I, Ishii R, Fujita S, Tamura Y (2005) Investigation of the melt flow on solidified structure by a levitation technique using alternative and static magnetic fields. ISIJ Int 45:991–996

50. Herlach DM, Eckler K, Karma A, Schwarz M (2001) Grain refinement through fragmentation of dendrites in undercooled melts. Mater Sci Eng A 304-306:20

51. Herlach DM (1994) Nonequilibrium solidification of undercooled metallic melts. Mater Sci Eng R 12:172

52. Lamb H (1932) Hydrodynamics. Cambridge University Press, Cambridge

53. Przyborowski M, Hibiya T, Eguchi M, Egry I (1995) Surface-tension measurement of molten silicon by the oscillating drop method using electromagnetic-levitation. J Cryst Growth 151:60

54. Fujii H, Matsumoto T, Nogi K (2000) Analysis of surface oscillation of droplet under microgravity for the determination of its surface tension. Acta Mater 48:2933–2939

55. Yasuda H, Tamura Y, Fujita D, Mizuguchi T, Nagira T, Ohnaka I (2005) Investigation of the melt flow on solidified structure by a levitation technique using alternative and static magnetic fields. ISIJ Int 45: 991–996

56. Farrell DE, Chandrasekhar BS, DeGuire MR, Fang MM, Kogan VG, Klem JR, Finnemore DK (1987) Superconducting properties of aligned crystalline grains of Y1Ba2Cu3O7-delta. Phys Rev B 36:4025–4027

57. Lusnikov A, Miller LL, McCallum RW, Mitra S, Lee WC, Johnson DC (1989) Mechanical and high-temperature (920-degrees-c) magnetic-field grain alignment of polycrystalline (Ho,Y)Ba2Cu3O7-delta. J Appl Phys 65:3136–3141

58. Tkaczyk JE, Lay KW (1990) Effect of grain alignment and processing temperature on critical currents in YBa2Cu3O7-delta sintered compacts. J Mater Res 7:1368–1379

59. de Rango P, Lees M, Lejay P, Sulpice A, Tournier R, Ingold M, Germi P, Pernet M (1991) Texturing of magnetic-materials at high-temperature by solidification in a magnetic-field. Nature 349:770–772

60. Suzuki TS, Sakka Y, Kitazawa K (2001) Orientation amplification of alumina by colloidal filtration in a strong magnetic field and sintering. Adv Eng Mater 3:490–492

61. Suzuki TS, Sakka Y (2002) Fabrication of textured titania by slip casting in a high magnetic field followed by heating. Jpn J Appl Phys 41:L1272–1274

62. Sakka Y, Suzuki TS, Tanabe N, Asai S, Kitazawa K (2001) Alignment of titania whisker by colloidal filtration in a high magnetic field. Jpn J Appl Phys 41:L1416–1418

63. Nakahira A, Konishi S, Honda Y, Yasuda H, Ohnaka I (2003) Sintering behaivors and microstructure of Apatite-based materials under a high magnetic field. Trans Mater Res Soc Jpn 28:287–290

64. Yasuda H, Nakahira A, Ohnaka I, Yamamoto Y, Kishio K (2003) Formation of crystallographically aligned grains during coarsening in a magnetic field. Mater Trans 43:2555–2562

65. Shimotomai M, Maruta K, Mine K, Matsui M (2003) Formation of aligned two-phase microstructures by applying a magnetic field during the austenite to ferrite transformation in steels. Acta Mater 51:2921–2932
66. Ohtsuka H, Xu Y, Wada H (2000) Alignment of ferrite grains during austenite to ferrite transformation in a high magnetic field. Mater Trans JIM 41:907–910
67. Ohtsuka H (2004) Effects of strong magnetic fields on Bainitic transformation. Curr Opin Solid State Mater Sci 8:279–284
68. Sheikh-Ali AD, Molodov DA, Garmestani H (2003) Magnetically controlled recrystallization texture in titanium. Scripta Materialia 48:483–488
69. Molodov DA, Sheikh-Ali AD (2004) Effect of magnetic field on texture evolution in titanium. Acta Materialia 52:4377–4383

Effect of AC Magnetic Fields on Free Surfaces

Yves Fautrelle[1], Alfred Sneyd[2], and Jacqueline Etay[1]

[1] EPM/CNRS/INPG, ENSHMG, BP 95, 38402 Saint Martin d'Hères cedex,
France(Yves.Fautrelle@hmg.inpg.fr, etay@grenoble.cnrs.fr)
[2] Department of Mathematics, University of Waikato, Private Bag 3105,
Hamilton, New Zealand(sneyd@waikato.ac.nz)

1 Introduction

When a liquid metal is submitted to an alternating magnetic field, electromagnetic forces, called Lorentz or Laplace forces, may be created in the metal
due to the interaction between the induced electric currents and the applied
magnetic field. When the magnetic field is pulsating and according to its
frequency f (which vanishes in the case of DC magnetic field), the electromagnetic forces generate various effects both on the bulk motion and at the
free surface of the liquid metal [1–3]. These effects have been applied to the
design of many industrial processes, and the use of AC magnetic fields is
now widespread. Formally, we may distinguish two kinds of phenomena: the
bulk flow hydrodynamics (the so-called electromagnetic stirring) and the free-
surface problem. Note that in practical applications the distinction is not so
strict. In the usual cases, the bulk hydrodynamics is decoupled from the electromagnetic aspects. The electromagnetic field and the electric currents are
calculated first as if the fluid is at rest. Then, the electromagnetic forces are
injected in the momentum equation to determine the bulk flow. However, concerning the free-surface problems, the situation is different. The free-surface
domain deformation affects the electric current path, and there appears a real
coupling between the electromagnetic and dynamical aspects.

Let us now focus on the behaviour of free surfaces submitted to AC electromagnetic fields. It is a fascinating boundary problem. For the sake of simplicity, two sub-cases may be distinguished: the case of static deformation and
the one of free-surface motion. These two cases will be considered in § 2.

2 Static deformation generated by AC magnetic fields

When the electric current is alternating, the electromagnetic forces $\mathbf{F}$ comprise
both a mean steady part $\langle \mathbf{F} \rangle$ and an alternating one, $\tilde{\mathbf{F}} \cos(2\omega t + \varphi)$, i.e.,

$$\mathbf{F} = \langle \mathbf{F} \rangle + \tilde{\mathbf{F}} \cos(2\omega t + \varphi), \tag{1}$$

where ω is the magnetic field pulsation and φ is a phase.

S. Molokov et al. (eds.), Magnetohydrodynamics – Historical Evolution and Trends,
345–355. © 2007 Springer.

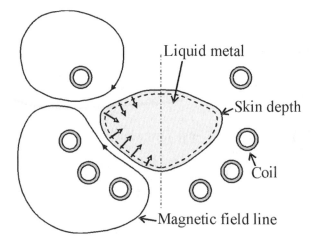

Fig. 1. Sketch of a liquid metal pool submitted to an AC magnetic field generated by a solenoidal coil supplied with alternating electric currents

The two parts usually are of the same order except in the very low frequency case [4]. However, in the medium frequency range because of the fluid inertia, the alternating part of the electromagnetic forces has no significant effect on the liquid metal free surface provided that the magnetic field frequency is much larger than the free surface natural frequencies. The mean part of the electromagnetic forces is responsible for various phenomena. Firstly, it is generally rotational and accordingly generates a significant bulk fluid flow [1], [4]. But one of the most striking effects is the free surface static deformations; thanks to the well-known repulsion effect. The latter phenomenon is caused by the so-called electromagnetic pressure. When the skin depth δ is much smaller than the pool dimension, the electromagnetic forces reduce to a gradient force perpendicular to the pool boundary. This is illustrated by the sketch in Fig. 1.

2.1 Electromagnetic levitation

Thanks to a special design of the coil, the repulsive electromagnetic forces are able to balance gravitational forces to levitate totally or partially a liquid metal blob located in a coil [5–7]. This has been first demonstrated experimentally by Okress [7]. An example of levitated drop experiment is shown in Fig. 2. Theoretical analysis have been carried out by Mestel [5] and then by Sneyd and Moffatt [6], who established that there exist a variational principle according to which the electrical energy input balances the gravitational energy and the superficial one. By means of a balance between the magnetic energy and the gravitational one, it may be easily shown that the typical height h_m of the levitated drop evolves as the square of the applied magnetic field according to the following relation:

Fig. 2. View of a liquid aluminium blob levitated by an AC magnetic field, pool diameter 30 mm (experiment performed in the EPM laboratory)

$$h_m = \frac{B_0^2}{2\mu\rho g},\tag{2}$$

μ, ρ, g, and B_0 being respectively the permeability of the vacuum, the liquid metal density, the gravity and the typical magnetic field strength (r.m.s. value).

2.2 Shaping and guiding

Another interesting use of the magnetic pressure is the possibility of shaping or guiding a liquid metal jet which crosses a coil [8–10]. An example of shaping is shown in Fig. 3. The liquid metal jet which crosses a Helmholtz-type coil tends to be aligned with the magnetic field lines [8]. Then, the jet cross section is flattened.

2.3 Dome effect

In a classical single-phase induction furnaces, the magnetic pressure is responsible for the dome formation which has been widely shown both experimentally and numerically [11]. Figure 4 shows a typical dome-shaped free surface obtained in a cold crucible induction furnace [12]. The magnetic field exerts a centring effect on the free surface and diminish the contact between the melt and the crucible. This property has been used to develop the concept of soft contact casting in continuous casting of steel.

2.4 Unsymmetric static free surfaces

The shape of the liquid free surface may not always remain axisymmetric even in axisymmetric coil configurations, especially for large magnetic field

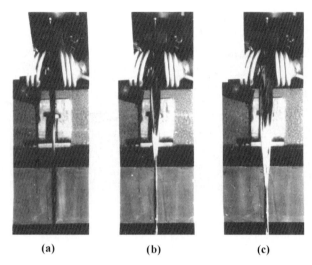

(a)   **(b)**   **(c)**

Fig. 3. Deformation of mercury jets in a Helmholtz-type coil. Flow rate: 25 cm^3/s; nozzle diameter : $\varnothing = 5$ mm (a) $B = 25.5$ mT $L/\varnothing = 1.13$; (b) $B = 54$ mT $L/\varnothing = 2.08$; (c) $B = 61.2$ mT $L/\varnothing = 2.73$

Fig. 4. Photographs of the static deformation of a liquid metal free surface under the effect of a AC magnetic field (a) classical stable dome shape on an aluminium pool for $f = 7.5$ kHz [12]

amplitude. A symmetry breaking may occur in some geometrical conditions. For example, for shallow horizontal liquid metal pools other types of static deformation may be obtained [13]. Highly non-symmetric steady shapes may be observed when the magnetic field amplitude increases. Figure 5 illustrates such phenomenon on a gallium layer in a 10 kHz AC magnetic field. According to the variational principle, the pool increases its surface energy (by increasing the drop perimeter) rather than the gravitational one. This phenomenon

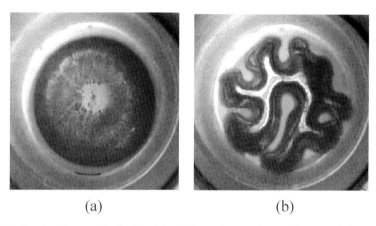

(a)   (b)

Fig. 5. Static shapes of the liquid gallium drop submitted to a AC magnetic field [13], pool diameter 60 mm, $f = 14$ kHz (a) the drop is at rest $B_{0\,\mathrm{max}} = 0$, (b) $B_{0\,\mathrm{max}} = 39$ mT

may also be interpreted in another way. Indeed, the free surface horizontal deformation modifies the induced electric current path giving birth to a kind of "edge instability". This instability is analogous to the pinch effect observed in another configuration by Mohring and Karcher [14]. In such a particular geometry, the system tends to maximize the electric current path which is confined near the pool edge. Accordingly, the global electric resistance of the system is higher, and the coupling between the melt and the inductor decreases.

3 Wave motion and instability under AC magnetic fields

The free-surface behaviour (shape, oscillation, instability) varies with the value of the magnetic field amplitude, as well as the applied frequency. Two kinds of situations may occur according to the value of the frequency of the magnetic field. They will be discussed in §§ 3.1 and 3.2.

3.1 Low frequency single-phase magnetic field

It has been shown that very low frequency AC magnetic fields could generate motions at the free surface of a liquid metal when the natural frequencies of the interface are of the same order as the magnetic field frequency [10, 12]. In the limit of low frequencies, i.e., large skin depth compared with the dimension of the pool, the oscillating part of the electromagnetic forces is dominant and is responsible for two types of actions [15–17]. Firstly, it generates forced standing free-surface waves. Such waves have the same degree of symmetry as that of the force [17]. For example, in an axisymmetric geometry, the wave

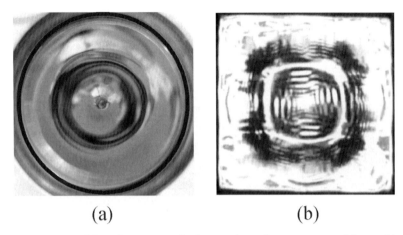

(a)   (b)

Fig. 6. Top view of forced waves on the free surface of a mercury pool located inside
a single phase coil [2], the oscillation frequency is $2f$, the magnetic field amplitude
and frequency are respectively 0.10 T and $f = 3.06$ Hz (a) circular tank of radius
100 mm, (b) rectangular tank of dimension 120 mm, the magnetic field amplitude
and frequency are respectively 0.15 T and 2.87 Hz

pattern is axisymmetric as well. Such types of forced waves are illustrated in
Figs. 6 and 9b. They occur in various geometries. In Fig. 6 the liquid metal is
contained in a cylindrical or square tank with a horizontal free surface, whilst
in Fig. 9 the liquid metal domain consists of a liquid drop set on a substrate.
An instability appears for higher amplitude magnetic fields, the interaction
parameter N is of the order of one, N being defined as

$$N = \frac{\sigma B_0^2}{2\pi\rho f},\qquad(3)$$

σ, ρ, B_0, and f being respectively the electrical conductivity of the liquid
metal, its density, the magnetic field amplitude, and its frequency. The free
surface becomes unstable and non-symmetric waves appear on the free surface. Such instabilities, which are illustrated in Figs. 7 and 9, mainly come
from "parametric" resonance effects due to the alternating part of the Lorentz
forces [15–19]. Note that, now, contrarily to the forced wave case, the free
surface oscillates at the applied magnetic field frequency. Two kinds of parametric instability may occur (see, e.g., Figs. 7 and 9c, e). Indeed, the type-I
parametric instability corresponds to a single-mode transition whilst type II
leads to a mode combination [17].

For large magnetic field values, the free surface becomes highly agitated.
Liquid metal ejections may be observed (Figs. 8 and 9). In some conditions an
emulsion of small droplets appears (see, e.g., Fig. 9f). The size of the droplets is
of the order of the gravito-capillary length. It is noteworthy that the emulsion
regime appears beyond a certain magnetic field threshold. The interpretation

Type I :
Sub-harmonic regime

$f = 3.90$ Hz, $I = 130$ A

Mode : (5, 0)
Eigenfrequency :
$fe = 3.91$ Hz

Type II :
Combination mode
regime

$f = 3.86$ Hz, $I = 110$ A

Modes : (3, 0) + (3, 1)
Eigenfrequencies :
$f1 = 2.76$ Hz
$f2 = 4.76$ Hz

Fig. 7. View of the various free-surface instabilities in an annular vessel submitted to a vertical uniform low frequency magnetic field [18]

Fig. 8. Free surface instabilities of a mercury pool located inside a single-phase coil [15], the oscillation frequency $f = 10.3$ Hz. The pool diameter is 200 mm

is as follows. When the magnetic energy increases, the surface energy must increase as well. This is achieved by a decrease of the width of the fingers. However, that width cannot be smaller than the capillary length. Thus, the only way to increase the surface energy of the drop is to break into droplets.

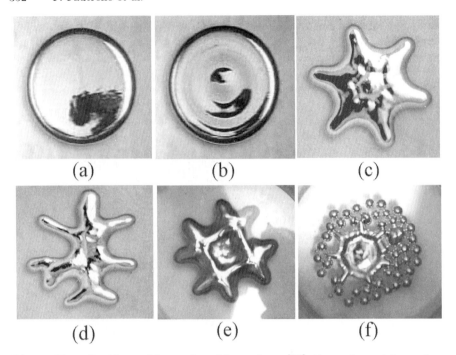

Fig. 9. Plane liquid metal layer viewed from above [19]; the various pictures show the regime according to both the magnetic field amplitude and the applied frequency, the typical drop diameter is 60 mm, the magnetic field amplitude varies from 0.1 to 0.2 T. (a) Mercury drop is at rest; (b) forced wave regime $f = 9.602$ Hz; (c) parametric regime, mercury drop $f = 2.10$ Hz; (d) unstructured regime with finger formation, mercury drop $f = 10$ Hz; (e) combination mode regime $(4 + 8)$, gallium drop $f = 4.01$ Hz; (f) emulsion of a gallium drop $f = 6.2$ Hz

3.2 Instability under medium- and high-frequency magnetic field

Let us consider now the case where the electromagnetic skin depth is smaller than the pool dimension. Both parts of the Lorentz forces are comparable. The idea to use AC magnetic field in order to stabilize the free surface of a liquid metal is quite old [14]. The action of AC magnetic fields still remains a pending question. It is not clear whether or not a magnetic field may stabilize a liquid metal free surface. The literature on the subject is quite controversial even in a simple geometry such as a planar free surface under the action of a parallel magnetic field. In the approximate quasi-steady analysis performed by Garnier and Moreau [20] the magnetic field is neutral. But, by means of a linear stability analysis, Mac Hale and Melcher [21] have found that a kind of electromechanical-electrothermal instability could occur for a mid-frequency. They confirmed their analysis by an experiment performed at a frequency of 2 kHz. Recent experiments on a liquid metal drop located in a coil also show that instability occurs for sufficiently strong magnetic fields in the frequency

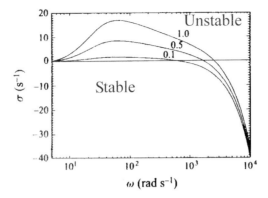

Fig. 10. Effect of a parallel AC magnetic field on a liquid metal free surface [24]; maximum growth rate σ of the wave perturbations versus the magnetic field pulsation $\omega = 2\pi f$ for three values of the non-dimensional magnetic field strength $M = B_0^2/2\mu\rho g\delta$; instability occurs when the growth rate is positive; instability disappears for the highest frequencies, but increasing the magnetic field strength M widens the unstable region

range of around 20 kHz [22]. Without using the quasi-steady approximation, Deepak and Evans [23] performed a stability analysis and showed also that a parallel magnetic field could be weakly destabilizing. Nevertheless, Iwai et al. [24] showed experimentally that forced waves were damped when a high frequency magnetic field (100 kHz) was applied parallel to a gallium free surface. A possible explanation of the controversy was given by Fautrelle and Sneyd [25] in the case of a uniform parallel single-frequency AC magnetic field. They showed that the mode parallel to the field might be unstable for moderate frequencies (the mode perpendicular to the magnetic field is neutral) [24]. However, for a given magnetic field increasing the frequency suppressed the instability, whilst increasing the magnetic field strength might enhance it. This effect is illustrated in Fig. 10 extracted from [25]. It may be seen from Fig. 10 that the higher the magnetic field strength, the easier the ability of the surface of the pool to be destabilized. In that analysis the most unstable wavelength were of the order of the electromagnetic skin depth. In conclusions, when a medium-frequency magnetic field is used, for example, to shape a liquid metal free surface or to levitate the liquid blob, the free surface may be subject to electromagnetic instabilities according to the frequency as well as the magnetic field amplitude.

Free-surface instabilities may exist even if the magnetic field frequency is not very low. Drop and pinch experiments discussed in § 3 clearly show that symmetry breaking occurs for sufficiently high magnetic fields. In that case the unstable modes are perpendicular to the magnetic field contrary to the previous one. Accordingly, another instability mechanism must be sought. One possible mechanism comes from the fact that the electric currents are significantly disturbed by the free-surface perturbations in that case (contrary

to the previous situation). The latter effect could contribute to enhance the
instability.

4 Conclusions

The use of AC magnetic fields to control or to levitate liquid metal free surface
is quite old, and the first attempts were made approximately 50 years ago. It is
an elegant means to fulfil various metallurgical aims. For example, levitation
or quasi-levitation avoids or minimizes the contact between the liquid metal
and the crucible. Surface agitation generated by low-frequency magnetic field
increases significantly mass transfers across a liquid–liquid interface in ladle
refining. Recent trends involve using complex magnetic fields such as two-
frequency fields [26–28], i.e., high and low frequencies (modulated magnetic
fields) or DC field ($f = 0$) with high frequency field. It is noticeable that
50 years later some fundamental questions concerning the stability of free
surface under AC magnetic fields are still pending even in the simplest planar
geometry.

References

1. Sneyd AD (1994) Theory of electromagnetic stirring by AC fields. IMA J Math
   Appl Bus Indust (invited review article) 5(2):87–113
2. Fautrelle Y, Perrier D, Etay J (2003) Free surface controlled by magnetic fields.
   Trans ISIJ Int 43(6):801–806
3. Moreau R (1990) Magnetohydrodynamics. Kluwer Academic, Dordrecht
4. Taberlet E, Fautrelle Y (1985) Turbulent stirring in a experimental induction
   furnace. J Fluid Mech 159:409–431
5. Mestel AJ (1982) Magnetic levitation of liquid metals. J Fluid Mech 117:27–43
6. Sneyd AD, Moffatt HK (1982) Fluid dynamical aspects of the levitation melting
   process. J Fluid Mech 117:45–70
7. Okress EC, Wroughton DM, Comenetz C, Brace PN, Kelly JCK (1952) Electromagnetic levitation of solid and molten metals. J Appl Phys 23:545–552
8. Brancher JP, Etay J, Sero-Guillaume O (1983) Formage de lames liquides, cal-
   culs et expériences. J de Mec Théor Appl 2(6):977–989
9. Etay J, Garnier M (1984) Some applications of high frequency magnetic field in
   metallurgical applications of magnetohydrodynamics. In: Proceedings of IUTAM
   Symposium. The Metals Society, London, pp 190–196
10. Shercliff JA (1981) Magnetic shaping of molten metals. Proc R Soc Lond
    375:455–473
11. Barbier JN, Fautrelle Y, Evans JW, Cremer P (1982) Simulation numérique des
    fours chauffés par induction. J de Mec Théor Appl 1(3):533–556
12. Leclerq I (1989) Ph.D. dissertation. Institut National Polytechnique de Greno-
    ble, France

13. Fautrelle Y, Perrier D, Etay J (2003) Free surface deformations of a liquid metal drop submitted to a middle-frequency AC magnetic field. In: Proceedings of the 4th International Conference on Electromagnetic Processing of Materials, Lyon, France, 14–17 October 2003, pp 279–282

14. Mohring J-U, Karcher Ch (2002) Electromagnetic pinch in an annulus: experimental investigation and analytical modelling. In: Proceedings of the 5th International PAMIR Conference on Fundamental and Applied MHD, Ramatuelle, France, 16–20 September 2002, I, pp 143–148

15. Galpin JM, Fautrelle Y (1992) Liquid metal flows induced by low frequency alternating fields. J Fluid Mech 239:383–408

16. Galpin JM, Fautrelle Y, Sneyd A (1992) Parametric instability in low frequency magnetic stirring. J Fluid Mech 239:409–427

17. Fautrelle Y, Sneyd A (2005) Surface waves created by low-frequency magnetic fields. Eur J Mech B/Fluids 24:91–112

18. Debray F, Fautrelle Y (1994) Free surface deformation frequencies of an electromagnetically excited mercury layer. Exp Fluids 16:316–322

19. Fautrelle Y, Etay J, Daugan S (2005) Free surface waves generated by low frequency alternating magnetic fields. J Fluid Mech 527:285–301

20. Garnier M, Moreau R (1983) Effect of finite conductivity on the inviscid stability of an interface. J Fluid Mech 127:365–377

21. Mac Hale EJ, Melcher JR (1982) Instability of a planar liquid layer in an alternating magnetic field. J Fluid Mech 114:27–40

22. Karcher Ch, Kocourek V, Schulze D (2003) Experimental investigations of electromagnetic instabilities of free surfaces in a liquid metal drop. In: Nacke B, Baake E (eds) Proceedings of International Scientific Colloquium "Modelling for Electromagnetic Processing", Institute for Electrothermal Processes, University of Hannover, Germany, pp 105–110

23. Deepak, Evans JW (1995) The stability of an interface between viscous fluids subjected to a high-frequency magnetic field and consequences for electromagnetic casting. J Fluid Mech 287:133–150

24. Iwai K, Suda M, Asai S (1994) Damping behaviour of surface wave motion on molten metals by imposing a high frequency magnetic field. In: Proceedings of International Symposium on Electromagnetic processing of Materials EPM '94, Nagoya, Japan, 25–28 October. Iron and Steel Institute of Japan, Tokyo, pp 127–131

25. Fautrelle Y, Sneyd A (1998) Instability of a plane conducting free surface submitted to an alternating magnetic field. J Fluid Mech 375:65–83

26. Li T, Sassa K, Asai S (1994) Dynamic meniscus behavior in continuous casting mold with intermittent high frequency magnetic field and surface quality of products. In: Proceedings of International Symposium On Electromagnetic Processing of Materials EPM '94, Nagoya, Japan, 25–28 October, The Iron and Steel Institute of Japan, Tokyo, pp 242–247

27. Takeuchi E, Miyazawa K (2000) Electromagnetic casting technology of steel. In: Proceedings of the 3rd International Symposium on Electromagnetic Processing of Materials EPM' 00, The Iron and Steel Institute of Japan, Tokyo, pp 20–27

28. Perrier D, Fautrelle Y, Etay J (2003) Experimental and theoretical studies of the motion generated by a two-frequency magnetic field at the free surface of a gallium pool. Metall Mat Trans B 34(5):669–678

Numerical Modelling for Electromagnetic Processing of Materials

Valdis Bojarevics and Koulis Pericleous

University of Greenwich, CMS, Park Row, London SE10 9LS, United Kingdom
(v.bojarevics@gre.ac.uk)

1 Introduction

Electromagnetic processing of materials (EPM) is one of the most widely practiced and fast growing applications of magnetic and electric forces to fluid flow. EPM is encountered in both industrial processes and laboratory investigations. Applications range in scale from nano-particle manipulation to tonnes of liquid metal treated in the presence of various configurations of magnetic fields. Some of these processes are specifically designed and made possible by the use of the electromagnetic force, like the magnetic levitation of liquid droplets, whilst others involve electric currents essential for electrothermal or electrochemical reasons, for instance, in electrolytic metal production and in induction melting. An insight for the range of established and novel EPM applications can be found in the review presented by Asai [1] in the EPM-2003 conference proceedings.

Due to the complex coupling between flow and electromagnetics, numerical modelling is the most economical way of analysing, optimising, and developing new EPM applications. Typically, numerical efforts are concerned with specific manufacturing processes, application conditions, and particular field configurations. This approach is encountered too often in numerical modelling, and we would like to quote Jaluria [2] in saying that "what is often missing is the link between the diverse processing techniques and the basic mechanisms that govern the flow". Bearing this in mind, we will restrict the present review – also because of the obvious space limitations – to three major applications of EPM, where numerical modelling brings out basic physical mechanisms and where validation can be obtained by analytical solutions and/or experimental measurements. We will consider (1) the magnetic levitation of liquid droplets, (2) the induction cold crucible melting, and (3) the magnetohydrodynamic (MHD) aspects of aluminium electrolysis cells. These three EPM processes have attracted considerable interest by modellers for many years and vast experience has been accumulated. Nevertheless, the following analysis will show how the link between the basic physical mechanisms is still

S. Molokov et al. (eds.), Magnetohydrodynamics – Historical Evolution and Trends,
357–374. © 2007 *Springer.*

a challenging task for the numerical modelling, especially where (a) the electromagnetic fields are coupled to the free surface and its time-dependent variation, (b) to the melting/solidification front position, and (c) the thermal field depends on the turbulent flow and the combination of magnetic fields.

2 Magnetic levitation of liquids

Since the earliest magnetic levitation experiments in AC fields [3], it became apparent that the levitated liquid metal is prone to oscillation and instability. A very intense internal fluid flow was visually observed, apparently in the turbulent regime for earthbound conditions. The visual observations indicated typical velocities of the order of 0.20–0.40 m/s in a 10 mm diameter droplet [4] corresponding to a Reynolds number of the order 10^3–10^4. Numerical modelling of the fluid flow was usually restricted to an assumed, fixed shape of the droplet [4–8]. The intense turbulent flow was therefore not included in the free-surface shape calculation or oscillation analysis because of numerical difficulties, until recently [9]. The common approach for predicting the free-surface behaviour of a magnetically levitated droplet influenced by an external high-frequency magnetic field is based on the idea of a thin skin-layer penetration depth, when the free-surface shape can be obtained independently of the internal fluid motion [5, 10, 11]. This approximation essentially removes the dependence on the electrical conductivity of the levitated material.

It is instructive to see how this assumption compares to the known analytical solution for a conducting sphere surrounded by a single/multiple coaxial AC loops [12]. Figure 1 shows that for typical levitation conditions the magnetic field still penetrates considerably the 8 mm aluminium sphere at 1 MHz AC, and there is an interesting feature of the secondary induced currents next to the skin layer. The same analytical solution can be used to validate numerical solutions for the levitated droplet (Fig. 2). The numerical solution method using the Green function representation [4, 7, 11, 13] is sensitive to the discretization grid in an unexpected way: it is more important to have a very fine mesh division along the skin layer, but not so much in the direction normal to the boundary. In addition to this, the self-induction contribution is essential for a realistic approximation.

With the assumption of an ideal fluid potential flow, droplet oscillations induced under such conditions were studied using small amplitude linear theory in [14], and the transient decay of the droplet oscillation with viscous corrections in [15]. The viscous decay of small amplitude oscillations for the relevant liquid metal was numerically simulated in [6], however, without a magnetic field and in the absence of gravity. The transient internal flow decay in the slow flow, Stokes approximation for a levitated droplet was theoretically investigated in [7]. The thin skin-layer assumption was recently tested for cases of typical experimentally used frequency and material property values, using an axisymmetric direct numerical solution based on a spectral method [16].

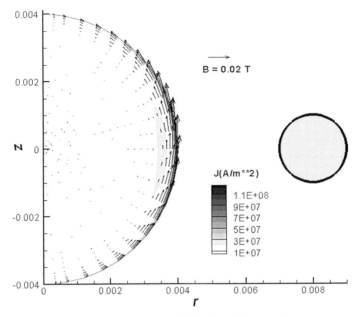

Fig. 1. Electric current and magnetic field induced in an aluminium sphere by a single current loop (195 A, 1 MHz, analytical solution)

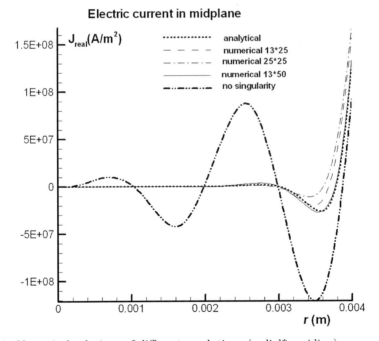

Fig. 2. Numerical solutions of different resolutions (radial*meridian) compared to the analytical for the real part of electric current amplitude induced in an aluminium sphere by a single current loop carrying 195 A at 1 MHz

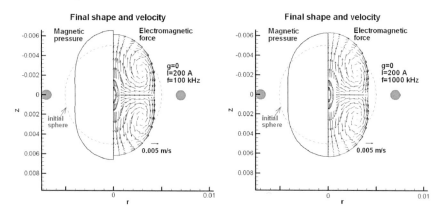

Fig. 3. High frequency AC solutions for the stationary velocity field and the surface of aluminium droplet in zero gravity for the full electromagnetic force and the magnetic pressure approximation

A significant difference was found between this and the asymptotic solution, for both the predicted deformation size and the velocity field pattern. In the asymptotic solution the magnetic force effects are restricted to the surface only (magnetic pressure approximation). Evidently, the magnetic pressure approximation is acceptable only for very high frequencies, i.e., about 1 MHz as in the 10 mm aluminium droplet, such as in the example shown in Fig. 3. The 10 kHz case, considered in [16], leads to a significant difference in shape; the difference is smaller at 100 kHz, and at 1 MHz there is no discernible difference between the full force and magnetic pressure cases (Fig. 3).

The three-dimensional (3D) finite element simulation of a magnetically levitated moving solid sphere was performed in [17], and a numerical gradual iterative shape change in 3D for levitated metal in [18]. Solid objects are much more prone to rotation and oscillation in levitated conditions owing to their usually irregular shape. However, even a spherical metallic specimen starts to rotate and oscillate when critical conditions are met [19]. The stability can be greatly enhanced by applying a relatively moderate DC magnetic field [19,20]. The combination of AC and DC magnetic fields was recently recognized as an efficient tool for the electromagnetic processing of materials [20, 21] and for contactless thermo-physical property measurements [22]. Even in a purely DC magnetic field, levitation is possible for paramagnetic and diamagnetic materials, and it can be used for advanced material research [23,24]. Numerical simulations in [9] predict long-lasting quasi-stationary surface shape oscillations, due to a dynamic change of the droplet position placed in the high-gradient magnetic force field.

An important test case for any numerical algorithm is that of an ideal fluid sphere, sustaining small amplitude oscillations, in the absence of gravity. In this case the flow is linear and potential. The Rayleigh capillary oscillation frequencies are known analytically [14]:

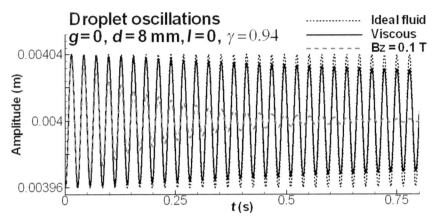

Fig. 4. Fluid droplet oscillation in absence of gravity: the top position change in time (...... for ideal fluid, ——— viscous, - - - - - in presence of DC magnetic field)

$$f_l = \frac{1}{2\pi}\sqrt{\gamma l(l-1)(l+2)/(\rho R_0^3)}; \qquad l = 2, 3, 4 \ldots,$$

where γ is the surface tension, ρ is the density, and R_0 is the droplet radius. The pseudo-spectral numerical solution in [9] obtains for zero viscosity the non-decaying oscillation shown in Fig. 4 corresponding to the four-digit accuracy to the above expression. When the laminar viscosity ($\nu = 10^{-6}\,\mathrm{m^2/s}$) is introduced, the oscillation is slowly damped (Fig. 4), yet it retains the same frequency. Adding the vertical uniform DC magnetic field B_z of moderate intensity 0.1 T, introduces quite a dramatic change in the local flow structure, and quite significant damping of the oscillations shown by the dashed line in Fig. 4. The flow is no longer irrotational even without viscosity. This can be easily ascertained by taking the curl of the magnetic damping force term, which is non-zero even for the uniform magnetic field case [9]. Remarkably, the oscillation frequency is almost unchanged.

AC levitation under normal gravity is often used for thermophysical property measurements [22, 25]. In a typical coil set-up [25], shown in Fig. 5, the electric current flows in the positive azimuthal direction in the bottom four turns of the coil and in the negative direction at the top two turns. The current frequency is 450 kHz; therefore one can expect a very small penetration depth for the electromagnetic field in a well conducting material like liquid aluminium. An external DC magnetic field can be added using a coaxial coil surrounding the AC coil as shown in Fig. 5. Passing a 200A DC current in the eight-turn external coil creates an almost uniform additional magnetic field inside the droplet.

When starting a numerical simulation (as with a physical experiment) great care is needed to position the droplet in the coil, so that the initial total electromagnetic force balances the weight of the droplet. However, there is an initial transient adjustment phase during which the droplet assumes

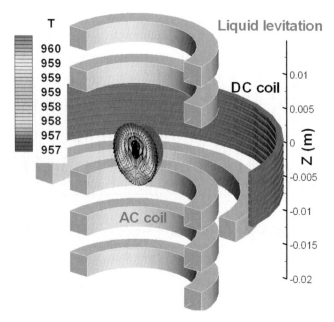

Fig. 5. The AC levitation coil for normal gravity conditions and the additional DC coil for stabilizing the oscillations [9]

the shape imposed by the force balance. The initial transient oscillations are soon damped, and a new non-decaying quasi-stationary oscillation pattern is established in 3–4 s time (Fig. 6). The centre of mass for the droplet is subject to a vertical cyclic motion which affects the shape-change oscillations. The final oscillation pattern does not show damping, which suggests a net energy transfer from the external field to the droplet mechanical motion. The generated oscillations depend on the material properties of the liquid metal. The motion of the droplet with a high surface tension is dominated by the centre of mass translational oscillation. The translational motion frequency at 8.76 Hz gives the first peak in the simulated spectra in Fig. 6. The $l = 2$ and $l = 3$ like modes are clearly present, yet the $l = 3$ peak is shifted from the pure Rayleigh frequency because of the non-linear interactions. The exact mechanism of the translational and normal mode oscillation interaction needs further analysis, the non-linearity being a clue as suggested by the close similarity of this effect to the translational motion generation as a result of two close normal mode interactions observed for a bubble in fluid oscillations [26].

Apart from the oscillation mode interaction, there is a considerable influence also of the intense circulation flow consisting of two vortices (Fig. 5), the intensity of which changes with the oscillation phase. The lower, smaller vortex is particularly affected by the bottom oscillation. The turbulent viscosity is mainly generated in this bottom part and then transported to the rest

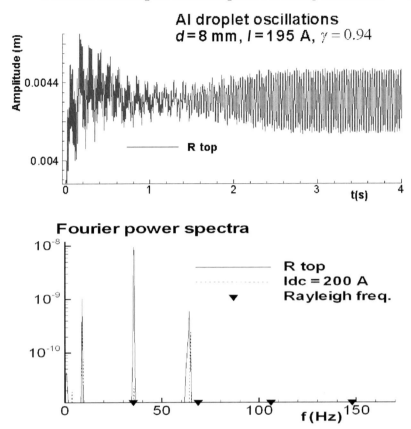

Fig. 6. The computed oscillations in the levitation coil under normal gravity and the Fourier power spectra with and without DC field

of the volume. The maximum magnitude for the time-dependent turbulent viscosity is about 15–20 times the laminar value, and it greatly enhances overall flow stability by limiting the velocity magnitude to below 0.3–0.4 m/s. In the absence of numerical diffusion, attempts to simulate the flow with only the laminar viscosity fail because the flow velocities start to increase continuously.

If a DC magnetic field is added to the aluminium droplet oscillation in the presence of the same AC coil, the droplet stability is greatly enhanced, and the resulting oscillation amplitude is significantly reduced. However, for the case of 200 A DC in the coil, the oscillation still reaches a quasi-stationary state and the power spectra in Fig. 6 exhibit essentially the same frequencies as without the DC field. The remaining oscillation can be completely suppressed when a 500 A current is supplied to the DC coil.

3 Melting in a cold crucible

The cold crucible technique is a process suitable for melting and preparing reactive metal alloys of high purity, prior to casting or gas atomization. This process is used to melt high-temperature materials like Ti, TiAl, Zr, Mo, and many others to produce near-net-shape cast components [27–29] and to investigate their material properties [30,31]. The water-cooled copper crucible is used to contain the metal charge melted by Joule heating from the induced current generated by an external medium to high-frequency AC coil (Fig. 7). The copper wall is made of electrically insulated segments so that the magnetic field can effectively penetrate through it; this penetration is achieved due to high density AC current loops induced within each individual segment (Fig. 8). These crucible currents incur relatively high energy losses removed by the cooling liquid circulating within the copper segments. In addition to the direct Joule losses, there are other, conductive and convective heat losses from the metal charge when in contact with the crucible walls.

The molten metal is normally held away from the sidewalls by the electromagnetic force. The shape and position of the liquid metal depends on the instantaneous balance of forces acting on it. Hence, the electromagnetic field and the associated force field are strongly coupled to the free-surface dynamics of the liquid metal, the turbulent fluid flow within it and the heat transfer. This complex problem has been studied extensively both experimentally and numerically. The early numerical simulation efforts mostly concentrated on the electrodynamics part of the problem [32–34] with the heat transfer mostly treated as a stationary problem. The liquid metal shape was obtained from a magnetostatic approximation, and the turbulence of the melt velocity field was considered within the stationary $k - \varepsilon$-type model range [28,34].

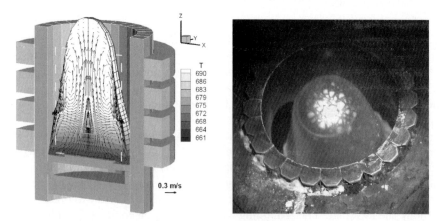

Fig. 7. Melting in the cold crucible and the corresponding numerical simulation for 2.8 kg Al alloy and 5560 A coil current [29]

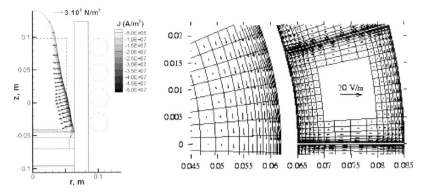

Fig. 8. The force and electric current distribution in the aluminium alloy melt at the final stage (*left*) and the calculated electric field distribution within the mid-section of the copper wall fingers (*right*), only one of the fingers from the total of 24 is shown

A new type of simulation was attempted in [35] using large eddy simulation (LES) technique for 3D, time-dependent flow within the domain of fixed shape, which in a time-average sense replicates the experimentally measured turbulent flow in the sodium experiment. The importance of the dynamic flow effects on the free surface was shown in related studies [36,37] of conventional induction crucibles.

At the present time only the axisymmetric model permits the simulation of the complete melting cycle with the coupled fluid flow, thermal and electromagnetic fields using the exact boundary conditions on the free surface. This technique was first applied to the closely related semi-levitation melting problem [13]. An appropriate turbulence model [38] for the time-dependent flow with free surface was implemented, and validated against detailed turbulent flow measurements within a "cold" liquid metal (In–Ga–Sn) AC field-driven experiment [39]. The fully coupled model for the cold crucible requires a custom-made combination of complementary sub-models and solution techniques: finite volume, integral equation, and pseudo-spectral methods combined to achieve the description of the dynamic melting process [29]. The predicted results match closely the experimental data of the temperature history of the melting at all stages and the heat losses in the various parts of the furnace (Fig. 9).

Visual observations and the height measurements of the free surface at various times compare to the numerically predicted surface shapes (Fig. 7). The detailed flow, turbulence intensity, and temperature fields from the dynamic numerical simulation provide explanations for the thermal efficiency loss at various stages of melting and the limitation of the resulting superheat of the melt. The process parameter dependence demonstrates the effect of changes in the melt weight, AC current frequency, induction coil position, and the effect of the heat loss reduction achieved using an additional, external DC electromagnetic field (Fig. 10).

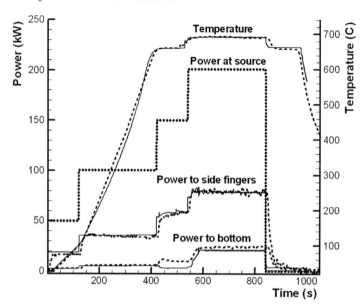

Fig. 9. Comparison of the measured temperature and power losses (*dashed lines*) with the computed data (*continuous lines*) when melting Al alloy showing the effect of stepwise increases in the supplied AC field power (*dotted line*)

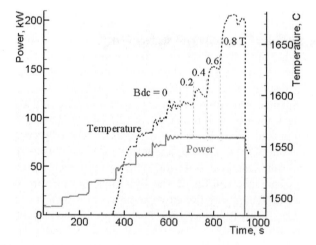

Fig. 10. The temperature at the top of the liquid TiAl melt and the total power released into the liquid metal. DC magnetic field is zero until 645 s, and then the DC magnetic field is increased in steps

In the presence of a strong DC magnetic field, an additional modification to the $k - \omega$ turbulence model [39] has been added by adapting Widlund's model [40, 41]. Using the pseudo-spectral representation, the computation follows in detail the time development of the melting front, free-surface evolution, and turbulent flow characteristics determined by the coupled non-linear transport equations. The DC field is shown to increase the melt superheat, a very desirable parameter for casting applications. This is in part due to turbulence damping in the whole melt volume and particularly close to the water-cooled walls and base where most of the losses occur.

4 Aluminium electrolysis cells

Large-scale industrial electrolysis cells used to produce primary aluminium, are sensitive to waves at the interface of liquid aluminium and electrolyte. An aluminium electrolysis cell is a part of a row of similar cells, each cell is connected in series to its neighbours by a complex arrangement of current-carrying busbars as shown in Fig. 11. The electric current to each individual cell is supplied from above through massive anode busbars made of solid aluminium, whence anode rods connect to the carbon anodes. The liquid electrolyte layer beneath the anode blocks is a relatively poor electrical conductor of a small depth (4–6 cm) if compared to its horizontal extension (2–4 × 6–20 m). The electrolyte density is only slightly different to that of liquid aluminium, which occupies a pool of depth 20–30 cm created as the result of

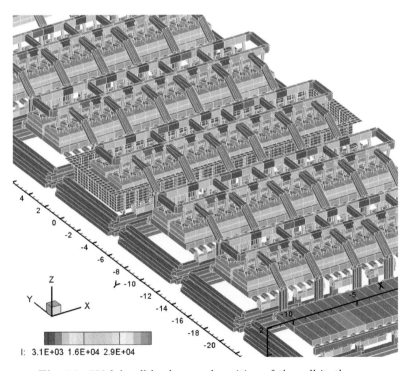

Fig. 11. 500 kA cell busbars and position of the cell in the row

electrolytic reaction at the bottom of the cell. The interfacial waves are similar in many aspects to stratified sea layers [42], but the penetrating electric current and the associated magnetic field are intricately involved in the oscillation process, which can lead to the unstable growth of the large amplitude waves. Even under quasi-stationary conditions these MHD modified waves have oscillation frequencies shifted from the purely hydrodynamic ones [43]. The resulting interface stability problem is of great financial and ecological importance, since aluminium production is a major electrical energy consumer, and it is linked to large-scale environmental pollution.

The first numerical investigation on the stability of the aluminium cells appeared in [44] and a short summary of the main developments was given in [43]. Important aspects of the multiple mode interaction were introduced in [45], and a widely used linear friction law to represent the governing bottom and top friction of the narrow layers was first applied in [46]. In [47] a systematic perturbation expansion of the fluid dynamics equations and electric current problem enabled reduction of the 3D physical problem to a two-dimensional (2D) mathematical one. The procedure, well known in oceanographic studies as the "shallow water approximation" can be extended to cover weakly non-linear and dispersive waves. The Boussinesq formulation allows generalization of the problem for non-unidirectionally propagating waves, accounting for sidewalls and for a two fluid layer interface [42]. Attempts to extend the electrolytic cell wave modelling to the weakly nonlinear case started in [48] where the basic equations were derived, including the non-linearity and linear dispersion terms. An alternative approach for the non-linear numerical simulation for an electrolysis cell wave evolution is described in [49] and references therein, yet this approximation omits the dispersion terms and treats the dissipation only at the linear friction law level. It can then only predict whether the waves are stable or not.

The self-sustained MHD modified interface waves, observed in the most of commercial cells under certain conditions [43], require numerical modelling of the coupled electric current, magnetic field, and fluid dynamics problem. The inclusion of the horizontal circulation-generated turbulence is essential in order to explain the small amplitude self-sustained oscillations of the liquid metal surface observed in real cells, known as "MHD noise". Reference [50] describes a very ambitious full model, where the coupled effects of: fluid dynamics, turbulent horizontal circulation, the non-linearity of the waves, and the extended electromagnetic field that covers the whole busbar circuit and ferromagnetic effects are accounted for.

It is instructive to analyse how a step by step inclusion of different physical coupling factors is affecting the wave development in the electrolysis cells. The general model in [50] permits the use of the full electromagnetic interactions between cells, making it suitable for realistic plant simulations: the magnetic field can be computed from up to six cells in the same row (Fig. 11) and five cells in the return row (not shown). The electric current distribution in the fluid layers (shown in Fig. 12 for the bottom aluminium layer) depends on the

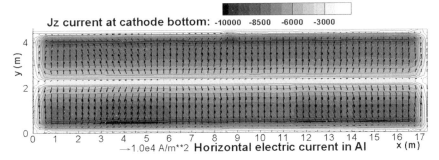

Fig. 12. The electric current distribution and the magnetic field in the liquid metal layer of the 500 kA cell

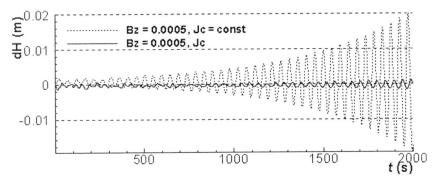

Fig. 13. Comparison of the interface oscillations for the cathode current J_c =const and for the computed J_c in the 500 kA cell (uniform $B_z = 0.0005$ T)

cathode collector connections to the full circuit of connecting bus elements of unequal lengths and cross sections. The horizontal current, even in the case of a flat aluminium surface without waves, arises because of geometrical differences in the anode and collector bars and the tendency for the cathode current to find the path of least resistivity (with the effect of the frozen ledge protecting the cell walls).

Most theoretical models for wave development do not account for this current distribution, instead assuming a uniform current density J_z at the bottom. Figure 13 illustrates the difference in the directly computed oscillation pattern for the two cases keeping the magnetic field uniform and fixed at $B_z = 0.0005$ T. The constant J_z assumption results in a fast-growing wave at a frequency shifted from pure gravitational waves but corresponding to the analytical prediction [47]. When the electric current is computed according to the actual electrical circuit the growth rate is significantly lower, and if a sufficient dissipation is included (due to friction say), does not lead to instability.

In addition to this effect, the velocity field is time-dependent and turbulent, thus ensuring higher dissipation rates for the waves than in the laminar

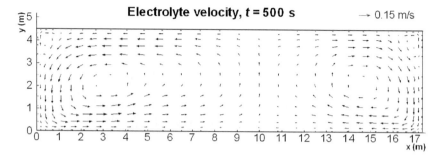

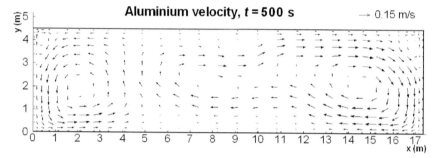

Fig. 14. Horizontal velocity in the two fluid layers for the 500 kA cell

case. The horizontal, depth-averaged circulation is different in each of the fluid layers because of the electric current distribution variation. The horizontal electric current in the aluminium is responsible for the variation in the nearly symmetric vortex structure, which is usually observed in the electrolyte layer (Fig. 14). The horizontal circulation vortices create a pressure gradient contributing to the deformation of the interface. Typically, an intense vortex in a single fluid layer with free surface is associated with a dip in its centre. For the two layers the effect on the common interface is in balance when two equal vortices are positioned one above the other, if the densities of the two fluids are not significantly different.

Instructive comparisons can be made for the interface at the same time moments if accounting for the horizontal circulation and without the effect. Fig. 15 (top) clearly shows the dips at the centre and the right side where the aluminium vortices are more intense and an oscillating wave crest on the left where the electrolyte circulation is more intense. There is nothing like this in the case when the horizontal circulation effect is set to zero, Fig. 15 (bottom).

The corresponding self-sustained oscillation pattern is shown in Fig. 16. The wave frequency is nearly the same in both cases, but the interface topology and the wave amplitudes are quite different. The inclusion of turbulent damping makes the cell stable for the appropriately designed bus network. The 500 kA cell considered in this example is absolutely stable if positioned

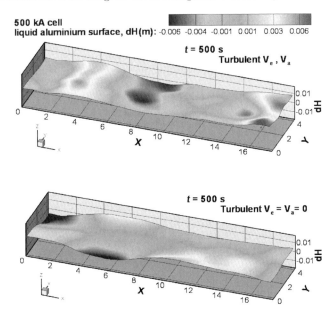

Fig. 15. Interface at fixed time 500 s: including horizontal velocity circulation effect (*top*) and without the horizontal circulation (*bottom*)

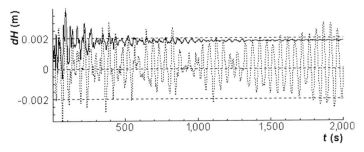

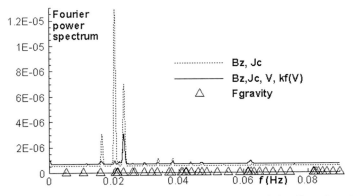

Fig. 16. Comparison of the interface oscillations in the 500 kA cell: with and without the horizontal circulation **v** effect

in the line of cells far from the end, but it gives rise to a very low self-sustained interface oscillation when positioned at the end of line.

The examples presented here show the importance of the fully coupled custom-made numerical models for predicting the behaviour of real electrolysis cells.

References

1. Asai S (2003) Challenging of EPM in economic mass production, nano-technology and environment protection. In: Proc 4th Int Conf Electodyn Proc Materials, Lyon:1–8
2. Jaluria Y (2001) Fluid flow phenomena in materials processing. J Fluids Eng 123:173–210
3. Okress E, Wroughton D, Comenetz G, Brace P, Kelly J (1952) Electromagnetic levitation of solid and molten metals. J Appl Phys 23:545–552
4. Schwartz E, Szekely J, Ilegbusi OJ, Zong J-H, Egry I (1991) The computation of the electromagnetic force fields and transport phenomena in levitated metallic droplets in the microgravity environment. In: MHD in process metallurgy, TMS, pp 81–87
5. Mestel AJ (1982) Magnetic levitation of liquid metals. J Fluid Mech 117:27–43
6. Szekely J, Schwartz E (1994) Perspectives on EM levitation in space experimentation. In: International Symposium on Electromagnetic Processing of Materials, ISIJ, Nagoya, pp 9–14
7. Li BQ (1994) The transient magnetohydrodynamic phenomena in electromagnetic levitation process. Int J Engng Sci 32:1315–1336
8. Hyers RW, Trapaga G, Abedian B (2003) Laminar-turbulent transition in an electromagnetically levitated droplet. Metall Materials Trans B 34:29
9. Bojarevics V, Pericleous K (2003) Modelling electromagnetically levitated liquid droplet oscillations. ISIJ Int 43(6):890–898
10. Sneyd AD, Moffatt HK (1982) Fluid dynamical aspects of the levitation melting process. J Fluid Mech 117:45–70
11. Gagnoud A, Brancher JP (1985) Modelling of coupled phenomena in electromagnetic levitation. IEEE Trans Magn 21:2424–2427
12. Smythe R (1989) Static and Dynamic Electricity. Hemisphere, New York
13. Bojarevics V, Pericleous K, Cross M (2000) Modelling the dynamics of the semi-levitation melting. Metall Materials Trans B 31:179–189
14. Cummings DL, Blackburn DA (1991) Oscillations of magnetically levitated aspherical droplets. J Fluid Mech 224:395–416
15. Bratz A, Egry I (1995) Surface oscillations of electromagnetically levitated viscous metal droplets. J Fluid Mech 298:341–359
16. Bojarevics V, Pericleous K (2001) Magnetic levitation fluid dynamics. Magnetohydrodynamics 37:93–102
17. Enokizono M, Todaka T, Yokoji K, Wada Y, Matsumoto I (1995) Three dimensional moving simulation of levitation melting method. IEEE Trans Magn 31:1869–1872
18. Winstead CH, Gazzerro PC, Hoburg JF (1998) Surface-coupled modeling of magnetically confined liquid metal in three-dimensional geometry. Metall Materials Trans B 29:275–281

19. Priede J, Gerbeth G, Mikelsons A, Gelfgat Y (2000) Instabilities of electromagnetically levitated bodies and their prevention. In: Proceedings of the 3rd International Symposium on Electromagnetic Processing Materials, ISIJ, Nagoya, pp 352–357

20. Yasuda H, Ohnaka I, Ninomiya Y, Ishii R, Fujita S, Kishio K (2003) Solidification behavior in the melt levitated by simultaneous imposition of alternative and high static magnetic fields. In: Proceedings of the 4th International Symposium on Electromagnetic Processing Materials, Lyon, pp 459–463

21. Gillon P (2000) Processing of materials with high DC magnetic field gradients. In: Proceedings of the 3rd International Symposium on Electromagnetic Processing Materials, ISIJ, Nagoya, pp 635–640

22. Egry I, Diefenbach A, Dreier W, Piller J (2001) Containerless processing in space - thermophysical property measurements using electromagnetic levitation. Int J Thermophys 22:569–578

23. Ikezoe Y, Hirota N, Nakgawa J, Kitazawa K (1998) Making water levitate. Nature 393:749–750

24. Motokawa M (2000) Orientation and levitation effects in high magnetic fields. In: Proceedings of the 3rd International Symposium on Electromagnetic Processing Materials, ISIJ, Nagoya, pp 612–617

25. Brooks RF, Day AP (1999) Observations of the effects of oxide skins on the oscillations of EM levitated metal droplets. Int J Thermophys 20:1041–1050

26. Feng ZC, Leal LG (1995) Translational instability of a bubble undergoing shape oscillations. Phys Fluids 7:1325–1336

27. Tadano H, Kainuma K, Take T, Shinokura T, Hayashi S (2000) Vacuum melting with cold crucible levitation melting furnaces. In: Proceedings of the 3rd International Symposium on Electromagnetic Processing Materials, ISIJ, Nagoya, pp 277–282

28. Bernier F, Vogt M, Muehlbauer A (2000) Numerical calculations of the thermal behaviour and the melt flow in induction furnace with cold crucible. In: Proceedings of the 3rd International Symposium on Electromagnetic Processing Materials, ISIJ, Nagoya, pp 283–288

29. Harding RA, Wickins M, Bojarevics V, Pericleous K (2004) The development and experimental validation of a numerical model of an induction skull melting furnace. Metall Materials Trans B 35:785–803

30. Gillon P (2000) Processing of materials with high DC magnetic field gradients. In: Proceedings of the 3rd International Symposium on Electromagnetic Processing Materials, ISIJ, Nagoya, pp 635–640

31. Toh T, Yamamura H, Wakoh M, Takeuchi E (2003) Inclusion behavior in cold crucible levitation melting and its applications to cleanliness evaluation. In: Proceedings of the 4th International Conference on Electromagnetic Processing Materials, Lyon, pp 226–231

32. Tanaka T, Kurita K, Kuroda A (1991) Mathematical modeling for electromagnetic field and shaping of melts in cold crucibles. Liquid Metal Flows ASME FED 115:49–54

33. Enokizono M, Todaka T, Matsumoto I, Wada Y (1993) Levitation melting apparatus with flux concentration cap. IEEE Magn 29 (6):2968–2970

34. Baake E, Muehlbauer A, Jakowitsch A, Andree W (1995) Extension of the $k - \varepsilon$ model for the numerical simulation of the melt flow in induction crucible furnaces. Metall Mater Trans B 26:529–536

35. Baake E, Umbrashko A, Nacke B, Jakovics A, Bojarevics A (2003) Experimental investigations and LES modelling of the turbulent melt flow and temperature distribution in the cold crucible induction furnace. In: Proceedings of the 4th International Conference on Electromagnetic Processing Materials, Lyon, pp 214–219
36. Fukumoto H, Hosokawa Y, Ayata K, Morishita M (1991) Numerical simulation of meniscus shape considering internal flow effects. MHD in Process Metallurgy. Miner, Met Mater Soc:21–26
37. Kageyama R, Evans JW (1998) A mathematical model for the dynamic behaviour of melts subjected to electromagnetic forces. Part 1. Metall Mater Trans B 29:919–928
38. Wilcox DC (1998) Turbulence Modelling for CFD. DCW Industries, La Canada, CA
39. Bojarevics A, Bojarevics V, Gelfgat J, Pericleous K (1999) Liquid metal turbulent flow dynamics in a cylindrical container with free surface: experiment and numerical analysis. Magnetohydrodynamics 35:258–277
40. Widlund O (2000) Modelling of magnetohydrodynamic turbulence. Ph.D. thesis. Royal Institute of Technology, Stockholm, Sweden, ISSN 0348-467X
41. Widlund O (2002) Draft of $K - \omega - \alpha$ closure for modeling of MHD turbulence (unpublished)
42. Mei CC (1989) Applied dynamics of ocean surface waves. World Scientific
43. Von Kaenel R, Antille JP (1996) Magnetohydrodynamic stability in alumina reduction cells. Travaux 23(27):285–297
44. Urata N, Mori K, Ikeuchi H (1976) Behavior of bath and molten metal in aluminium electrolytic cell. Keikinzoku 26(11):573–600
45. Sneyd AD, Wang A (1994) Interfacial instability due to MHD mode coupling in aluminium reduction cells. J Fluid Mech 263:343–359
46. Moreau R, Ewans JW (1984) An analysis of the hydrodynamics of aluminium reduction cells. J Electrochemical Society 131(10):2251–2259
47. Bojarevics V, Romerio MV (1994) Long waves instability of liquid metal-electrolyte interface in aluminium electrolysis cells: a generalization of Sele's criterion. Eur J Mech B/Fluids 13:33–56
48. Bojarevics V (1998) Nonlinear waves with electromagnetic interaction in aluminium electrolysis cells. In: Progress Fluid Flow Research: Turbulence and Applied MHD. AIAA Chapter 58, pp 833–848
49. Sun H, Zikanov O, Finlayson BA, Ziegler DP (2005) The influence of the basic flow and interface deformation on stability of Hall-Herault cells. Light Metals 2005, TMS, pp 437–441
50. Dupuis M, Bojarevics V (2005) Weakly coupled thermo-electric and MHD mathematical models of an aluminium electrolysis cell. Light Metals 2005, TMS, pp 449–454

Magnetic Fields in Semiconductor Crystal Growth

Hiroyuki Ozoe[1], Janusz S Szmyd[2], and Toshio Tagawa[3]

[1] Institute for Materials Chemistry and Engineering, Kyushu University, Kasuga Koen 6-1, Kasuga, Fukuoka 816-8580, Japan (hozoe@osu.bbiq.jp)
[2] AGH, University of Science and Technology, 30 Mickiewicz Ave., 30059 Krakow, Poland (janusz@uci.agh.edu.pl)
[3] Tokyo Metropolitan University, Asahigaoka 6-6, Hino, Tokyo 191-0065, Japan (ttagawa@cc.tmit.ac.jp)

1 Introduction to semiconductor crystal growth

We may define three main categories of crystal growth techniques: growth from solid, vapour, and melt. These three main categories of crystal growth methods need careful control of the phase change. We may introduce a subcategory, growth from the solution, which is strictly already included in the above definitions, and which represents crystal growth processes of solute from an impure melt.

Figure 1 shows techniques commonly used for the crystal growth from the melt. All of these growth techniques can be referred to two main categories: meniscus-controlled crystal growth systems and confined crystal growth systems. In meniscus-controlled crystal growth systems (Czochralski technique, floating zone technique) there is a three-phase boundary at which crystal, melt and gaseous phase coexist. In confined crystal growth systems both crystal and melt are confined within a solid container. Such techniques can be divided into normal freezing method (in which the whole charge is melted initially and then progressively crystallized), and zone-melting method (in which a molten zone is established and traversed along an ingot). In those techniques a crystal–melt interface moves vertically or horizontally. The vertical directional solidification technique is commonly known as the Bridgman technique, while the horizontal directional solidification technique as the Chalmers technique. Zone-melting techniques are designed vertically or horizontally [1].

1.1 Czochralski method

Currently, the most important technique for the growth of bulk crystals (Czochralski method) uses pulling from the melt and has its origin in the problem of crystallization velocities [2]. This technique [3] and various modifications

S. Molokov et al. (eds.), Magnetohydrodynamics – Historical Evolution and Trends,
375–390. © 2007 *Springer.*

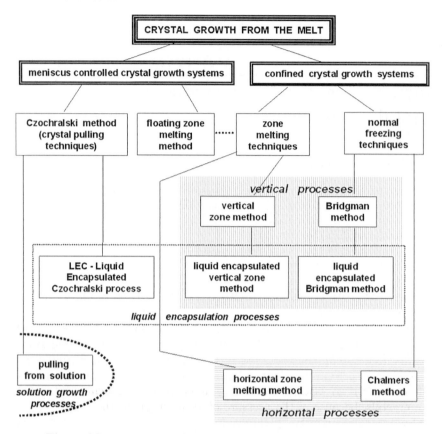

Fig. 1. Main categories of crystal growth techniques from the melt

have become the dominant process used in industry today for the production of semiconductor and oxide single crystals. The melt (molten charge) is in the crucible, which is heated (by resistive heating system or by frequency induction heating system). The pull rod with a single crystal ("seed crystal") is positioned axially above the crucible and is lowered. The temperature of the melt is then adjusted so that the centre of the liquid is at its freezing point. The seed crystal is dipped into the melt. The pull rod is rotated and lifted. During the process of crystal growth by the Czochralski method, the crystal is rotated to produce homogeneity near the melt–crystal interface. Since buoyancy and surface tension forces are very strong and produce complex flow structures (oscillatory and/or turbulent), forced convection is superimposed by rotating crucible. Application of a magnetic field for the melt in Czochralski crystal growth system is also considered as one of the effective tools for suppressing the melt convection. The main effort in the growth of silicon (Si) single crystals is to control impurities and imperfections. Molten silicon reacts slowly with the crucible material (SiO_2). The oxygen from the crucible has an appreciable solubility in molten silicon and therefore acts as impurity in the final silicon crystal. In addition to oxygen, other impurities in the crucible can be dissolved into the molten silicon and can affect the electrical properties of grown silicon single crystal. These impurity effects can be minimized through pressure control and proper stirring of the melt. Crystal lattice imperfections, such as dislocation, is controlled and eliminated by using a technique [4], which requires the reduction in diameter of the seed followed by growth at the reduced diameter before enlarging the crystal to the final diameter. The current industry standard for Si growth is 200 mm diameter, and 300 mm large diameters Si single crystals can be produced.

The liquid-encapsulated Czochralski (LEC) technique and high-pressure liquid-encapsulated Czochralski (HPLEC) technique have been developed for the production of single crystals of group III – group V compounds (GaAs, InP, GaP). These techniques (LEC, HPLEC) have been developed to overcome one of the main material limitations of the crystal pulling, that the material to be grown should have a relatively low vapour pressure (in GaP, InP, and GaAs compounds, the dissociation pressure of phosphorus or arsenic at melting point is greater than 1 atm). An encapsulant layer (generally B_2O_3) is placed over the compound melt to prohibit the escape of the volatile component. Crystals of GaAs can be grown at about atmospheric pressure, because the equilibrium vapour pressure at the melting temperature is about 0.9 atm. For the InP melt crystal pulling chamber is a pressure chamber, because the vapour pressure of the volatile component in the melt is high (about 27.5 atm for the phosphorus in the InP melt). HPLEC system is very different from "Czochralski furnace" for Si crystals, because of the pressure of an inert gas in the crystal growth furnace. The gas convection plays an important role in a HPLEC furnace [5].

The flux pulling method has been developed to overcome another limitation in the Czochralski technique, namely that the material should not decompose upon or before melting. The flux pulling method consists of the pulling of a crystal as in the conventional Czochralski technique from a melt, which contains the desired compound in solution. The crystal growth process in these techniques is dependent upon the mass transfer of the compound from the solvent to the crystal interface. The bottom of the crucible is usually hotter than the top of the melt, so that the nutrient dissolves at the bottom and passes to the top of the melt by convection and diffusion [6]. A detailed discussion on crystal growth using the Czochralski method (crystal pulling technique) may be found in several books and review articles [7–11].

1.2 Floating zone method

The crucible-free zone melting process (floating zone technique) was developed by Theuerer [12], Keck and Golay [13] to obtain high-purity crystals, especially to avoid impurities from the crucible material. In the floating zone system, a molten pool (molten zone) is formed by a circumferentially allocated heat source. This molten zone (like a drop between two parts of the rod) separates a melting polycrystalline feed rod and a solidifying crystal. The molten zone is moved through the rod over its whole length by the motion of the heater or the rod. A single crystal can be generated by spontaneous nucleation or by using "seed crystal" as the initial part of the rod, (seed crystal is kept to be un-molten). For higher stability of the process, the molten zone is usually moved upwards. The growing crystals and sometimes also the melting rods are rotated mostly with different rotation rates or counterrotation. The floating zone technique is used for the growth of high-purity semiconductor materials, such as silicon and germanium. This technique is also applied for the growth of materials with high (or very high) melting points for which no crucible materials are available. In small-scale, molten zone is performed with conventional resistance heaters. Moreover, floating zone melting can also be performed using electron bombardment, electric arc or plasma, thermal or light radiation. For large-diameter industrial floating zone systems heating is achieved by using radio-frequency induction heating elements shaped so that the induction coil has smaller diameter than the growing crystal. The stability of the molten zone is determined by surface tension, gravity, centrifugal forces (due to crystal rotation), Marangoni convection, buoyancy-driven convection and electrodynamic forces (due to the high-frequency field of the heater coil). The techniques related to the crucible-free zone melting processes (floating zone techniques) are discussed in the books by Pfann [14] and Bohm et al. [15].

1.3 Bridgman method and confined crystal growth systems

In confined crystal growth systems the material is loaded into an ampoule (a solid container) and then melted and solidified by varying the temperature field: by changing the heat power of the furnace (the gradient freeze technique) or by translating the ampoule through the furnace (the Bridgman technique). After solidification, the crystal is removed from the ampoule. The growth of a single crystal in a confined system from the melt can be "seeded" or "un-seeded". The confined crystal growth techniques are very widely used and the range of material grown by these techniques is enormous, because of the simplicity and the ease of construction of equipment. The vertical directional solidification technique is commonly known as the Bridgman technique or as the "vertical Bridgman" (VB) technique. Horizontal directional solidification technique has also been widely employed. In this technique, the ampoule is laid horizontally with respect to the gravity and the temperature gradient in the melt is changed by translating heater or by variation of the power of heater. The melt is contained in the ampoule or open ampoule (the composition of the melt can be equilibrated with the surrounding ambient). The horizontal directional solidification technique is known today as the Chalmers technique or as the "horizontal Bridgman" (HB) technique.

Zone melting set-up (in which a molten zone is established and traversed along an ingot) was invented by Kapitza [16] to grow bismuth crystals in a glass tube. Repeated zone melting can produce a substantial purifying effect. Pfann [17, 18] applied zone-melting technique for the purification of germanium. Zone-melting techniques are designed vertically or horizontally. The confined crystal growth systems have also been employed in space. Microgravity experiments of crystal growth in confined systems have been intensively carried out in recent years in order to reduce buoyant convection and to grow high quality crystals.

1.4 Industrial requirements of crystal growth processes

From the industrial requirements the crystal should be: large size (demand for reduction of operation costs), perfect quality ("perfect crystal", i.e., crystal without dislocations, vacancies, impurities; concentration of the dopants should be regular and in the case of a pseudobinary alloy the composition of the crystal should be uniform). However, this is never completely obtained because of the complexity of the phenomena taking place in the melt and in the vicinity of the solidification interface. The quality and the size of the grown crystal depend on many physical effects (macroscopic and microscopic). The solidification front between crystal and melt plays a major role in the process. The difficulty of growing high-quality crystals is the dependence of interface shape and segregation on the melt flow pattern. Generally, the melt flow is recirculating, three-dimensional (3D), time-dependent and is caused by a combination of natural, forced, and Marangoni convection. The natural convection is introduced by buoyancy forces, which are produced by thermal boundary conditions (heating and cooling of the sidewalls, phase change at the crystal–melt interface, radiative heat transfer from the free surface). The buoyancy caused by concentration difference can play an important role in the melt growth of semiconductor-mixed crystals. The mode of natural convection plays a significant role in all crystal growth techniques from the melt. The Marangoni convection due to the variation of surface tension is caused by gradients of temperature and concentration. The mode of Marangoni convection plays a significant role in the crystal growth techniques containing free surface of the melt. The forced convection is introduced by crystal rotation, crucible rotation, crystal pulling, and by the reduction of the melt height (in a Cz system). The growing crystal and sometimes also the melting rod are rotated mostly with different rotation rates during the crucible-free zone-melting process. The mode of forced convection plays a significant role in these growth techniques. In semiconductor crystal growth, the electric current induced in the melt may be important. For instance, in Czochralski crystal growth system, the meniscus shape can depend on the electromagnetic field if a significant Lorentz force is exerted on the surface of the melt, due to the surface tension effect. Since buoyancy and surface tension forces are very strong and produce complex flow structures (oscillatory and/or turbulent), application of a magnetic field is considered to be one of the effective tools for suppressing melt flow oscillations. The use of magnetic fields can stabilize the melt convection [19].

2 Fundamentals prerequisite to crystal growth in magnetic fields

As mentioned in the previous subsection, semiconductor crystal growth is closely relevant to various fundamental phenomena, such as buoyancy effect caused by temperature or concentration difference, rotation of crystal or enclosure, free surface, solid–liquid interface, etc., as well as the type of an applied magnetic field. Therefore, in the following, each fundamental phenomenon subjected to the magnetic field is discussed and reviewed.

2.1 Buoyancy effect

In the presence of temperature (or concentration) inequality, the fluid density varies depending on the local temperature, and buoyant, or natural convection arises. The melt of silicon and/or gallium arsenide semiconductors is an electrically conducting, low-Prandtl number fluid, and its fluid motion exhibits oscillatory or turbulent convection. In order to avoid such unwanted oscillatory melt flow, a static magnetic field is applied to stabilize and to damp out the convection by means of the Lorentz force. There have been many researches related to natural convection in an enclosure in the presence of a magnetic field in order to investigate fundamentals prerequisite to crystal growth [20–22]. The buoyancy effect is usually studied by employing the Boussinesq approximation. The electric current density is governed by the Ohm's law, since the magnetic Prandtl number, Pr_{m}, is much less than unity in most cases of semiconductor crystal growth. The temperature field can be estimated from the energy equation, in which the viscous dissipation and the Joule heat are usually neglected.

The thermal natural convection in an electrically conducting fluid in the presence of a magnetic field is governed by three independent non-dimensional parameters: the Rayleigh number, the Prandtl number, and the Hartmann number. The other parameters such as the Grashof number, Gr, the Reynolds number, Re, and the interaction parameter, N, can be estimated from combinations of the three parameters. The Rayleigh number, Ra, characterizes the strength of the natural convection when the inertial effect is neglected. The Prandtl number, Pr, represents a fluid property and characterizes the ratio of kinematic viscosity to thermal diffusivity of the fluid. For decreasing Pr and for a constant Ra, the inertial effect prevails and therefore natural convection of low-Pr fluids exhibits unstable oscillatory or turbulent behaviour. The Hartmann number, Ha, represents the square root of the ratio of the Lorentz to the viscous force. For an increasing Ha, the Lorentz force becomes dominant with respect to the viscous one, and therefore the viscous effect is confined to thin boundary layers. Besides, as Ha increases, the Lorentz force is also dominant to the inertial force ($N \gg 1$) and the effect of Pr becomes negligible. In such an inertialess case, the governing parameters are Ra and Ha. The Nusselt number, Nu, is used to reflect the non-dimensional rate of heat transfer.

When the effect of solutal buoyancy in crystal growth is considered, an additional buoyancy term should be included into the momentum equation, and the diffusion equation for mass transfer must be included into the analysis.

2.2 The effect of the magnetic field on the flow

The pioneering study by Hartmann on what is now known as the Hartmann flow, indicates that the Lorentz force acts to decelerate the core flow far from two parallel walls, and to accelerate boundary flows in the vicinity of walls perpendicular to the magnetic field if walls perpendicular to the electric current are electrically insulating. In general, when a magnetic field is imposed on the electrically conducting melt flow, the flow tends to be flat along the applied magnetic field except for the boundary layers near the walls perpendicular to the field. This type of the boundary layer is called the Hartmann layer. The importance of the Hartmann layer depends on the flow symmetry, direction of magnetic field, electrical conductivity of the wall, and its thickness [23,24].

One of the other types of magnetic fields applied to the crystal growth may be the rotating magnetic field. In case of a horizontal magnetic field, which rotates at a constant angular velocity around the central axis of the stationary crucible, the electrically conducting melt is rotated at the same speed as that of the field in the interior region. However, near the sidewall, a boundary layer exists, which becomes thinner as the Hartmann number increases. This stirring effect is due to the principle of the temporal variation of the magnetic field. The rotating magnetic field may be used to stir and to change the melt flow mode as an alternative method to the rotation of the crucible in the Czochralski method. The references on the use of the rotating magnetic field can be found in [25–27].

2.3 Solid–liquid interface

During the crystal growth, wall–melt interfaces and a crystal–melt interface exist. The effect of the electric conductance of a wall is as significant as the direction of the applied magnetic field. Usually as the electric conductance of a wall increases, the flow damping effect due to the Lorentz force becomes remarkable. Such studies may be found in [28,29].

The electrical conductivity of semiconductor crystal is much smaller than that of melt, and the crystal–melt interface is usually assumed to be electrically insulating. However, it should be mentioned that when the crystal volume is large, a certain amount of electric current passes in the crystal and this may change the electric current field in the melt. As a consequence, the change in the electric current in the melt causes the change in the flow mode itself in the presence of a magnetic field.

Another important factor during the crystal growth under the magnetic field is attributed to the Seebeck effect. When two different metals (conducting media) are jointed in the presence of a temperature gradient, a thermoelectric current is induced in the media owing to the difference of the absolute thermoelectric power, which is called the Seebeck effect. The absolute thermoelectric power is a function of temperature. It shows steep change between solid and liquid phase with respect to the melting temperature. Hence, additional electric current due to the Seebeck effect would occur and affect the flow field as well in a magnetic field. There are some references related to the thermoelectric MHD [30–32].

2.4 Rotation effect

During the process of Czochralski method, the crystal and the crucible are rotated to produce crystals as homogeneous as possible. The rotation of crucible causes both the centrifugal force and the Coriolis force, which change the melt flow mode caused by buoyancy. The centrifugal force is conservative, similar to gravity, when the fluid is homogeneous. However, the melt has local variation of fluid density, which may cause an additional buoyancy force. The Coriolis force makes the flow tend to be uniform along the axis of rotation by the principle of the Taylor–Proudman theory and may suppress the meridional flow.

Concerning the combined convection due to buoyancy and rotation in the Czochralski method, the flow inside the crucible is combined by both natural and forced convection. When a uniform axial magnetic field is imposed on the Czochralski system, the radial flow caused by buoyancy is efficiently damped out by the Lorentz force, while the azimuthal flow is not because of the existence of electric current within the Hartmann layer at the bottom of the rotating crucible. On the other hand, when a horizontal transverse magnetic field is applied to the system, it is difficult to describe this phenomenon clearly, since the flow is 3D and the phenomenon depends on the balance of intensities of the natural convection, forced convection, and magnetic field. Such a detailed numerical analysis for the transverse magnetic field is given, for instance, in [33].

2.5 Free surface

At the liquid–gas interface, heat is mainly transferred by radiation from the melt surface, and interfacial tension takes place. The combination of the two at the interface causes Marangoni convection. As mentioned in many references, the Marangoni convection arises even in the absence of gravity. By applying a magnetic field, the Lorentz force acts to affect the convection and indirectly the temperature field. The reference related to magnetohydrodynamic (MHD) Marangoni convection can be found, for instance, in [34].

3 Semiconductor crystal growth in various magnetic fields

Magnetic fields have recently been employed in manufacturing single crystals by the Czochralski crystal growth, the floating zone, and the Bridgman methods, and others. Magnetic fields have also been employed in continuous steel casting processes to improve the quality of products and the safety of operation of the system. In these processes, magnetic fields are applied to liquid metals (electrically conducting fluids) to be solidified as described in previous chapters and they are expected to suppress the disturbance of convection by means of the Lorentz force. The convection of liquid metal is known to be oscillatory but suppressible by a magnetic field, as demonstrated by Chandrasekhar [35] theoretically and by Nakagawa [36] experimentally. Details are given in Chandrasekhar's book [37]. The Prandtl number of a liquid metal is of the order of 0.01, and convection becomes oscillatory. This characteristic has been employed for material processing as described in this chapter.

Chedzey and Hurle [38] applied a magnetic field of 0.25 T to a tellurium-doped InSb melt and observed a decay of temperature fluctuation of 5°C amplitude. Utech and Fleming [39] reported the application of a magnetic field of 0.175 T in the crystal growth of InSb with a Bridgman boat system. The amplitude of the temperature oscillation was suppressed. Witt et al. [40] appear to be the first to have reported an application of a transverse magnetic field to the Czochralski crystal growth process of InSb for a field of 0.4 T. The amplitude of temperature oscillation decreased but with additional higher order of fluctuation. These pioneering works appear to have proved the effectiveness of a magnetic field in calming the oscillatory convection in crystal growing systems. The first industrial application of a magnetic field to Czochralski crystal growth of silicon was carried out by Hoshi et al. [41, 42]. They needed good quality silicon for manufacturing charge-coupled device (CCD) cameras. They succeeded in controlling oxygen concentration, defect formation, and crystal growth rate with a transverse magnetic field. However, the magnetic Czochralski system did not develop as expected after this. The reasons stated at crystal growing factories included the fact that the strong magnetic field for practical growth process requires expensive superconducting magnets, while the floating zone crystal growing process could anyway provide an oxygen-free silicon crystal rod, since it does not employ a SiO_2 crucible. For various reasons, whether stated or not, the practical use of a magnetic field appears to have been delayed until recently. The production of large silicon crystal rods of 16 in. or so in diameter, however, appears to need magnetic control. The required large-scale crucible induces turbulent natural convection of molten silicon, since the Grashof number increases in proportion to the cube of geometrical size. For such a crucible the melt surface may become wavy and therefore inhibit the initial seeding process and various subsequent operations. The following is a rather limited review of the application of a magnetic field to the crystal growth processes.

3.1 Transverse magnetic field applied to Czochralski crystal growth

Suzuki et al. [43] reported that with a transverse magnetic field of 0.15 T, thermal convection was suppressed to give crystals with oxygen concentration from 10^{18} to 10^{17} atoms/cm^3 and resistivity of about 380 Ω-cm. Kobayashi [44] performed theoretical analysis on the solute distribution in Czochralski crystals grown in either an axial or a transverse magnetic field. Unlike the axial magnetic field, which was predicted to suppress the forced convection in the radial direction resulting from the rotation of the crystal rod, the transverse magnetic field does not affect the centrifugal forced convection. The melt motion in Czochralski crucible involves an intrinsically 3D convection, and the effect on it of a weak transverse magnetic field was studied by Williams et al. [45] employing the axisymmetric base solution.

More recently, Ozoe and Iwamoto [46] carried out fully 3D numerical analysis with a rotating crucible and a transverse magnetic field. They employed a reasonable treatment developed by Ozoe and Toh [47] to obtain the radial velocity component at the cylinder axis, rather than assuming it was zero, as did Michelcic et al. [48]. Akamatsu et al. [33] obtained an elliptic temperature profile at the top of a crucible with its major axis parallel and its minor axis perpendicular to the transverse magnetic field in the absence of any rotation of the crystal or the crucible. This was supported by the elliptic cross-sectional crystal rod grown up for LEC GaAs by Kajigaya et al. [49]. Krauze et al. [50] recently reported similar shapes of isotherms using 3D and turbulent models.

The number of papers published since 1999 in the *Journal of Crystal Growth* are listed in Table 1 to show the general trends. Fully 3D numerical analyses have become widespread, with or without turbulence models. Comparison of axial and transverse magnetic fields has also been studied.

3.2 Axial magnetic field applied to Czochralski crystal growth

Hoshikawa [51] proposed using an axial (vertical) magnetic field for Czochralski crystal growth, in contrast to the previously discussed transverse one. He mentioned that the problem with the transverse magnet is its large size and heavy weight, and that there is thermal asymmetry at the growth interface, which causes difficulty in crystal shape control, and the generation of periodic rotational striations of impurities. An axial magnetic field of 0.1 T was

Table 1. Number of papers published in JCG since 1999

Transverse magnetic field in Cz	11
Axial magnetic field in Cz	8
Cusp-shaped magnetic field in Cz	11
Bridgman crystal growth system	30
Floating zone crystal growth with magnetic field	10
Strong magnetic field and others for growth of protein	16

provided by an 80 kg solenoid to suppress convection. Growth striations and microscopic phosphorous dopant heterogeneity in the silicon rod decreased below the limit of detection. This solenoid did not require the rearrangement of the Cz furnace. This axial magnetic field allows rigorous axially symmetric, two-dimensional (2D) modeling of the melt flow, and many reports can be found. For example, Langlois and Lee [52] pointed out that the induced magnetic field can be neglected and that the electric field is irrotational, and they computed streamlines of the meridional circulation. Hurle and Series [53] reported the effective distribution coefficient in a steady axial magnetic field. Organ [54] mentioned that the effect of an axial magnetic field is to enhance the azimuthal motion in the lower part of the crucible, and that the stirring effect due to crystal rotation penetrates to a much greater depth in the melt. Hjellming and Walker [55] performed linear analysis of the melt motion with an axial magnetic field and found the effect of electrical conductivity of a crystal rod in determining the flow. Series [56] measured the effects of an axial magnetic field on the incorporation of interstitial oxygen, carbon and phosphorous into silicon crystals. Hicks and Riley [57] examined the boundary layer flow at the crystal–melt interface under either an axial or a radial magnetic field. More recently, Fukui et al. [58] studied the effect of an axial magnetic field and found that the fluid column under a rotating crystal rod starts to rotate with the application of an axial magnetic field for a static crucible. On the other hand, the rotating fluid column under a rotating rod in a reversely rotating crucible stops rotating with the application of a magnetic field. By numerical analysis these effects were found to be due to the Lorentz force.

3.3 Cusp-shaped magnetic field applied to Czochralski crystal growth

Hirata and Hoshikawa [59,60] proposed a cusp-shaped magnetic field as shown in Fig. 2c for the Czochralski crystal growth process to overcome various shortcomings of the axial magnetic field. The axial magnetic field is perpendicular to the crystal–melt interface and suppresses the desirable rotational forced convection. Furthermore, the previous know-how for the operation of crystal rotation can not be employed [61]. The diffusion layer near the melt-free surface is thickened by the vertical magnetic field, causing the oxygen concentration in the bulk melt to increase toward the saturation level. These difficulties forced them to adopt the cusp-shaped magnetic field employed in nuclear engineering. They employed two superconducting magnets with counter-directional electric currents placed coaxially to the crucible to produce a cusp-shaped magnetic field with a free surface near the centre of the cusp. This cusp-shaped magnetic field is axially symmetric and orthogonal to the whole melt–crucible interface, and it suppresses the boundary layer flow along the crucible wall and decreases the dissolution rate of oxygen from the wall. It also results in the absence of the magnetic field near the crystal

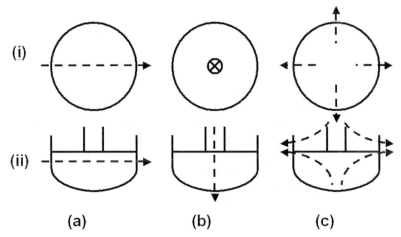

(i) Top view of a crucible,   (ii) Side view of a crucible
(a) Transverse magnetic field, (b) Axial magnetic field,
(c) cusp-shaped magnetic field.

Fig. 2. Three types of magnetic field applied in Czochralski crystal growth systems

growing interface, which allows free flow of the melt near the interface. The oxygen concentration could be controlled from 10^{18} to 2×10^{17} atoms/cm^3 in the crystal rod. The radial and axial oxygen distribution became homogeneous. Series [62] reported the use of a shaped magnetic field for similar reasons. He called it symmetrical internal split pair Czochralski (SISP-CZ). His system appears to be almost the same as the cusp-shaped magnetic system reported by Hirata and Hoshikawa [59, 60]. Furthermore, according to Hoshikawa and Hirata [61], this cusp-shaped magnetic field system has been patented by Nanishi et al. [63] for the growth of GaAs.

3.4 Bridgman crystal growth system with a magnetic field

Kim [64] reported the suppression of thermal convection by a transverse magnetic field for the melt of InSb in a vertical column of 1.1 cm in diameter and 4 cm in length. Temperature of the InSb melt was measured. At magnetic induction $B = 0$, temperature fluctuated extensively with amplitude of 15°C or so. From $B = 0.09$ to 0.166 T, the temperature fluctuated in a regular oscillatory manner with amplitude from 6 to 1°C. At $B = 0.169$ T, the oscillation was suppressed and striation-free crystal growth was achieved. Baumgartl and Muller [65] compared three different complexities of MHD models for zone-melting crystal growth and reported a flow mode map of Ha versus Ra_w from the experimental results. They concluded that for the steady state or for slowly varying flow, the MHD2 model is enough to ensure that the fluid flow does not disturb the externally applied magnetic field, while the

Lorentz force still acts. Bridgman type crystal growth has been extensively studied, as the number of reports listed in Table 1 shows. The magnetic fields are either static, rotating, traveling, alternating and/or time dependent. The direction of the magnetic field is either transverse or axial. Recently, a strong magnetic field has been applied for crystal growth of compound materials and proteins as well as levitation and chemical vapour deposition.

3.5 Floating zone crystal growth system with a magnetic field

This system was reported, for example, by Danilewsky et al. [66]. Heterogeneity of the dopant due to the time-dependent natural convection was suppressed by an axial magnetic field. The floating zone crystal growth system has been widely studied with static axial, transverse, or rotating magnetic field.

References

1. Szmyd JS, Suzuki K (2000) Introduction to crystal growth processes from the melt. In: Szmyd JS, Suzuki K (eds) Modelling of Transport Phenomena in Crystal Growth. WITPress, Southampton, pp 1–18
2. Czochralski J (1917) Ein neues Verfahren zur Messung der Kristallisationsegeschwindigkeit der Metalle. Z Phys Chem 92:219–221
3. Teal GH, Little JB (1950) The growth of germanium single crystals. Phys Rev 78:647
4. Dash WC (1959) Growth of silicon crystals free from dislocations. J Appl Phys 30:459–474
5. Prasad V, Zhang H (1997) Transport phenomena in Czochralski crystal growth processes. Advances in Heat Transfer, Vol 30, Academic Press, New York, pp 313–435
6. Szmyd JS, Suzuki K (2000) Transport phenomena during growth of superconducting materials by Czochralski method. In: Szmyd JS, Suzuki K (eds) Modelling of Transport Phenomena in Crystal Growth. WITPress, Southampton, pp 279–321
7. Brown RA (1988) Theory of transport processes in single crystal growth from the melt. AIChE J 34:881–911
8. Shimura F (1989) Semiconductor Silicon Crystal Technology. Academic Press, Orlando
9. Hurle DTJ (1993) Crystal Pulling from the Melt. Springer-Verlag, Berlin
10. Hurle DTJ, Cockayne B (1994) Czochralski growth. In: Hurle DTJ (ed) Handbook of Crystal Growth. Elsevier Science BV, Amsterdam 2a:99–211
11. Dupret F, van den Bogaert N (1994) Modelling Bridgman and Czochralski growth. In: Hurle DTJ (ed) Handbook of Crystal Growth Vol 2b. Elsevier Science BV, Amsterdam, pp 875–1010
12. Theuerer HC (1952) US patent No 3 060 123
13. Keck PH, Golay MJE (1953) Crystallization of silicon from a floating liquid zone. Phys Rev 89:1297
14. Pfann WG (1966) Zone Melting. Wiley, New York

15. Bohm J, Lüdge A, Schröder W (1994) Crystal growth by floating zone melting. In: Hurle DTJ (ed) Handbook of Crystal Growth Vol 2a. Elsevier Science BV, Amsterdam, pp 214–257

16. Kapitza P (1928) The study of the specific resistance of bismuth crystals and its change in strong magnetic fields and some allied problems. Proc Roy Soc A 119:358–443

17. Pfann WG (1952) Principle of zone-melting. Trans AIME 194:747–753

18. Pfann WG (1962) Zone melting. Science 135:1101–1109

19. Ozoe H (2000) Effect of a magnetic field in Czochralski silicon crystal growth. In: Szmyd JS, Suzuki K (eds) Modelling of Transport Phenomena in Crystal Growth. WITPress, Southampton, pp 201–237

20. Ozoe H, Okada K (1989) The effect of the direction of the external magnetic field on the three-dimensional natural convection in a cubical enclosure. Int J Heat Mass Transfer 32:1939–1954

21. Okada K, Ozoe H (1992) Experimental heat transfer rates of natural convection of molten gallium suppressed under an external magnetic field in either the x, y, or z direction. J Heat Transfer 114:107–114

22. Garandet JP, Alboussière T, Moreau R (1992) Buoyancy driven convection in a rectangular enclosure with a transverse magnetic field. Int J Heat Mass Transfer 35(4):741–748

23. Alboussière T, Garandet JP, Moreau R (1996) Asymptotic analysis and symmetry in MHD convection. Phys Fluids 8(8):2215–2226

24. Bühler L (1998) Laminar buoyant magnetohydrodynamic flow in vertical rectangular ducts. Phys Fluids 10(1):223–235

25. Barz RU, Gerbeth G, Wunderwald U, Buhrg E, Gelfgat YuM (1997) Modelling of the isothermal melt flow due to rotating magnetic fields in crystal growth. J Crystal Growth 180:410–421

26. Ben Hadid H, Vaux S, Kaddeche S (2001) Three-dimensional flow transitions under a rotating magnetic field. J Crystal Growth 230:57–62

27. Walker JS, Martin Witkowski L, Houchens BC (2003) Effects of a rotating magnetic field on the thermocapillary instability in the floating zone process. J Crystal Growth 252:413–423

28. Tagawa T, Ozoe H (1998) The natural convection of liquid metal in a cubical enclosure with various electro-conductivities of the wall under the magnetic field. Int J Heat Mass Transfer 41:1917–1928

29. Molokov S, Bühler L (2003) Three-dimensional buoyant convection in a rectangular box with thin conducting walls in a strong horizontal magnetic field. Forschungszentrum Karlsruhe Report FZKA 6817

30. Shercliff JA (1979) Thermoelectric magnetohydrodynamics. J Fluid Mech 91(2):231–251

31. Moreau R, Lasker O, Tanaka M (1996) Thermoelectric and magnetohydrodynamic effects on solidifying alloys. Magnetohydrodynamics 32(2):173–177

32. Kaneda M, Tagawa T, Ozoe H (2002) Natural convection of liquid metal in a cube with Seebeck effect under a magnetic field. Int J Transport Phenom 4(3):181–191

33. Akamatsu M, Higano M, Ozoe H (2001) Elliptic temperature contours under a transverse magnetic field computed for a Czochralski melt. Int J Heat and Mass Transfer 44 3253–3264

34. Maekawa T, Tanasawa I (1988) Natural convection driven by buoyancy and surface tension forces under external magnetic filed. Adv Space Res 8(12): 215–218

35. Chandrasekhar S (1952) XLVI. On the inhibition of convection by a magnetic field. Phil Mag 7 43:501–532

36. Nakagawa Y (1955) An experiment on the inhibition of thermal convection by a magnetic field. Nature 175:417–419

37. Chandrasekhar S (1961) Hydrodynamic and Hydromagnetic Stability. Oxford University Press, Oxford

38. Chedzey HA, Hurle DTJ (1966) Avoidance of growth-striae in semiconductor and metal crystals grown by zone melting techniques. Nature 210:933

39. Utech HP, Flemings MC (1966) Elimination of solute banding in indium antimonide crystals by growth in a magnetic field. J Appl Phys 37(5):2021–2024

40. Witt AF, Herman CJ, Gatos HC (1970) Czochralski-type crystal growth in transverse magnetic fields. J Mater Sci 5:822–824

41. Hoshi K, Suzuki T, Okubo Y, Isawa N (1980) Cz silicon crystal grown in transverse magnetic fields. Electrochem Soc Ext Abstr, St. Louis, 324:811–813

42. Hoshi K, Isawa N, Suzuki T (1984) Growth of silicon monocrystals in a magnetic field. Oyobutsuri (Applied Physics, in Japanese) 53(1):38–41

43. Suzuki T, Isawa N, Okubo Y, Hoshi K (1981) Cz silicon crystals grown in a transverse magnetic field. Semiconductor Silicon, Electrochemical Society, Pennington, pp 90–100

44. Kobayashi S (1986) Effects of an external magnetic field on solute distribution in Czochralski grown crystals-A theoretical analysis. J Crystal Growth 75:301–308

45. Williams MG, Walker JS, Langlois WE (1990) Melt motion in a Czochralski puller with a weak transverse magnetic field. J Crystal Growth 100:233–253

46. Ozoe H, Iwamoto M (1994) Combined effects of crucible rotation and horizontal magnetic field on dopant concentration in a Czochralski melt. J Crystal Growth 142:236–244

47. Ozoe H, Toh K (1998) A technique to circumvent a singularity at a radial center with application for a three-dimensional cylindrical system. Numer Heat Transf B 33:355–365

48. Mihelcic M, Wingerath K, Pirron Chr (1984) Three-dimensional simulations of the Czochralski bulk flow. J Crystal Growth 69:473–488

49. Kajigaya T, Kimura T, Kadota Y (1991) Effect of the magnetic flux direction on LEC GaAs growth under magnetic field. J Crystal Growth 112:123–128

50. Krauze A, Muiznieks A, Mühlbauer A, Wetzel Th, Tomzig E, Gorbunov L, Pedchenko A, Virbulis J (2004) Numerical 3D modeling of turbulent melt flow in a large CZ system with horizontal DC magnetic field. II. Comparison with measurements. J Crystal Growth 265:14–27

51. Hoshikawa K (1982) Czochralski silicon crystal growth in the vertical magnetic field. Japanese J Appl Phys 21(9):L545–L547

52. Langlois WE, Lee K-J (1983) Czochralski crystal growth in an axial magnetic field: Effects of joule heating. J Crystal Growth 62:481–486

53. Hurle DTJ, Series RW (1985) Effective distribution coefficient in magnetic Czochralski growth. J Crystal Growth 73:1–9

54. Organ AE (1985) Flow patterns in a magnetic Czochralski crystal growth system. J Crystal Growth 73:571–582

55. Hjellming LN, Walker JS (1986) Melt motion in a Czochralski crystal puller with an axial magnetic field: isothermal motion. J Fluid Mech 164:237–273

390   H. Ozoe et al.

56. Series RW (1989) Czochralski growth of silicon under an axial magnetic field. J Crystal Growth 97:85–91
57. Hicks TW, Riley N (1989) Boundary layers in magnetic Czochralski crystal growth. J Crystal Growth 96:957–968
58. Fukui H, Kakimoto K, Ozoe H (1998) The convection under an axial magnetic field in a Czochralski configuration. Adv Comput Method Heat Transfer V: 135–144
59. Hirata H, Hoshikawa K (1989) Silicon crystal growth in a cusp magnetic field. J Crystal Growth 96:747–755
60. Hirata H, Hoshikawa K (1989) Homogeneous increase in oxygen concentration in Czochralski silicon crystals by a cusp magnetic field. J Crystal Growth 98: 777–781
61. Hoshikawa K, Hirata H (1991) Oxygen concentration control and magnetic field application technique in Czochralski silicon crystal growth. Oyobutsuri (Applied Physics, in Japanese) 60(8):808–812
62. Series RW (1989) Effect of a shaped magnetic field on Czochralski silicon growth. J Crystal Growth 97:92–98
63. Nanishi Y, Tada K, Nakai T (1983) Japanese Patent Syowa 58–217493
64. Kim KM (1982) Suppression of thermal convection by transverse magnetic field. J Electrochem Soc 129(2):427–429
65. Baumgartl J, Müller G (1992) Calculation of the effects of magnetic field damping on fluid flow-Comparison of magnetohydrodynamic models of different complexity. In: Proceedings of the 8th European Symposium on Materials and Fluid Sciences in Microgravity, Brussels, ESA SP, pp 333
66. Danilewsky AN, Dold P, Benz KW (1992) The influence of axial magnetic fields on the growth of III–V semiconductors from metallic solutions. J Crystal Growth 121:305–314

An Outline of Magnetoelectrochemistry

Antoine Alemany[1] and Jean-Paul Chopart[2]

[1] Pamir team, LEGI, 1025 rue de la Piscine, Domaine universitaire de Saint-Martin d'Hères, France (`Antoine.Alemany@hmg.inpg.fr`)

[2] DTI, UFR Sciences, BP 1039, REIMS Cedex 2, France (`jp.chopart@univ-reims.fr`)

1 Introduction

Magnetoelectrochemistry (MEC) is electrochemistry in the presence of an imposed magnetic field. This relatively new branch of electrochemistry has seen rapid development during the last years [1], the potential applications being very promising even if not industrially realized up to now. Several studies have been performed with the objective to elucidate the effect of a magnetic field on the electrolyte properties, on the mass transfer processes and, at a smaller scale, on the electrochemical kinetics and on the structure and quality of the deposit.

Fahidhy [2] and Ulrich and Steiner [3] summarise the progress starting from the first observations of Faraday. Our intent here is to give an overview of the most important points developed in previous studies, to propose a simplified way to approach the mass transfer calculation essentially for the case when a supporting electrolyte is used assuring the electrical conductivity of the bath, and to discuss the main objectives for future research.

Roughly speaking, the role of a magnetic field on the electrochemical phenomena can be divided into three main categories. The *first* one concerns the effect of an external field on the main properties of the bath. The *second* one, studied most extensively, is focused on the influence of an external magnetic field on the mass transfer phenomena. The *third* category at a smaller scale includes the influence of the magnetic field on both the electrochemical kinetics and the structure and quality of deposits.

2 Influence of the magnetic field on the electrolyte properties

An electrolyte is a solution more or less dissociated in different charged particles, the ions. When these particles are submitted to an *electric field* $\mathbf{E}$, they

S. Molokov et al. (eds.), Magnetohydrodynamics – Historical Evolution and Trends,
391–407. © 2007 *Springer.*

experience the action of an electrostatic force in the form:

$$\mathbf{F} = q\mathbf{E}, \tag{1}$$

where q represents the electric charge of the particle. When a *magnetic field* $\mathbf{B}$ is superimposed to the electric field, the particles are subjected, to a new force in the form:

$$\mathbf{F} = q(\mathbf{E} + \mathbf{v}_p \times \mathbf{B}), \tag{2}$$

where $\mathbf{v}_p$ is the velocity of the particle. The second term on the right-hand side of Eq. (2), representing the deviation from the linear trajectory, can be associated with the Hall effect. As the trajectory of the particles deviates, which results from the interaction between the two fields, the properties of the electrolyte are bound to be modified.

2.1 Conductivity

The electrical conductivity of an electrolyte under an applied magnetic field is difficult to identify. Depending on the magnetic field intensity, the measurements show that the electrical conductivity increases by a factor of 1.04–1.2 with respect to the conventional one [2].

Later, using a method analogous to the kinetic gas theory, Tronel-Peyroz and Olivier [4] have calculated the transport coefficient in an ionic solution in the steady state in the presence of electric and magnetic fields. Using the Boltzmann equation they have demonstrated that the ionic mobility and the diffusion coefficients become tensor quantities, which depend on the magnetic field. The matrix that represents the ionic mobility of the species is:

$$\bar{u} = \begin{pmatrix} u_T & u_H & 0 \\ -u_H & u_T & 0 \\ 0 & 0 & u_\| \end{pmatrix}, \tag{3}$$

where $u_T = \frac{u}{1+u^2\Gamma^2B^2}$ is the transverse mobility, $u_H = \frac{u^2\Gamma B}{1+u^2\Gamma^2B^2}$ is the Hall mobility, $u_\| = u$ is the conventional mobility in the absence of the magnetic field, and Γ is a factor depending on the kinetic energy of the ion [4].

The matrix of diffusion can be deduced from the mobility by the following relation:

$$\bar{D} = \frac{KT}{q}\bar{u}, \tag{4}$$

where K represents the Boltzmann constant and T the bath temperature. The authors estimate that for a KCl solution containing the ferro-ferricyanide, and for a magnetic induction less that 1 T, and for $\Gamma \approx 10^3$ and $u \approx 10^{-7}\mathrm{m}^2/\mathrm{s}$, one gets $u_H \approx u^2\Gamma B$ and $u_T \approx u$. This gives, neglecting the diffusion, the following expression for the current induced by the electric field (the migration current):

$$\mathbf{J} = \sigma(\mathbf{E} + u\Gamma\mathbf{E} \times \mathbf{B}). \tag{5}$$

2.2 Diffusivity

Concerning the diffusivity, very few measurements devoted to the influence of the magnetic field are accessible. Nevertheless, it appears that its variation is relatively small and depends on the considered species.

2.3 Viscosity, temperature

In the same way, the relative increase of the viscosity measured under magnetic field of 1 T decreases when the species concentration increases. Some thermal effects have also been observed by Tronel-Peyroz and Olivier [5]. A local modification of the bath temperature has been measured when the electrolyte is subject to both an electric and a magnetic field perpendicular to each other.

3 Influence of the DC magnetic field on the mass transport

The rate of mass transfer inside an electrolyte can be significantly modified by an external magnetic field. This is generally attributed to the effect of the electromagnetic force, which modifies the existing flow (e.g., in the case of continuous electrodeposition), or induces a fluid motion in an otherwise quiescent fluid.

Generally (but this has to be considered carefully), the magnetic field effect on mass transport consists in increasing the limiting current. The existence of a critical value of the magnetic field intensity beyond which the diffusional limit disappears has also been observed.

The explanation about the increase in the limiting current when the magnetic field increases is of the same nature as for the phenomena observed on the rotating electrode when the rotation increases.

Let us consider the mass transfer on a plane surface, a rotating electrode for example (Fig. 1). By improving the voltage between the anode and the cathode, the electric current intensity J, and consequently the mass transfer, increases. For a fixed voltage the rate of mass transfer can be evaluated using the Nernst approximation, which gives:

$$J \approx \alpha \frac{C_b - C_w}{\delta_d}, \tag{6}$$

where α is a constant, which depends on the nature of the electrolyte, C_b and C_w are concentrations of the species in the bulk of the solution and at the electrode wall, respectively, and δ_d is the diffusion layer thickness. In the case of a high Schmidt number, which usually takes a value greater than 1,000 in the electrolyte, δ_d is generally controlled by the thickness of the hydrodynamic boundary layer, δ_h (it is the case, for example, on the rotating electrode), the typical ratio between both being: $\delta_h/\delta_d = Sc^{1/3}$.

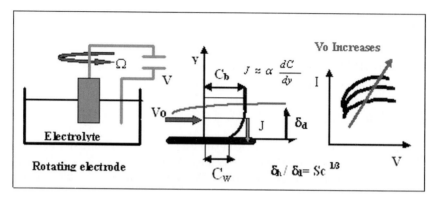

Fig. 1. The rotating electrode experiment

Considering a given rotation velocity with typical value V_0, when the voltage increases the intensity increases also and if the hydrodynamics is fixed (δ_d is imposed), the only possibility to follow this increase consists in a decrease of the concentration of the electroactive species, C_w. When C_w reaches 0, the current is fixed at a value that depends on the hydrodynamics of the bath: this is the diffusion limit.

For a fixed voltage, when the velocity increases, the hydrodynamic boundary layer, and consequently the diffusion layer, decrease corresponding to an improvement of the diffusion limit. At the first order in the case of diffusion control, according to Levich [6], the rate of mass transfer on the rotating electrode, characterised by the Sherwood number, Sh, depends on the Reynolds Re and Schmidt Sc numbers, and is governed by the classical correlation:

$$Sh = K Re^{1/2} Sc^{1/3}, \tag{7}$$

where the constant K depends on the flow configuration.

Let us now consider the situation where a magnetic field is superimposed on the process. It is supposed that initially the electrolyte is at rest and the mass transfer is governed only by diffusion. The electromagnetic force, which results from the interaction between the magnetic field and the electric current, gives rise to a fluid motion and to hydrodynamic boundary layers, which control the mass transfer. The same mechanism as that explaining classical electrochemistry can be invoked to explain the evolution of the current density versus the magnetic field intensity: increasing this intensity results in an increase of the fluid velocity and consequently in an increase of the limiting current (Fig. 2). If the magnetic field intensity becomes very high, the flow can reach the turbulent condition.

Of course the efficiency depends on the orientation of the magnetic field. The most important effect corresponds to an applied magnetic field being parallel to the electrode (perpendicular to the electric current).

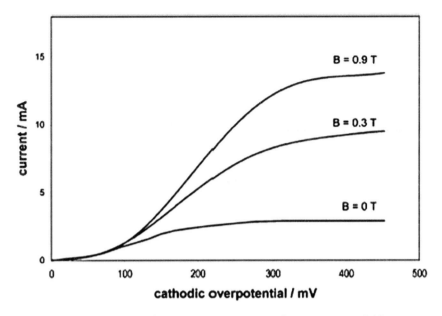

Fig. 2. Increase of the limiting current under a magnetic field

3.1 Simple theoretical model

Two main types of problems can be considered depending on the origin of the flow motion. It can be imposed by an external device (forced convection problem) or can result from the natural convection. In such a case it is the variation of the concentration of electroactive species that generates natural convection controlled by the electrochemical coefficient of expansion β [7]:

$$\beta = \frac{d\rho}{dC_{\text{EA}}}, \tag{8}$$

where subscript EA stands for electro-active species and ρ stands for the solution density. Finally, the flow motion controls the mass transfer processes.

In both cases the superimposed electromagnetic forces are able to completely change the structure of the flows that can be partially or mainly controlled by the magnetic field effect for sufficiently high values of the magnetic field.

The explanation of the phenomena and the governing equations are based on the hypothesis that the electrical conductivity of the bath is imposed by a supporting electrolyte, which is the case in the industrial cells. In this case the distribution of the electric current (Ohm's law in electrochemistry) inside the bath takes the following simplified expression [8]:

$$\mathbf{J} = (zFD)_{\text{EA}}\nabla C_{\text{EA}} - \sigma\nabla\Phi_b, \tag{9}$$

where z is the number of electrons exchanged with the electrode, F is the Faraday constant, D is the diffusion coefficient, σ is the conductivity of the bath, and Φ_b is the electric potential of the bath. This expression supposes that (i) the term $\mathbf{V} \times \mathbf{B}_0$ ($\mathbf{V}$ being local velocity) of the classical Ohm's law in magnetohydrodynamics (MHD) is negligible compared to the imposed voltage drop inside the bath; this is the case for most of the applications for which the ratio of the two terms, $|\mathbf{V} \times \mathbf{B}|/|\nabla \Phi_b|$, is of order 10^{-2}, (ii) the electrical conductivity is imposed by the supporting electrolyte, (iii) at the vicinity of the electrode the current is mainly controlled by the concentration gradient of the electroactive species, (iv) the Hall effect is neglected, (v) the magnetic Reynolds number is supposed to be very small so that the induced magnetic field is negligibly small.

Under these approximations the following set of non-dimensional equations can be used for describing the mass transfer processes in MEC [8]:

$$\nabla \cdot \mathbf{V} = 0, \tag{10}$$

$$\frac{D\mathbf{V}}{dt} = -\nabla P^* + \frac{1}{Re}\nabla^2 \mathbf{V} - \frac{M_d}{Re^2 Sc}\nabla f \times \mathbf{B}_0, \tag{11}$$

$$\frac{DC}{dt} = \frac{1}{ReSc}\nabla^2 C, \tag{12}$$

$$\Delta f = 0. \tag{13}$$

These are continuity, Navier–Stokes and transport equations (in which C represents the concentration of the electroactive species), and the equation controlling the current density, respectively. The above equations are written in dimensionless variables using a proper choice of the kinetic and dynamic variables. The term P^* includes gravity and pressure, and the dimensionless function f represents the apparent potential, which depends on the concentration distribution of the electroactive species and on the electric potential. The expression for function f is:

$$f = \frac{zFD}{\sigma V_0 B_0 L_0}C + \frac{\Phi_b}{V_0 B_0 L_0}. \tag{14}$$

In Eq. (14), L_0 and B_0 are typical scales of the length and the induction of the magnetic field, respectively.

The dimensionless electric current distribution can easily be deduced form this expression as follows:

$$\mathbf{J} = \nabla f. \tag{15}$$

The continuity of the electric current reduces the distribution of the apparent potential f to the solution of the Laplace equation. When the mass transfer is controlled by the diffusion layer, the concentration gradient of electroactive species becomes very high in the vicinity of the electrode, and the current density is mainly controlled by this term. To the contrary, it is the imposed electric field that controls the current density in the bath.

It can be observed that the presence of the magnetic field in the governing equations introduces a strong coupling between the Navier–Stokes equation, which depends on the concentration field by the way of the current density distribution, and the convection–diffusion equation that depends on the velocity distribution. This coupling disappears in classical electrochemistry for which the Navier–Stokes equation is independent of the concentration equation.

The non-dimensional number, the so-called *magneto-diffusion parameter*, M_d, which appears in Eq. (1), is:

$$M_d = \frac{zFC_bB_0L_0^2}{\rho\nu}.$$

It can be noticed that $\frac{M_d}{ReSc}$ represents the ratio of the electromagnetic to the viscous forces. It is analogous to the square of the Hartmann number in classical MHD, while $\frac{M_d}{Re^2Sc}$ is similar to the interaction parameter.

It can be also noted that if the motion occurs in the presence of a magnetic field, the typical velocity, V_0, can be estimated from the equilibrium between the electromagnetic and inertia forces. This gives:

$$V_0 \approx \left[\frac{zFDC_bB_0}{\rho}\right]^{1/2}, \tag{16}$$

and consequently:

$$Re = \sqrt{\frac{zFDC_bB_0L_0^2}{\rho\nu^2}}. \tag{17}$$

If the difference of concentration ΔC between the bath and the electrodes generates a density variation, which controls the fluid motion, then the Reynolds number must be replaced by the electrochemical Grashof number:

$$Gr = \sqrt{\frac{g\beta\Delta CL_0^3}{\nu^2}}. \tag{18}$$

The above theoretical formulation of MEC has been tested in two typical experiments performed in Japan.

3.2 Experimental validation

The first experiments to identify the influence of an imposed magnetic field on the mass transfer processes have been performed by Ryoichi Aogaki [9] and Shigeru Mori [10]. Aogaki experimented with forced convection in the presence of the electromagnetic forces, while Mori tested the influence of the magnetic field on the natural convection.

The Aogaki experiment (Fig. 3) can be viewed as an electrochemical conducting pump. It involved a horizontal channel of rectangular cross section immersed inside an electrochemical bath. Without the magnetic field the mass

a: Working and counter
electrodes, b: lugging
capillarity, c: electrodes for
measuring main flow rate
d: streamlines

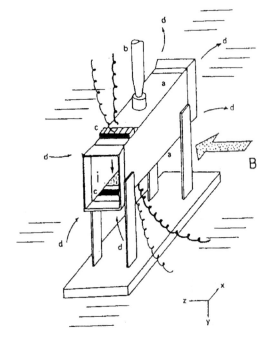

Fig. 3. The Aogaki experiment [9]

transfer is controlled by diffusion corresponding to a linear variation of electroactive species between the anode and the cathode. The electrochemical mechanism is consequently very slow. On the contrary, when the magnetic field is applied, a fluid motion induced by the electromagnetic forces takes place, which controls the mass transfer processes. Two types of channels were tested, both using the same electrolytic solution composed of 1 mol/L of sulfuric acid (H_2SO_4) and a varying concentration of copper sulfate ($CuSO_4$).

The first channel was small, of length 5 cm, width 0.1 cm, and height 1 cm. The second one was larger, of length 5 cm, width 2.5 cm, and height 2.5 cm.

The results obtained by numerical simulation [9] are partially given in Figs. 4 and 5. For the small channel (Fig. 4), they can be regrouped to be presented in the form of a linear dependence of the Sherwood number proportional to the limiting current versus the product of $C_b^{3/2} B_0^{1/2}$, corresponding to a linear dependence of the flow rate versus the product $C_b^{3/2} B_0^{3/2}$.

It can be seen that the theoretical results obtained by numerical simulation [8] present a good agreement with the Aogaki data, even if the values are greater than the experimental ones. This can be attributed to the numerical procedure based on the assumption of a two-dimensional (2D) flow, and which was not the case in the experiment.

For the large channel the agreement between the numerical simulation and the Aogaki results was almost perfect. It is shown in Fig. 5.

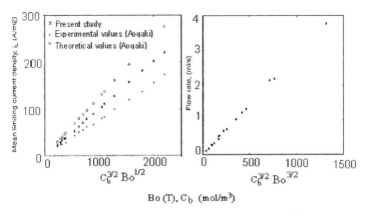

Fig. 4. Comparison between experimental [9] and numerical [8] data for the small channel

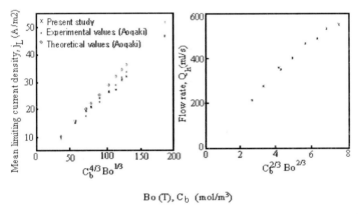

Fig. 5. Comparison between experimental data [9] and numerical one [8] for the large channel

It can be noticed that there is a difference with the previous case. The limiting current (the Sherwood number) is proportional to the product $C_b^{4/3} B_0^{1/3}$, corresponding to the flow rate, which varies as $C_b^{2/3} B_0^{2/3}$. This difference can easily be explained. In the first, narrow channel the two boundary layers, developing along the anode and cathode join at a short distance from the inlet. Thus, in most of the channel the flow is hydrodynamically established. The phenomenological analysis of the situation performed by Ngo Boum and Alemany [8] fits well with both numerical and experimental results. For the second, wide channel the two boundary layers are separate over the whole length. This leads to a different dependence between the magnetic field and mass transfer, which is also phenomenologically explained in [8]. Concerning comparison between simulation and experiment, the agreement is better for

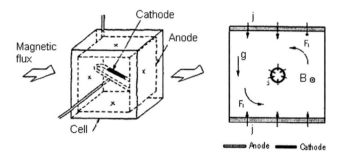

Fig. 6. The Mori experiment and the geometry of the associated numerical simulation [10]

the large channel than for the small one, as has already been noted above. This can be explained by the fact that the two-dimensionality of the flow is more pronounced in the large channel, where the boundary layers are smaller that the typical size of the channel.

The Mori experiment [10] has been performed to test the influence of the magnetic field on natural convection.

The schematic diagram of the Mori cells is shown in Fig. 6. The experimental set-up consists of a cylindrical cathode immersed inside a square box with two walls being the two anodes. The axis of the cathode can take different angles with respect to the direction of the horizontal magnetic field. The numerical simulation has been performed only for a cathode orientation parallel to the magnetic field (Fig. 6).

The comparison between the experimental results by Mori, and the numerical ones by Ngo Boum [9] shows a relatively good agreement for a weak magnetic field. The choice of the dimensionless parameter $M_d Ra^{1/2}/Sc$ (see Fig. 7), where Ra is the electrochemical Rayleigh number, has been justified by Mori by the fact that it combines the parameter used to characterise the natural convection with the dimensionless expression for the electromagnetic forces due to the diffusion current. This is the reason why the agreement between the experiment and the simulation is not too bad for small and moderate values of this parameter. It can also be noted that the numerical simulation agrees very well with the experimental Sherwood number obtained for no magnetic field. Moreover, when the magnetic field increases, the flow is completely controlled by forced convection induced by the Lorentz (Laplace) forces, which are tangential to the cathode. In this case, and because there is no renewal of the electroactive species, the diffusion layer is not controlled by the hydrodynamic boundary layer, and thus becomes very thick. In such a case, the mass transfer decreases when the magnetic field increases, explaining the form of the curve obtained numerically.

Another approach, based on the work by Mollet et al. [11], has been taken by Olivier and co-workers [12] with the aim to develop a method to study

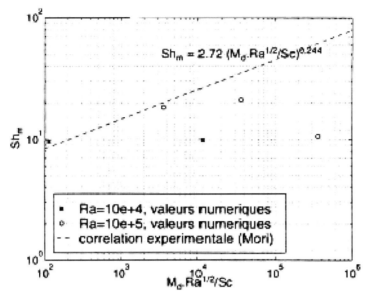

Fig. 7. Comparison between numerical and experimental results for the Mori experiment [10]

the Lorentz (Laplace) force effect. For this case, they have deduced a relation between the magnitude of the magnetic field and the current.

For limited diffusion at a circular microelectrode the current I_{lim} is:

$$I_{\text{lim}} = 0.678 F D^{2/3} C_b d^{5/3} \alpha^{1/3}, \tag{19}$$

where d is the diameter of the electrode and α is the velocity gradient at the wall. As the Lorentz (Laplace) force is proportional to B, the velocity gradient α is also proportional to B, while the current is proportional to $B^{1/3}$ [12]. It has been shown that α also depends on the nature of the electroactive species and the supporting electrolyte [13].

4 Magneto-electrolysis under AC magnetic field

To investigate any process, non-stationary perturbations can be applied during the investigated phenomenon. Such a dynamic method has been undertaken by two different methods. Aogaki and co-workers used cyclic magnetammetry (CM), which consists of a periodically changing magnetic field applied to an electrode under MHD control [14]. The current is measured during linear sweep of the magnetic field. The authors examine specific magnetic field effects on the reaction and diffusion processes excluding the usual MHD effect. They have shown typical differences between a totally mass-transport-controlled system (i.e., ferri-ferrocyanide system) and a mixed electrochemical response (i.e., copper electrodeposition) wherein a hysteresis effect of the magnetic field and the periodic current response have been observed.

Olivier et al. [15] perturbed a constant magnetic field by a small sinusoidal amplitude superimposed by varying magnetic fields. The sinusoidal modulation of the magnetic field ΔB generates a flow velocity modulation Δv_x, which creates a hydrodynamic perturbation and thus an electroactive species concentration gradient perturbation $\Delta(dC/dx)_{x=0}$. The latter involves an electrolytic current response ΔI, i.e.,

$$\Delta B \rightarrow \Delta v_x \rightarrow \Delta\alpha_y \rightarrow \Delta(dC/dx)_{x=0} \rightarrow \Delta I.$$

By measuring the current response at a constant applied potential they obtained a new MHD potentiostatic transfer function $(\Delta I/\Delta B)_E$ [15].

This transfer function is analogous to the classical electrochemical impedance. It allows in-depth mass transport analysis to be made. By this technique, it has been possible to elucidate the hydrogen evolution mechanism in nickel deposition from a Watts bath [16] and the diffusion process of zincate species in zinc electrodeposition from a basic medium [17].

5 Turbulence

The examples given in the previous paragraph are devoted to the laminar situation. Even in this case the problems are difficult to analyse numerically, taking into account that for high Schmidt numbers the diffusion layer is confined to an extremely small region at the walls. This necessitates the use of a very fine mesh at the vicinity of the electrode.

Concerning turbulence, the problem to be solved is extremely difficult for two main reasons.

The *first* one is also related to high values of the Schmidt number. In this case, small variations of velocity inside the viscous sub-layer, which in ordinary turbulent flows are neglected as far as the transfer of momentum is concerned, become dominant for mass transfer processes. The reason is that the diffusivity is generally 1,000 times lower than the kinematic viscosity. Consequently, the problem is very difficult to analyse both numerically and experimentally. The only numerical study [18] dealt with the problem using large eddy simulation.

The *second* reason is related to the influence of the electromagnetic forces, which are able to modify the characteristics of turbulence, and consequently the mass transfer processes. As usual, the few attainable results reveal an improvement in the transfer rate when a strong magnetic field is applied. However, up to now no interpretation of the magnetic field influence have been proposed for turbulent flow.

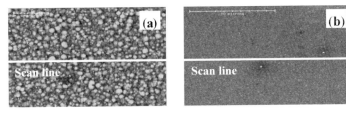

Fig. 8. Ni–Fe alloy electrodeposited at –1.5 V/ESS: (a) without magnetic field, (b) with a 0.9 T magnetic field parallel to the electrode [27]

The consequence is that the prediction of the mass transfer in magneto-electrochemistry in a turbulent regime is a completely open problem.

6 Influence of a magnetic field on electrochemical kinetics and on the structure and the quality of the deposit

The influence of the magnetic field effects on electrochemical kinetics is still subject to controversy. Many magnetic phenomena in magnetic fields of up to 1 T have been proclaimed relevant to such effects, only to be proved later to be induced by MHD convection [19, 20]. Few experiments have produced certain results that can be claimed as an evidence for magnetic effect on the electronic transfer kinetics. However, this question remains open and requires a much deeper insight before a definite conclusion is made.

In contrast, when metal electrodeposition is undertaken with a magnetic field superimposed on the electrochemical cell, the obtained materials may exhibit significant texture and surface morphology modifications. Many examples have been given in recent reviews [21, 22]. Electrochemical investigations for copper and cobalt electrodeposition under high magnetic fields have shown complex convective effects depending on the magnetic field orientation with respect to the electrode [23, 24]. During the electrodeposition of iron and cobalt, a magnetic field superimposed on a thin-layer cell induces typical arborescences that cannot be explained by classical MHD convection only [25, 26]. Physical insights by x-ray diffraction (XRD) and Scanning electron microscopy (SEM) analyses on nickel–iron alloys have highlighted modifications of the composition and the smoothness of the deposit [27] (Fig. 8).

For cobalt–iron alloy, superimposition of the magnetic field leads to a change in texture orientation without any modification of the composition of the deposit [28] (Fig. 9).

Finally, it has to be noted that certain effects during non-metallic compound electrodeposition, and chirality modifications have been observed during polyaniline electropolymerization under high magnetic fields [29] (Fig. 10).

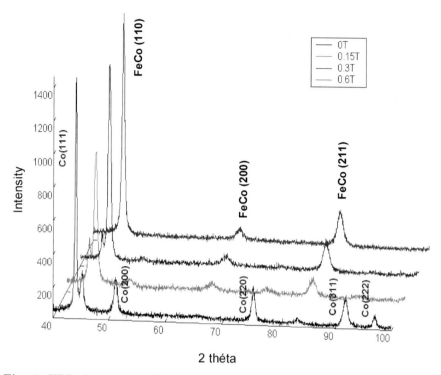

Fig. 9. XRD diagrams for Co–Fe alloys electrodeposited at –1.4 V/SSE: $CoSO_4$ 0.2 M, $FeSO_4$ 0.025 M, Na_2SO_4 0.5 M, NaCl 0.3 M, H_3BO_3 0.4 M for pH = 3, $T = 25°C$. The magnetic field is parallel to the horizontal electode

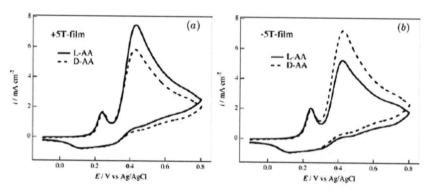

Fig. 10. Cyclic voltammograms of L- and D-ascorbic acids in a 0.5 M H_2SO_4 aqueous solution on polyaniline film electrodeposited under (a) +5 T and (b) −5 T magnetic field superimposition. The potential sweep rate is 50 mV/s [30]

7 Concluding remarks

A new field of interest arises in the electrolysis under a magnetic field, when the field of force that acts on the electrolyte has not an electromagnetic origin, but a purely magnetic origin taking advantage of the magnetic susceptibility of the electroactive species [31–33]. This is a new and promising field, which presents high potential for applications to the deposition of paramagnetic materials. For this case, main equations governing the flows have to be reconsidered, more specifically the distribution of the electric current. This should take into account the magnetic forces acting on the ions, which are able to modify their trajectories.

A general conclusion is that MEC is a new domain of MHD, almost completely open for research with a large range of possible applications. For example, we can refer to the numerical simulation by Olivas et al. [34] concerning the improvement of both the rate of transfer and the homogeneity of the gold electrodeposition around cylindrical connectors using a low frequency magnetic field. The main difficulty for the applications results in the possibility to apply the magnetic field with significant intensity only onto some specific zone.

References

1. A bibliographic investigation (CAPLUS) on the effect of the magnetic field in electrochemistry leads to 20 articles published before 1940. The annual article average number is 7 for the seventies and the eighties, 20 for the nineties and more than 30 for the first years of the new century.
2. Fahidy TZ (1983) Magnetoelectrolysis. J Appl Electrochem 13:553–563
3. Ulrich T, Steiner E (1989) Magnetic field effects in chemical kinetics and related phenomena. Chem Rev 89:51–147
4. Tronel-Peyroz E, Olivier A (1982) Application of the Boltzman equation to the study of electrolytic solution in the presence of electric and magnetic fields. Physico-Chemical Hydrodynamics 3:251–265
5. Tronel-Peyroz E, Olivier A, Fahidy TZ, Laforgue-Kantzer D (1976) Effet thermo-magnétoélectrique en solution électrolytiques. Electrochimica Acta 19:835–840
6. Levich VG (1974) Physicochemical Hydrodynamics. Prentice Hall, Englewood Cliffs, NY
7. Ngo Boum G (1998) Etude numérique du transport de matière au sein d'un électrolyte: Effet d'un champ magnétique. Thèse INPG, France
8. Ngo Boum GB, Alemany A (1999) Numerical simulation of electrochemical mass transfer in electromagnetically forced channel flows. Elecrochimica Acta 44:1749–1760
9. Aogaki R, Fueki K, Mukaibo T (1976) Diffusion process in viscous flow of electrolyte solution in magnetohydrodynamic pumps electrodes. Denki Kagaku 44:89–94

10. Mori S, Satoh K, Takeushi M (1994) Electrolytic mass transfer around inclined cylinder in static magnetic field. Electrochimica Acta 39:2789–2794

11. Mollet L, Dumargue P, Daguenet M, Bodiot D (1974) Calcul du flux limite de diffusion sur une microélectrode de section circulaire – équivalence avec une électrode de section rectangulaire. Vérification expérimentale dans le cas du disque tournant en régime laminaire. Electrochimica Acta 19:841–844

12. Aaboubi A, Chopart JP, Douglade J, Olivier A, Gabrielli C, Tribollet B (1990) Magnetic field effects on mass transport. J Electrochem Soc 137:1796–1804

13. Olivier A, Chopart JP, Amblard J, Merienne E, Aaboubi O (2000) Direct and indirect electrokinetic effect inducing a forced convection. EKHD and MHD transfer functions. ACH – Models in chemistry 137:213–224

14. Sugiyama A, Morisaki S, Mogi I, Aogaki R (2000) Application of cyclic magnetammetry to the analysis of electrochemical reaction in a high magnetic field. Electrochemistry 68:771–778

15. Olivier A, Chopart JP, Douglade J, Gabrielli C, Tribollet B (1987) Frequency response of the limiting diffusion current to a magnetic field perturbation. J Electroanal Chem 227:275–279

16. Devos O, Aaboubi A, Chopart JP, Merienne E, Olivier A, Amblard J (1998) Magnetic field effects on nickel electrodeposition. J Electrochem Soc 145:4136–4139

17. Devos O, Aaboubi A, Chopart JP, Merienne E, Olivier (1999) Magnetic impedance method: the MHD transfer function. Electrochemistry 67:181–187

18. Gurniki F, Brak FH, Zahrai S (2000) Large-eddy simulation of electrochemical mass transfer. In: Proceedings of the pamir International Conference, Giens, France, pp 1:327–332

19. Devos O, Aaboubi O, Chopart JP, Olivier A (2000) Is there a magnetic field effect on electrochemical kinetics. J Phys Chem A104:1544–1548

20. Fricoteaux P, Jonvel B, Chopart JP (2003) Magnetic effect during copper electrodeposition: diffusion process considerations. J Phys Chem B107:9459–9464

21. Fahidy TZ (2001) Characteristics of surfaces produced via magnetoelectrolytic deposition. Prog Surf Sci 68:155–188

22. Coey JMD, Hinds G (2001) Magnetic electrodeposition. J Alloys Compd 326:238–245

23. Uhlemann M, Schlörb H, Msellak K, Chopart JP (2004) Electrochemical deposition of Cu under superimposition of high magnetic fields. J Electrochem Soc 151:C598–C603

24. Krause A, Hamann C, Uhlemann M, Gebert A, Schultz L (2005) Influence of a magnetic field on the morphology of electrodeposited cobalt. J Magn Magn Mater 290–291:261–264

25. Bodea S, Ballou R, Molho P (2004) Electrochemical growth of iron and cobalt arborescences under a magnetic field. Phys Rev E69:021605/1-021605/12

26. Heresanu V, Ballou R, Molho P (2001) Magnetic properties of Fe arborescences grown by electrodeposition. J Magn Magn Mater 226–230:1978–1980

27. Msellak K, Chopart JP, Jbara O, Aaboubi O, Amblard J (2004) Magnetic field effects on Ni-Fe alloys codeposition. J Magn Magn Mater 281:295–304

28. Harrach A, Douglade J, Dupuis M, Amblard J, Chopart JP (2005) Characterisation of Co-Fe alloys electrodeposited with magnetic field superimposition. In: Proceedings of the Joint 15th and 6th Pamir International Conference, Riga 2: pp 155–158

29. Mogi I, Watanabe K (2005) Chirality of magnetoelectropolymerized polyaniline electrodes. Jpn J Appl Phys 44:L199–L201
30. Mogi I, Watanabe K (2005) Chiral electrodes of magneto-electropolymerized polyanilines films. In: Proceedings of the Joint 15th and 6th Pamir International Conference, Riga 2: pp 127–130
31. Hinds G Coey JMD, Lyons MEG (2001) Influence of magnetic forces on electrochemical mass transport. Electrochem Com 3:215–218
32. Rabah LK, Chopart JP, Schloerb H, Saulnier S, Aaboubi O, Uhlemann M, Elmi D, Amblard J (2004) Analysis of the magnetic force effect on paramagnetic species. J Electroanal Chem 571:85–91
33. Leventis N, Dass A (2005) Demonstration of the elusive concentration-gradient paramagnetic force. J Am Chem Soc 127:4988–4989
34. Olivas P, Alemany A, Bark F (2004) Electromagnetic control of electroplating of a cylinder in forced convection. J Appl Electrochem 34:19–30